VPNs:
A Beginner's Guide

ABOUT THE AUTHOR

John Mairs, CCNA, CCDA, MCSE, MCP+I, MCP, is a Systems Engineer for GE Capital IT Solutions. He lives and works in San Francisco and is the co-author of the *Cisco Router Handbook, Second Edition* (Osborne/McGraw-Hill). He has more than 16 years of experience in the computer industry, working with corporate networks.

VPNs:
A Beginner's Guide

JOHN **MAIRS**

McGraw-Hill/Osborne

New York Chicago San Francisco
Lisbon London Madrid Mexico City Milan
New Delhi San Juan Seoul Singapore Sydney Toronto

McGraw-Hill/Osborne
2600 Tenth Street
Berkeley, California 94710
U.S.A.

To arrange bulk purchase discounts for sales promotions, premiums, or fund-raisers, please contact **McGraw-Hill**/Osborne at the above address. For information on translations or book distributors outside the U.S.A., please see the International Contact Information page immediately following the index of this book.

VPNs: A Beginner's Guide

34567890 FGR FGR 019876543

ISBN 0-07-219181-3

Publisher
   Brandon A. Nordin
Vice President and Associate Publisher
   Scott Rogers
Acquisitions Editor
   Tracy Dunkelberger
Project Editor
   Lisa Wolters-Broder
Acquisitions Coordinator
   Alexander Corona
Technical Editors
   Mark Newcomb
   Ariya Parsamanesh
Developmental Editor
   Martin Minner
Copy Editor
   Jan Jue

Proofreader
   Susie Elkind
Indexer
   Valerie Perry
Computer Designers
   Carie Abrew
   Lauren McCarthy
Illustrators
   Beth Young
   Jackie Sieben
   Lyssa Wald
   Michael Mueller
Cover Design
   Amparo Del Rio
Series Design
   Peter F. Hancik

This book was composed with Corel VENTURA™ Publisher.

CONTENTS

Part I

Networks and Security

v

Part II
Virtual Private Networks

Part III

VPN Protocols

Part VI

MPLS

ACKNOWLEDGMENTS

When I started out in this effort I did so in the interest of learning. When I say "learning," I mean that my intention was to impart whatever knowledge I had in the subjects of virtual private networking and security in the hope that someone could learn from, and refer to, this book to further their understanding in the broad subject of computer communications.

As I proceeded, I eventually discovered that it was I who was "learning" and it turned out to be the most important thing I have ever learned in my life. I learned the true meaning of friendship, loyalty, and love.

Sometimes, in the quest of looking for happiness, and the many other things that one could ever hope to possess, you find that they are all right under your nose. All to often, and in many cases, that is the last place we tend to look. Fortunately, I was lucky enough to look under mine. I point this out because I would like to dedicate this book to the very things I found when I glanced under my nose.

I want to dedicate this book to Carlyn.

Carlyn is, and has been, my inspiration ever since I found her, all the while unaware that she was there the entire time. I am not an easy person to live with and no woman has ever loved me so unconditionally and accepted me so completely through all of my faults and idiosyncrasies as she has. She is my constant companion, my most loyal ally, my biggest fan, and above all else, my best friend. I love you so much, Carlyn. I wake up every morning and go to sleep every night with butterflies in my stomach every day knowing that I will get to spend the rest of my life with you.

I would also like to dedicate this to my children.

To my daughters Nicole and Jessica Mairs. I want to thank you guys for always loving me through thick and thin (and there was a lot of thin at times) and putting up with the weirdest dad in the world. You are both very beautiful and I am so proud to be the father of the two girls with the brightest smiles on earth. I love you both very much.

I would like take to this opportunity to make a very special dedication to a very special son, John "Jake" Mairs. I want to thank Jake, my namesake and pal, for all he put up with during the writing of both of my books. Jake graciously sacrificed a lot of his time and literally donated that time to these books. I would *never* have been able to complete either of these works had he not sacrificed that time—time that I know is very precious to him. Jake, I want you to know how much I love you and how very proud I am that you are my son. Jake has often said to me and others that when he grows up he wants to be like his dad. The truth is, Jake, that as I grow up I would be happy and extremely lucky if I could just manage to be half the man you are now. You are a very special person and soul and the world is lucky you are in it.

I would also like to dedicate this to Susan, the mother of my three children, for doing such a good job raising them. I want her to know that I appreciate and love her more than she thinks. She has been infinitely patient with me throughout my life (anyone who knows me is aware that I unfortunately require "a tad more than infinity") and a good friend ever since that day on the basketball court (she beat me…hey! be quiet she'd beat you, too). With respect to Susan's chosen (or thrust upon) profession, one she excels so well at, I want everyone to know and I want to recognize all of the thankless hours of work and heartache that comes with the most difficult job on earth…motherhood. My greatest hope is that someday you can fully come to recognize and accept your own considerable worth.

I would like to thank the Stolec family. John Jr. (*Yohn Yohnson*, the best friend a person could ever hope to have), Christine, Natalie (not at first—wink), my godson John Ibahko (mighty Thor) and my very special adopted mother, Anna, for always being there for me unconditionally and at times when no one else would. I am proud to consider myself a member of your family.

I would like to thank Arlene Uyemura for being such a good friend to such a nitwit. Thank you for helping mend a very broken fence in our family and under such heavy fire. I will never forget all you have done for our family and, above all, me.

I would like to thank the others that made this book possible. I would like to thank Steven Elliot for having the confidence in me to offer the opportunity. I would also like to thank Steve for his constant encouragement, patience, and faith (when it counted) in writing this book. I wish you the best of luck in whatever you do. I would also like to thank Tracy Dunkelberger for her help and patience in getting this book finished and promoted.

I would like to thank Mark J. Newcomb and Ariya Parsamanesh for their vast knowledge and expertise in the technical editing. Thanks for catching all of my errors and glaring omissions throughout the project. Your input was very valuable and the book is better for it.

I would like to thank Alexander Corona his help and never loosing faith in me the entire time. I would also like to thank Lisa Wolters-Broder, Jan Jue, Martin Minner, and the other editors who worked on this and for having to put up with such a greenhorn. All your input have been made me a way more gooder writer, thank yu!

I would like to thank Carlyn and Sean Manly for all of their help with many of the illustrations and Frances Auman for her help with…well, she knows. Thanks for being such good friends to me through what was easily the most interesting period of my existence. Life is funny, well, hilarious in our case. I mean, man! Who would have ever thought….wow! I would like to thank Robert Monsanto and little Bobby (my other godson) for all their help toward the end of the book and keeping me motivated.

Finally I want to thank my all four of my parents for their individual contributions to my very odd personality and for not having me aborted (even after birth, although I am sure it was considered).

INTRODUCTION

Virtual private networking has become almost a household term these days. It is also one of the most misused and beat-up terms. The idea of VPNs has been around for some time, but it's definition has undergone many changes over the years. Many companies have been offering many different VPN solutions for prospective customers. Unfortunately, all of "their" definitions vary just as much as the products themselves do. One of the goals of this book is to dispel the confusion that comes with those definitions. We will break down all of the different protocols and technologies used in the creation of VPNs, and by the time you finish this book, you will be able to easily make a decision about which implementations might work best for any application you may encounter. This book makes no assumptions as to how much you know or what level of experience you have.

How this book is organized

This book is divided into six parts. Part I is comprised of seven chapters and is entitled *Networks and Security*:

- ▼ Chapter 1, *Layering Architectures and the OSI Model,* talks about layering principles and the Open Systems Interconnection Model.

- ■ Chapter 2, *Network Architectures,* talks about the different networking architectures, as well as Local Area Networks (LANs), Wide Area Networks (WANs), and their respective protocols.

- ■ Chapter 3, *The TCP/IP Protocol Stack,* talks about the Department of Defense (DOD) model and all of the protocols that make up the TCP/IP stack.

■ Chapter 4, *Security,* talks a bout general security concepts and how to plan a comprehensive security policy.

■ Chapter 5, *Threats and Attack Methods,* details the different methods used by malicious hackers to break into or disable your network. This includes things like packet sniffing, spoofing techniques, denial of service, password attacks, and application layer attacks.

■ Chapter 6, *Intrusion Detection,* goes into Intrusion Detection Systems (IDS) and Honeypots.

▲ Chapter 7, *Firewalls,* talks about the firewalls and the different architectures used. We also talk about Network Address Translation (NAT) and Port Address Translation (PAT).

Part II is entitled *Virtual Private Networks* and is comprised of two chapters:

▼ Chapter 8, *VPNs and Tunneling,* discusses tunneling types, tunneling protocols, and internets, intranets, and extranets. This is where we talk about what a VPN is, as well as the history behind its creation.

▲ Chapter 9, *VPN Architectures,* defines the different types of VPNs and their architectures, such as Access VPNs, Intranet VPNs, and Extranet VPNs.

Part III is entitled *VPN Protocols* and is comprised of two chapters:

▼ Chapter 10, *Generic Routing Encapsulation (GRE) and Point to Point Tunneling Protocol (PPTP),* discusses GRE and PPTP in great detail. It talks about the history behind the creation of these protocols and how these protocols work, as well as their individual packet structures.

▲ Chapter 11, *Layer 2 Forwarding (L2F) and Layer 2 Tunneling Protocol (L2TP),* discusses L2F and L2TP in great detail. It talks about the history behind the creation of these protocols and how these protocols work, as well as their individual packet structures.

Part IV is entitled *Secure Communication* and is comprised of four chapters:

▼ Chapter 12, *Cryptography,* talks about the history and concepts of cryptography, ciphers, codes, and randomness.

■ Chapter 13, *Cryptographic Algorithms,* discusses hash algorithms and Public-Key and Private-Key Algorithms.

■ Chapter 14, *Data Integrity,* details digital signatures certificates Public-Key Infrastructure (PKI).

▲ Chapter 15, *Authentication,* talks about Access Control Services (ACS) and the different authentication protocols used in secure communication.

Part V is entitled *IPSec* and is comprised of four chapters:

▼   Chapter 16, *IPSec Components,* talks about the background behind IPSec as well as the different concepts and components that make up this protocol.

■   Chapter 17, *Key Management Concepts and Overview,* talks about the Internet Security Association Key Management (ISAKMP) Protocol and the Internet Key Exchange (IKE) Protocol.

■   Chapter 18, *Mechanics of Key Exchange,* talks In-band and Out of Band exchanges and Diffie Hellman, SKIP, Photuris, SKEME, and Oakley, and how these protocols accomplish key exchange.

▲   Chapter 19, *IPSec Architecture and Implementation,* talks about archictectures and implementation specifics, such as IPSec and NAT, tunnel and transport modes, managing security associations (SAs), and security association and security policy databases.

Part VI is entitled *MPLS* and is comprised of four chapters:

▼   Chapter 20, *Quality of Service,* talks about what QoS is and why it is needed. This is also where we talk about QoS frameworks and protocols.

■   Chapter 21, *Traffic Engineering–Movement of Data,* details how traffic can be controlled throughout the network from end to end. We also talk about routing and switching, and IP overlay networks over Asynchronous Transfer Mode (ATM).

■   Chapter 22, *MPLS Background,* talks about what Multi-Protocol Label Switching is and the history behind MPLS.

▲   Chapter 23, *MPLS Concepts and VPNs,* talks about the components of MPLS and how MPLS works. We also talk about label distribution and how MPLS is used to create VPNs.

PART I

Networks and Security

CHAPTER 1

Layering Architecture and the OSI Model

When the Internet was established, it was not done with security in mind. To understand how VPNs and firewalls work and where security needs to be applied, you must have a basic understanding of the protocols being used and the methods most likely to be used to subvert them. Networking protocols are based on a principle called *layering*. Attacks can come at any one of these layers. In this chapter, we talk about each one of these layers and how an attack might be pursued. We'll start with the OSI model.

LAYERING PRINCIPLES

Approaching any problem methodically helps reduce its complexities. Whether it is something as simple as cleaning your house or as complicated as navigating the federal income tax system, breaking the task into sections, or layers, makes the task more achievable. This is especially true when it comes to working with communications and networking protocol suites or stacks. (Protocol suites are often referred to as *stacks* because the layers are stacked on top of each other.)

Discrete functions are performed in each layer of a protocol stack, and specific protocols have been written for each layer in order to accomplish these functions. The goal is to keep each layer as independent as possible from the other layers and to hide the complexities of each layer from the other layers. Besides speeding protocol development, this simplifies updates and changes because modifications to one layer do not affect the whole stack. Prior to the use of layered protocols, simple changes such as adding one terminal type to the list of those supported by a nonlayered architecture often required changes to all of the software in the protocol stack.

OPEN SYSTEMS INTERCONNECTION (OSI) MODEL

As local area networks (LANs) became popular, people attempted to connect different LANs together. These attempts were frequently unsuccessful because each company used its own proprietary standards of communication over the network.

As one of the largest and most sought customers in the world, the U.S. government has tremendous power to dictate technical requirements to its vendors. In the early 1990s, the U.S. government decided that it would buy products and services only from vendors that supported the Open Standards Interconnection model (otherwise known as the OSI model). This model was proposed by the International Standards Organization (ISO) in 1984 and is sometimes referred to as the *ISO-OSI model*. It was created to provide a common conceptual framework for the development of standards that would allow different systems to be connected. Although a fully operative OSI communications protocol suite exists, the OSI model is used as more of a guideline than a strict set of rules that must be adhered to. However, it is important to understand how this model works because it is frequently referred to when describing the actions of any given protocol suite. The ISO-OSI reference model defines seven discrete layers, as shown in Figure 1-1.

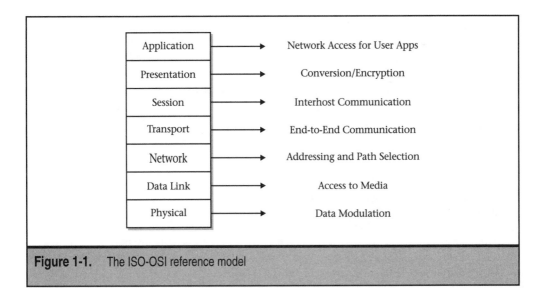

Figure 1-1.   The ISO-OSI reference model

The Application Layer (Layer 7)

The Application layer is the interface that applications use to access network resources. For example, the application may be a word processor attempting to open a file that is stored on a file server volume. In this case, a basic local function is redirected to the Application layer and then to the rest of the OSI stack to access remote resources. Or, the application may exist solely to provide an interface to a network resource, functioning like an executable protocol such as Telnet or FTP. In either case and others, including e-mail access, database access, and file and network management, the Application layer provides the tools needed to access those network resources. The Application layer interacts directly with and issues requests to the Presentation layer through the Presentation Service Access Point (PSAP). It differs from the other layers in that it does not provide services to any of the other OSI layers, but rather to application processes above the OSI model. The Application layer identifies and establishes the availability of intended communication partners, synchronizes cooperating applications, and establishes agreement on procedures for error recovery and control of data integrity. The Application layer also determines whether sufficient resources for the intended communication exist.

Examples of protocols that occupy this layer are File Transfer Protocol (FTP), Hyper-Text Transfer Protocol (HTTP), Telnet, Network File System (NFS), Simple Mail Transfer Protocol (SMTP), Simple Network Management Protocol (SNMP), and File Transfer Access Management (FTAM).

The Presentation Layer (Layer 6)

The Presentation layer makes sure that information passed down from the Application layer of one system is readable by the Application layer of the system it is transmitting to.

The Presentation layer is also responsible for data encryption, data compression, and graphics compression, as well as character conversion between disparate systems such as American Standard Code for Information Interchange (ASCII) to IBM's Extended Binary Coded Decimal Interchange Code (EBCDIC).

Examples of protocols that occupy this layer are Graphics Interchange Format (GIF), Joint Photographic Experts Group (JPEG), Tagged Image File Format (TIFF), PICT, ASCII, EBCDIC, Moving Picture Experts Group (MPEG), Musical Instrument Digital Interface (MIDI), and HyperText Markup Language (HTML).

The Session Layer (Layer 5)

The Session layer controls the creation, management, and termination of connections between two machines and enables two Application layer or Presentation layer entities to synchronize and manage their data exchange. While the connection is maintained, the machines might communicate either unidirectionally, which is called *half duplex*, or bidirectionally, which is called *full duplex*.

The Session layer not only allows ordinary data transport, as does the Transport layer, but it also provides enhanced services such as remote logins and remote file transfers.

Another session service is token management. For some protocols, it is essential that both sides do not attempt the same operation at the same time. To prevent this, the Session layer provides tokens that can be exchanged. Only the side holding the token can perform the critical operation.

Another session service is synchronization. Consider the problems that might occur when trying to perform a two-hour file transfer between two machines on a network with a one-hour mean time between crashes. After each transfer is aborted, the whole transfer will have to start over and will probably fail again with the next network crash. To eliminate this problem, the Session layer provides a way to insert checkpoints into the data stream, so that after a crash, only the data following the last checkpoint has to be repeated.

Sometimes the Session layer is referred to as the "port layer" because computer services communicate with each other through ports. A good example of this would be HTTP, which occupies the well-known port 80. Firewalls and packet filters sometimes restrict or allow certain protocols based on their port numbers. Examples of protocols that occupy this layer are Remote Procedure Call (RPC), Structured Query Language (SQL), NetBIOS, and AppleTalk Session Protocol (ASP).

The Session layer, Presentation layer, and Application layer collectively handle the network functionality that users can see. The remaining layers handle network operations that are transparent to the user.

The Transport Layer (Layer 4)

The Transport layer is involved in reliable (*connection oriented*) and unreliable (*connectionless*) end-to-end data transport. (I will discuss these topics later in the chapter.) The transported data is broken down into smaller units called *segments*. Data transport services are transparent to the upper layers and generally involve flow control, multiplexing, error checking, and recovery.

▼ *Flow control* manages data transmission between devices, ensuring that the receiving device is not overwhelmed with more data than it can handle at any given time.

■ *Multiplexing* alternates or interleaves segments from multiple applications, allowing for what appears to be simultaneous transmission over a single interface.

■ *Error checking* detects transmission errors as they occur.

▲ *Recovery* resolves data transmission errors by, for example, requesting the retransmission of lost segments.

Examples of protocols that occupy this layer are Transmission Control Protocol (TCP), User Datagram Protocol (UDP), and Sequence Packet Exchange (SPX).

The Network Layer (Layer 3)

The Network layer is responsible for delivering messages with *logical addressing;* that is, messages that are transported from end to end (for example, from the host in San Francisco to the host in New York City). Logical addressing is different from *physical addressing,* which handles communications from machine to machine (for instance, from the host in San Francisco to the router in San Francisco). For example, when you buy an airline ticket from San Francisco to New York City, you (like the logical addressing at the Network layer) are only concerned that you get from point A to point B. The airline, however, must concern itself with physical details such as stopovers in Phoenix and Dallas, plane changes, seat numbers, and so on. The Network layer supports both connection-oriented and connectionless services from higher-layer protocols. Message units in this layer are generally referred to as *packets* or *datagrams.*

Examples of protocols that occupy this layer are Internet Protocol (IP), Internetwork Packet Exchange (IPX), Border Gateway Protocol (BGP), Open Shortest Path First (OSPF), and Routing Information Protocol (RIP).

The Data Link Layer (Layer 2)

The Data Link layer handles physical addressing and is concerned with the delivery of packets on a given wire segment. Using the airline example, the pilot's job is to get the plane from San Francisco to Phoenix. He or she is not concerned whether the passengers continue on to Dallas, New York, Chicago, or London for that matter. Message units in the Data Link layer are generally referred to as *frames.*

Additional Data Link layer tasks involve topology definition, error control, frame sequencing, and flow control.

▼ Topology, which is defined by the Data Link layer, determines how devices are to be logically connected, such as in a bus, star, or ring.

- ■ Error control involves alerting upper-layer protocols that a transmission error has occurred.

- ■ Frame sequencing takes frames that are transmitted out of sequence and rearranges them into their proper order.

- ▲ Flow control ensures that the receiving device is not overwhelmed with more data than it can handle at any given time.

Examples of protocols that occupy this layer are Frame Relay, High-Level Data Link Control (HDLC), Point to Point Protocol (PPP), Fiber Distributed Data Interface (FDDI), Asynchronous Transfer Mode (ATM), IEEE 802.3 (for Ethernet), and IEEE 802.5 (for Token Ring).

The Physical Layer (Layer 1)

The Physical layer is the actual medium through which *bits* are transmitted from one location to another. It is comprised of every component that is needed to realize the connection, including all connectors, hubs, transceivers, network interfaces, and transmission mediums or cables. It also includes the method of signaling used to transmit bits to a remote location. The connecting medium between two network stations may be copper or some other electrically conductive medium such as cable, fiber optic, radio signals, microwaves, lasers, or infrared. The OSI model makes no distinctions concerning the actual hardware involved.

Many specifications can be combined to define the components of the Physical layer. For example, the RJ-45 specification defines the shape of the connector and number of wires in the cable as well as the specific pinouts for the connector being used. Ethernet and 802.3 define the use of wires to pins 1, 2, 3, and 6. These are combinations of different Physical layer specifications for use with a twisted-pair Ethernet scheme.

Examples of specifications that occupy this layer are EIA/TIA-232, EIA/TIA-449, V.35, V.24, RJ-45, Ethernet, 802.3, 802.5, FDDI, NRZI, NRZ, and B8ZS.

PEER COMMUNICATIONS AND ENCAPSULATION

When communicating with another entity, it is important that the same protocol be used on both sides of the conversation—just as it is in everyday life. After all, there is no point in writing a letter in English to a person who only understands French. This is where the concept of peer communication becomes important. Take a look at Figure 1-2 to get an idea of what we mean by this.

When two machines communicate, both computers must be running the same protocol stack so each layer of the protocol stack can communicate with the corresponding layer on the other machine, that is, with its *peer*. For example, the Application layer on Host 1 needs to communicate directly with the Application layer on Host 2. This is done logically through the protocol of the specific layer in question. Communication between

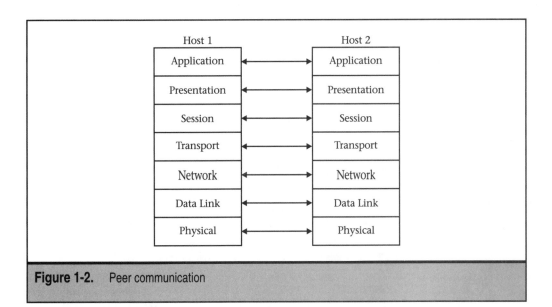

Figure 1-2.   Peer communication

peers is done in Protocol Data Units (PDUs), which are usually referred to as *packets*. Even though this communication takes place logically between peers, each layer depends upon the layers below it for the actual delivery of data.

NOTE   As you may have noticed earlier in this chapter, different terms are used for different layers. For example, the Transport layer communicates via *segments*, the Network layer via *packets* or *datagrams*, the Data Link layer via *frames*, and the Physical layer via *bits*.

Each layer passes its PDU to the layer beneath. The underlying layer then adds a header, creating its own PDU. The "data payload" for the layer waiting below is the PDU of the layer immediately above. The message starts from the highest layer and travels down its own protocol stack until it reaches the network medium (usually a wire). After traveling across this wire to the other computer, the message travels from the receiving computer's lowest layer and on up to the top layer. This process is called *encapsulation*, as shown in Figure 1-3. Encapsulation can be compared to a mail delivery system. Consider for a moment what it takes to deliver a piece of mail from one building to another. At the top of the mail delivery model, the letter is composed. Then it is placed in an envelope. The envelope is addressed, postage is added, and then it is placed into a mail carrier's bag. The bag is placed into a delivery truck and driven to the recipient's building. At the destination building, the truck door is opened and the bag is removed. Then the bag is opened and the letter given to the appropriate mail carrier. The carrier brings the envelope to the recipient, who opens the letter and reads the contents. The encapsulating/de-encapsulating process can be thought of in the same way.

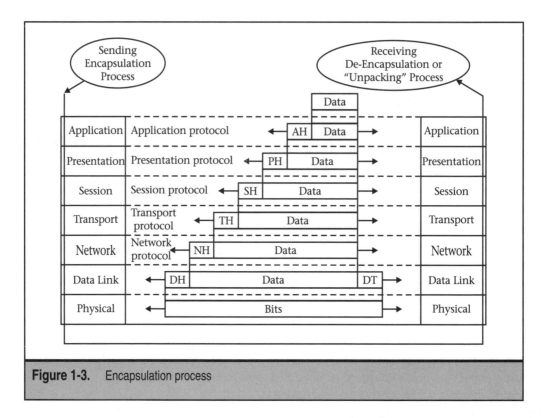

Figure 1-3. Encapsulation process

LAYER INTERACTIONS

Before getting into the actual mechanics of layer interactions, it is important that you familiarize yourself with the components of the protocol stack layers. First, let us define the components of the protocol stack itself: layers, entities, and protocols.

Layers, Entities, and Protocols

As you've already learned, every protocol stack consists of layers. The notations N, N+1, N+2, and so on, are used to identify a layer and the layers that are related to it. The purpose of these layers is to either use or provide some kind of service to the layer adjacent to it, depending on its hierarchy in the protocol stack. This means that each N layer *provides* services to the layer above it (referred to as *N+1*) and *uses* the services provided by the layer below it (referred to as *N–1*). (For more information about N notation and other naming conventions, see the sidebar.)

N Notation and Other Naming Conventions

When referenced individually, each layer in a protocol stack is labeled the *N layer*. The adjacent layer above the N layer is the *N+1 layer*. The adjacent layer below the N layer is the *N–1 layer*. For example, if the Transport layer is layer N, the Physical layer is N–3, and the Presentation layer is N+2. When discussing the OSI model, N always has a value from 1 through 7.

This notation enables writers to refer to other layers without having to write out their names every time. It also makes flowcharts and diagrams of interactions between layers a little easier to follow. The terms N+1 and N–1 are commonly used in both the OSI model and TCP. You may also run across situations in which OSI standards are referred to by the first letter of the layer in question or by the number of the layer. This can be confusing, because, for example, "S-entity," "5-entity," and "layer 5" all refer to the Session layer.

Entities are the active elements within each layer. An entity of any given layer can only communicate with:

▼ *Adjacent layer entities* directly above and/or below it, by using a service interface

▲ *Peer layer entities,* by using a protocol common to the entity on the other machine

Each layer also contains one or more *protocols,* which are the sets of rules and conventions used by the entity to communicate with its peer entity in the same layer on the other machine, as shown in Figure 1-4.

Besides the components that make up the protocol stack itself, there are components inside of and between the individual layers. Figure 1-5 and the definitions that follow will help you understand the processes and concepts discussed in the upcoming text.

The following list describes the objects illustrated in Figure 1- 5:

▼ **Service interface**   This is where the borders of each layer meet.

■ **Service Access Point (SAP)**   This is the address or logical location where each service transaction between layers takes place.

■ **Layer N+1**   This entity is the service *user* for layer N below.

■ **Layer N**   This entity is the service *provider* for layer N+1.

■ **Interface Control Information (ICI)**   This control information is transferred down through the SAP to coordinate data transfer at the interface, and then it is analyzed and discarded.

■ **Interface Data Unit (IDU)**   This is the data unit that is passed through the SAP from layer N+1. From this perspective, it is considered the ICI + PDU. It is then passed down to layer N, where it is seen as the ICI + SDU.

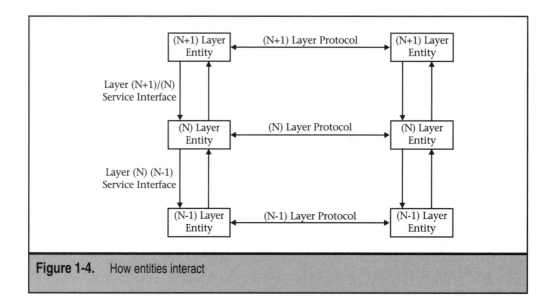

Figure 1-4. How entities interact

■ **Protocol Control Information (PCI)** This control information is exchanged between the local N entity and its remote peer N entity using the services of the N–1 connection to coordinate their joint operation.

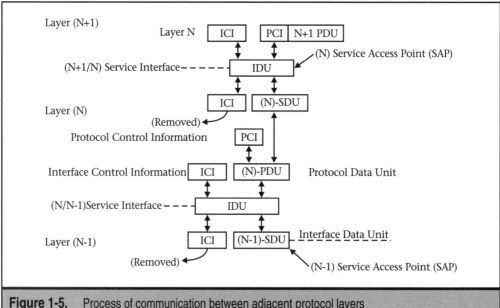

Figure 1-5. Process of communication between adjacent protocol layers

■ **Protocol Data Unit (PDU)**   This information is the combination of the PCI of the current layer and the SDU from the layer above (N+1).

▲ **Service Data Unit (SDU)**   The layer N+1 PDU that passes through the SAP becomes the SDU layer N below.

PDUs, SDUs, and SAPs

N layers communicate with each other through Service Access Points (SAPs). For example, say I am sending a letter to someone in France (I represent the service user). My ultimate goal is to have that person read my letter. However, I also need a way to communicate with the mail carrier (who represents my service provider) who comes to pick up the letter. I do that by placing the letter in the mailbox outside of my house and lifting the flag on the side of it. (In this case, the mailbox represents the interface or SAP.) So an N Service Access Point, which is often referred to as an *N-SAP*, is where an N service is provided to an N+1 entity (me) by the N service provider (the mail carrier). The N service Protocol Data Unit (PDU) is the packet of data exchanged at an N-SAP (the letter represents the PDU). My letter, or PDU, is considered to be the *logical* data flow with the person in France. (The recipient's mail carrier is the other service provider.) The address on the letter represents the Service Data Unit (SDU). The SDU is the *physical* representation of the data flow because it determines the actual path that the letter (or PDU) must follow to get there. Figure 1-6 illustrates PDU communication, that is, the logical flow of the data.

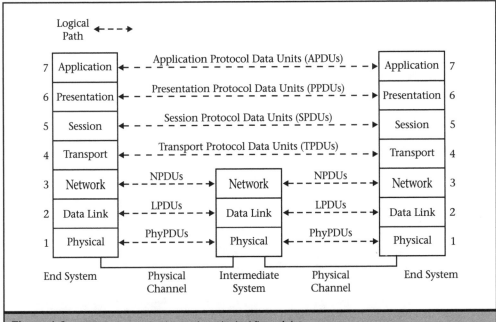

Figure 1-6.   PDU communication path, or logical flow of data

N Service Data Units or N-SDUs are the individual units of data that are exchanged at an N-SAP; therefore, N service data is made up of N-SDUs. Figure 1-7 shows how SDUs, which represent the actual path of the data, communicate.

Encapsulation Mechanics

As described earlier in the "Peer Communications and Encapsulation" section, encapsulation is the process of encoding data from one layer so that it can be transmitted to the next layer. This may involve adding information such as a header or translating the data into another form. For example, to transport an SDU, the current layer encapsulates the SDU by appending a Protocol Control Information (PCI) header. Figure 1-8 illustrates the encapsulation process.

Upon arrival, the receiving station on the network reverses the process, as seen in Figure 1-9.

Figure 1-10 displays the headers associated with each of the seven layers of the OSI reference model. When a unit of data is passed from whatever program the user is running to the Application layer, the Application layer adds its protocol header and passes the packet along with the header to the layer below. This layer in turn processes the data and adds its own header. Each layer treats the unit of information from higher layers as

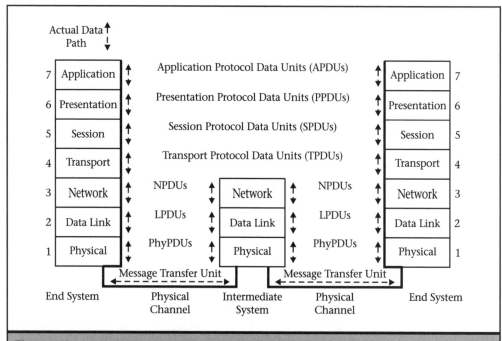

Figure 1-7. SDU communication path, or actual flow of data

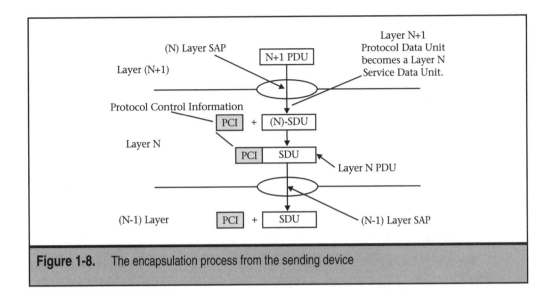

Figure 1-8.   The encapsulation process from the sending device

data and does not concern itself with the contents therein. This process continues until the packet reaches the Data Link layer, which applies both a header and a trailer (mainly for error control). At this point, the packet becomes a frame. It is then passed to the Physical layer, which converts the Data Link PDU into a series of bits that are sent across the transmission medium (such as a wire) to the intended receiver.

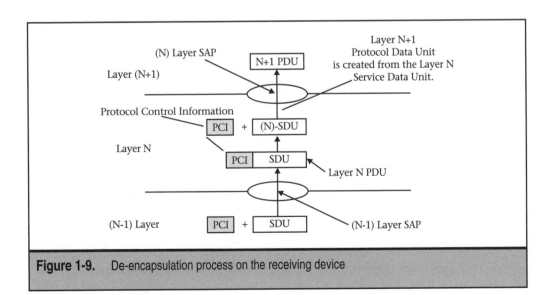

Figure 1-9.   De-encapsulation process on the receiving device

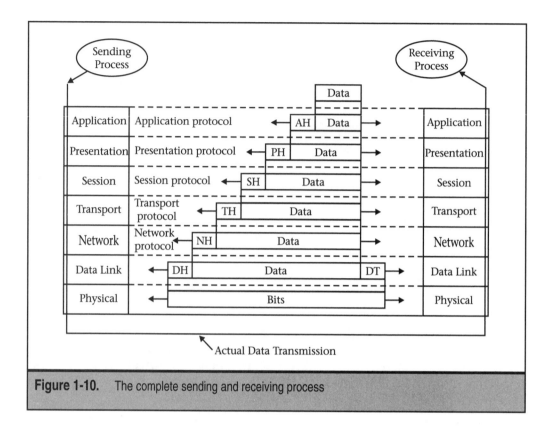

Figure 1-10.   The complete sending and receiving process

At the receiving end, the remote system reassembles the series of bits back into a frame, which is then forwarded and processed by the Data Link layer. Then the Data Link header and trailer are removed, and the packet (or datagram) is passed to the Network layer. The Network layer then removes its header from the packet and passes the data up to the Transport layer. This process continues until the original packet's data unit is sent to the remote machine's application program.

Services and Service Primitives

All the separate entities and protocols in each autonomous layer need a way to communicate with the entities in adjacent layers.

Communication between adjacent layers is done using function calls, called *service primitives*. Service primitives are the interactions that take place between the service user in the N+1 layer and the service provider in the N layer. All service primitives occur at the SAP in the form of SDUs at the service interface boundaries that separate the layers in the protocol stack.

The OSI model defines four classes of service primitives: request, indication, response, and confirm, as shown in Figure 1-11.

▼ **Request**   This primitive is issued by the service user entity in the upper layer to the service provider entity in the layer below to do some task.

■ **Indication**   This primitive is issued by the service provider entity in the layer below to the service user entity above to indicate that some event has taken place in the peer service of the remote protocol stack.

■ **Response**   This primitive is issued by the service user entity in response to the event from the indication primitive received.

▲ **Confirm**   This primitive is the response to an earlier request that has come back.

Connection-Oriented and Connectionless Communications and Service Primitives

The following process is called *confirmed service* because the applications wait for confirmation that communications are established and ready.

The first application sends a request primitive to the service provider of the Session layer and waits. The Session layer's service provider removes the request primitive from the inbound queue of the first application and sends an indication primitive to the second application's inbound queue.

The second application takes the indication primitive from its queue to the Session layer service provider and decides to accept the request for connection by sending a positive response primitive back through its queue to the Session layer. This is received by the Session layer service provider, and a confirmation primitive is sent to the first application in the Presentation layer.

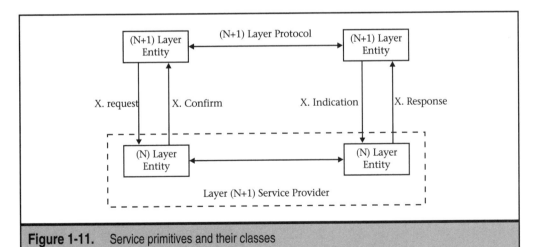

Figure 1-11.   Service primitives and their classes

CHAPTER 2

Network Architectures

To understand networking protocols, it is useful to know a little about networks in general. This chapter looks at the most common network architectures and the relationships between their physical and logical designs. We will discuss local area networks (LANs), wide area networks (WANs), as well as the protocols and technologies most commonly used by each.

NETWORKS AND NETWORK TOPOLOGIES

The term *network* usually means a set of computers and peripherals that are connected by some medium. The different devices on the network communicate with each other through a predefined set of rules known as *protocols*.

The devices on a network can be in the same room or scattered throughout a building. They can be connected over many miles through the use of dedicated telephone lines or microwave transmission. They can even be scattered around the world, connected by a long-distance communications medium or satellite systems. The layout of the actual devices and the manner in which they are connected to each other is called the *network topology*.

A network topology defines the way in which network devices are organized. The four most common topologies are bus, ring, star, and mesh, shown in Figures 2-1, 2-2, 2-3, and 2-4, respectively.

▼ **Bus topology**   A bus topology is where all of the devices connect to one common cable usually referred to as a *backbone*. The devices all use this backbone by using media access methods that share this medium by waiting until there is no activity on the network before transmitting data. The term is most commonly used in reference to the original Ethernet specification (which will be discussed shortly), where all of the devices attach or *tap* into the *bus cable* with a specialized interface connector. The nature of this design is that should there be a fault at any one point in the network, all network communication would be disabled.

■ **Ring topology**   In a ring topology all devices have two neighbors: one "upstream" neighbor and one "downstream" neighbor. All data communication travels around the ring in one direction. That direction can be either clockwise or counterclockwise. These devices all use this ring by using media access methods that share this medium by taking turns transmitting data. Should there be a fault at any one point in the ring, all network communication would be disabled. In some cases, two rings are used, as you will see shortly, where the data travels in both directions simultaneously for redundancy. Then, should there be a fault in the primary ring, the secondary ring would be brought into play for backup.

■ **Star topology**   A star topology is one in which each device is connected to a central device, usually a hub. Star topologies can be nested or "cascaded" to create hierarchical network topologies. In this case, should there be a fault on any one or more connections, the network will still remain active. If the central device or hub fails, then everything connected to it would be affected.

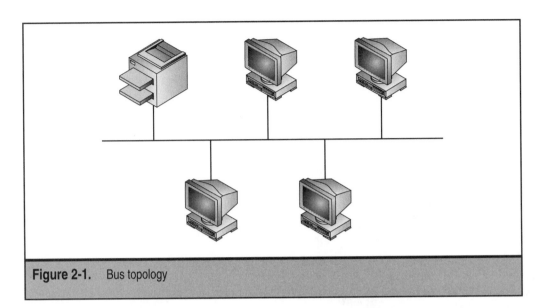

Figure 2-1.   Bus topology

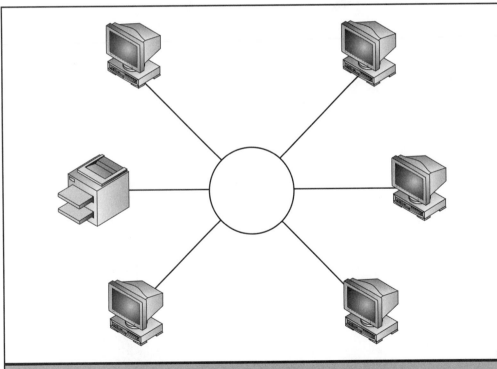

Figure 2-2.   Ring topology

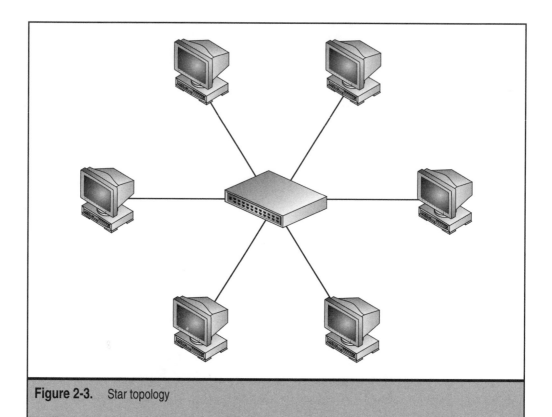

Figure 2-3.   Star topology

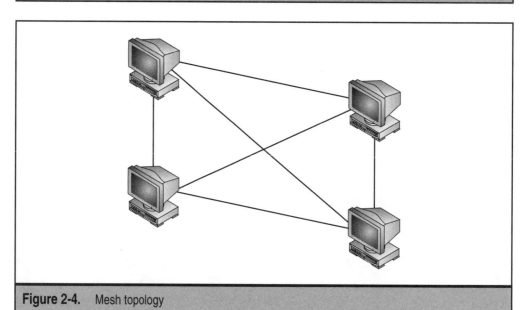

Figure 2-4.   Mesh topology

▲ **Mesh topology**   Every device in a mesh is connected to every other device in the network. This type of topology is extremely fault tolerant. Should there be a fault in one of the connections, one or more other connections between the devices remain. This type of topology can become difficult to install or to reconfigure due to the fact that all the devices are interconnected. Every time a new device is added to the network, additional connections are installed to all the existing devices. Every time a device is physically relocated, all the connections must be moved to the new location. Topologies are referred to either in a physical or logical sense. The distinction is based on whether you view them from the Physical layer or the Data Link layer.

LOCAL AREA NETWORKS (LANS)

Usually, if the devices on a network are in a single location or in a relatively small geographic area, they are considered to be a local area network (LAN). The connections between the individual nodes on a LAN range from just a few feet to several thousand feet. All the devices on a LAN are usually connected by a single type of network cable. LANs are generally used to connect workstations, personal computers, printers, and other devices.

LAN Protocols

LAN protocols occupy the lowest two layers of the OSI reference model: the Data Link layer and the Physical layer. The three most common LAN protocols are

▼ Ethernet, or IEEE 802.3

■ Token Ring, or IEEE 802.5

▲ Fiber Distributed Data Interface (FDDI)

Figure 2-5 illustrates how these protocol specifications map to the OSI reference model.
Before we get started talking about the 802.3 and 802.5 protocols, let's first talk about what sits between them and Layer 3, the Network layer. That very important thing that sits between them is the LLC sublayer, IEEE 802.2.

The LLC Sublayer, or IEEE 802.2

As shown in Figure 2-5, the Data Link layer is divided into two sublayers. The upper portion of the sublayer is called the *Logical Link Control* (LLC) sublayer. The lower portion is called the *Media Access Control* (MAC) sublayer.

The IEEE created the Logical Link Control standard, called IEEE 802.2, to create a programming interface between the part of the communications software that controls the network interface card and the rest of the protocols in the layers above it.

The connection between the network interface card and the rest of the communications protocols is through the Service Access Point (SAP), which was discussed in Chapter 1 and is shown in Figure 2-6.

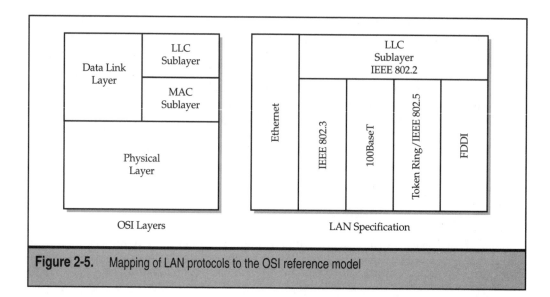

Figure 2-5. Mapping of LAN protocols to the OSI reference model

The SAP *identifier* is used to differentiate the different communications protocols. There are SAP identifiers for IP, NetBIOS, SNA, NetWare, and so on. These identifiers are located in the LLC sub-frame, as shown in Figure 2-7.

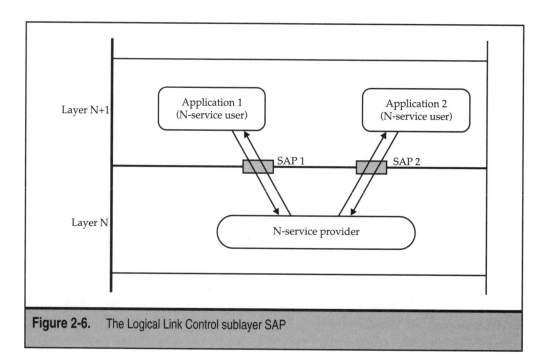

Figure 2-6. The Logical Link Control sublayer SAP

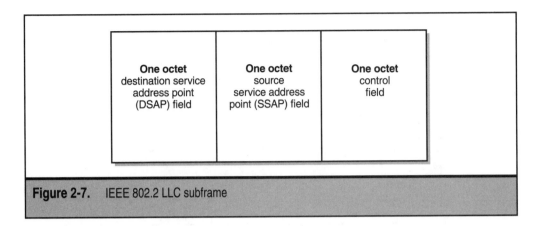

Figure 2-7.   IEEE 802.2 LLC subframe

The LLC layer is capable of providing two types of data link control operations, *LLC1* for connectionless and *LLC2* for connection-oriented (called *Type 1* and *Type 2,* respectively).

If the programmer chooses a Type 2 LLC, the frames (units of data at the Logical Link layer) are given sequence numbers as they pass through the sending SAP. This enables the 802.2 LLC layer at the receiving SAP to ensure that all frames have been received; once it receives the frames, it sends an acknowledgement. This creates a connection-oriented (reliable) data transfer mechanism at the Data Link layer.

For Type 1 (LLC1) services no connection is required; we only need a "data request" to hand the frame off to the LLC from the Network layer for transmission and a "data indication" to pass the frame up to the Network layer. There is no sequence and acknowledgement process because the data transfer is connectionless. Figure 2-8 illustrates the LLC Service Access Point concept.

Ethernet and the IEEE 802.3 Standard

Ethernet was originally developed by Digital, Intel, and Xerox in the early 1970s and was designed as a broadcast system. That simply means that stations on the network can send messages whenever and wherever they choose. The original format for Ethernet was developed at Xerox's Palo Alto Research Center (PARC) in 1972. The two men credited with the invention are Robert Metcalf and David Boggs. The original implementation was capable of transmitting data at the whopping rate of 2.94 megabytes per second (Mbps) and could support up to 256 devices over cable stretching up to 1 kilometer (km). Modern advancements have increased both the speed and distance limitations considerably. Current Ethernet technologies work at 10–1,000 Mbps and can span many kilometers. Today, Ethernet is the most popular and most widely used network technology in the world.

There are many kinds of Ethernet implementations, and the standard has grown to encompass many new technologies. But the basic mechanics of Ethernet operation still stem from the original design, which revolves around a technology called Carrier Sense Multiple Access with Collision Detection (CSMA/CD). CSMA/CD is discussed later in this chapter.

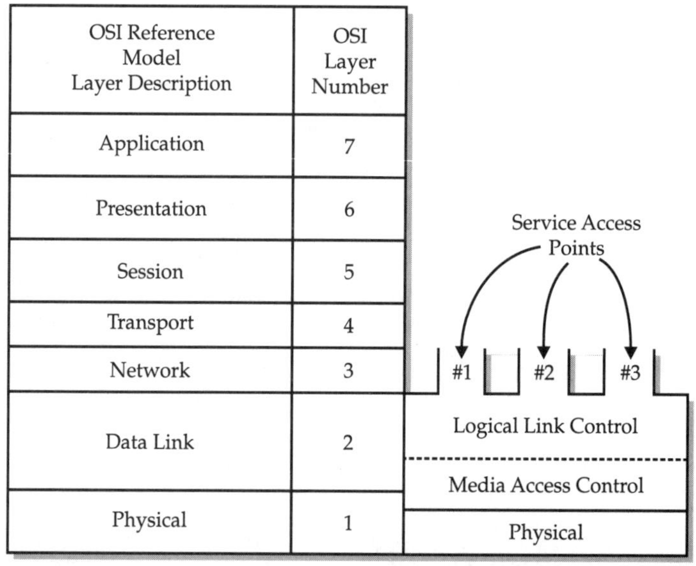

OSI Reference Model Layer Description	OSI Layer Number
Application	7
Presentation	6
Session	5
Transport	4
Network	3
Data Link	2
Physical	1

Service Access Points

#1 #2 #3

Logical Link Control

Media Access Control

Physical

Figure 2-8. LLC Service Access Points

Ethernet and 802.3 Frame Formats   There are two basic frame formats used in Ethernet: one is called Ethernet; the other is called IEEE 802.3. The 802.3 specification is the most common today.

As shown in Figure 2-9, the Ethernet and IEEE 802.3 specifications have the following components:

▼ **Preamble**   The alternating pattern of ones and zeros tells receiving stations that an Ethernet or IEEE 802.3 frame is coming. The Ethernet frame includes an additional byte that is the equivalent of the Start of Frame field specified in the IEEE 802.3 frame.

■ **Start of Frame (SOF) (IEEE 802.3 only)**   The IEEE 802.3 delimiter byte ends with two consecutive 1's, which serve to synchronize the frame-reception portions of all stations on the LAN. SOF is explicitly specified in Ethernet.

■ **Destination and Source Addresses**   The IEEE specifies the first three bytes of the destination and source addresses on a vendor-specific basis. The Ethernet or IEEE 802.3 vendor specifies the last three bytes. The source address is always a *unicast* (single-node) address. The destination address can be unicast, multicast (group), or broadcast (all nodes).

- **Type (Ethernet only)**   The type specifies the upper layer protocol that will receive the data after Ethernet processing is completed.

- **Length (IEEE 802.3 only)**   The length indicates the number of bytes of data that follow this field.

- **Data (Ethernet)**   After Physical layer and Data Link layer processing is complete, the data contained in the frame is sent to an upper-layer protocol, which is identified in the Type field. Although Ethernet Version 2 does not specify any padding (in contrast to IEEE 802.3), Ethernet expects at least 46 bytes of data. (To allow for collision detection, Ethernet requires a minimum packet size of 64 bytes. If the message being sent contains less data than the minimum, the remainder of the packet must be padded with zeros. Hence, the term *padding*.)

- **Data (IEEE 802.3)**   After Physical layer and Data Link layer processing is complete, the data is sent to an upper-layer protocol, which must be defined within the data portion of the frame, if at all. If data in the frame is insufficient to fill the frame to its minimum 64-byte size, padding bytes are inserted to ensure at least a 64-byte frame.

- ▲ **Frame Check Sequence (FCS)**   This sequence contains a 4-byte cyclic redundancy check (CRC) value, which is created by the sending device and is recalculated by the receiving device to check for damaged frames.

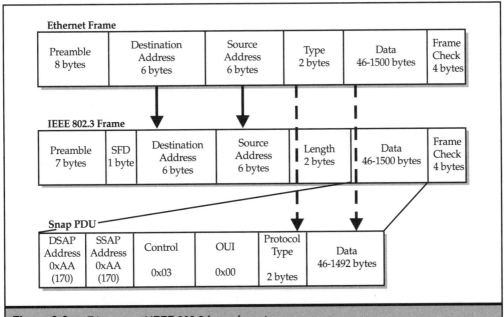

Figure 2-9.   Ethernet and IEEE 802.3 frame formats

LAN Specification Notation   Ethernet LAN specifications are written in a "shorthand" notation that is based on approximations. The notation includes the following information:

▼   LAN data rate (in Mbps)

■   LAN signal type (discussed in the next section)

■   Maximum segment length (in meters multiplied by 100)

▲   Media type, which may be one or a combination of the following types: twisted-pair cable, coaxial cable, or fiber-optic cable

For example, Thicknet Ethernet uses the notation "10Base5," where "10" refers to the data rate, "Base" refers to the signal type (baseband), and "5" refers to the maximum segment length. I'll explain this in more detail in the section "10Base5."

Signal Types   The signal type refers to the method used to transmit data across the medium (usually a wire) and how the data is multiplexed. The following terms are used to describe signal type:

▼   **Multiplexing**   This describes the method used to share the medium, such as Time Division Multiplexing (TDM) or Frequency Division Multiplexing (FDM).

■   **FDM**   Frequency Division Multiplexing uses multiple frequencies to transmit data simultaneously over the same wire. Television or cable TV, where the frequencies are called channels, is an example of FDM.

■   **TDM**   Time Division Multiplexing makes each signal take turns using the wire, meaning only one signal can be on the wire at any time.

■   **Broadband**   Broadband uses multiple frequencies to transmit data simultaneously over the same wire. Television or cable TV, where the frequencies are called channels, is an example of broadband. ("Broadband" is a term that is widely abused by companies marketing different versions of Internet access. The true definition is synonymous with FDM.)

▲   **Baseband**   A baseband network provides a single channel for communications across the medium, allowing only one device to transmit at a time. Devices on a baseband network are permitted to use all the available bandwidth for transmission, and the signals they transmit do not need to be multiplexed onto a carrier frequency.

10Base5   The original Ethernet was designed to use a "thick" coaxial cable and was called *Thicknet*. Thicknet is usually 50-Ohm, RG-8 coaxial cable (but not always and may have no RG label at all). It is about a half-inch thick and has a yellow shield. As mentioned earlier, the notation for Thicknet is "10Base5," where "10" denotes a 10Mbps-Mbps data rate; "Base" is short for "Baseband," which is the signal type; and "5" (multiplied by 100, remember) denotes a maximum cable length of 500 meters. The minimum length between stations is 2.5 meters.

The cable is run in one continuous length and terminated with 50-Ohm resistors, forming a physical bus topology. Transceivers attach to the cable using an inline N-type connector that "taps" into the cable (often referred to as a *Vampire Tap*). This has a 15-pin, or DB-15, connector called the *Attachment Unit Interface (AUI)* connection, which is used as the "drop cable" connection having a maximum length of 50 meters. The drop cable runs from the transceiver to the network device, for example, a computer or printer.

5-4-3 Rule A single Ethernet segment or bus can be appended with up to four repeaters (*hubs*). This means that five segments (with a total length of 2,460 meters) can be connected together. Of the five segments, only three can have devices attached (no more than 100 per segment). Therefore, a total of 300 devices can be attached on a 10Base5 (Thicknet) broadcast domain. Figure 2-10 illustrates this concept.

The preceding specifications have been summarized in the 5-4-3 rule, which states that:

▼ There may be no more than five segments between any two devices.

■ There may be no more than four repeaters or concentrators between any two devices.

▲ There may be no more than three populated segments (in other words, segments having any network devices attached). The remaining two segments are unpopulated and contain one repeater or concentrator at each end; they are used for increasing distance.

10Base2 Thin Ethernet (Thinnet) uses RG-58 cable and is called 10Base2, where "10" denotes a 10Mbps-Mbps data rate; "Base" is short for "Baseband," which is the signal type; and "2" (multiplied by 100) denotes an approximate maximum cable length of 200 meters. The actual maximum is 185 meters but remember, the "2" in the notation is approximate. The minimum length between stations is 0.5 meter.

Instead of using transceivers and cable drops as Thicknet does, each station connects to the Thinnet using a network interface card (NIC), which has a BNC connector. "BNC" stands for "British Naval Connector" or "Bayonet Nut Connector," depending on whom you ask. (If for some reason you really need to get to the bottom of that, don't ask the British, because I am pretty sure I know which one they'll tell you.) At each station, the Thinnet terminates at a T-piece. A 50-Ohm terminator is required at each end of the Thinnet run to absorb stray signals so that noise isn't reflected back into the line. (Plus you don't want a bunch of 1's and 0's leaking out of the ends all over the carpet.)

The 5-4-3 rule will always apply to Ethernet; therefore, a 10Base2 segment can be appended using up to four repeaters, or five segments in total. Of these five segments, only three can be populated. This means that two segments cannot be populated by any nodes; these segments are only used to extend the length of the broadcast domain to approximately 1000 meters (925 meters exactly, if you must know). The three populated segments can have a maximum of 30 stations each, for a total of 90 devices on a Thinnet broadcast domain. These, again, are the theoretical maximums in a pure situation. There are many ways of increasing those numbers using various combinations of repeaters (hubs), routers, and other devices.

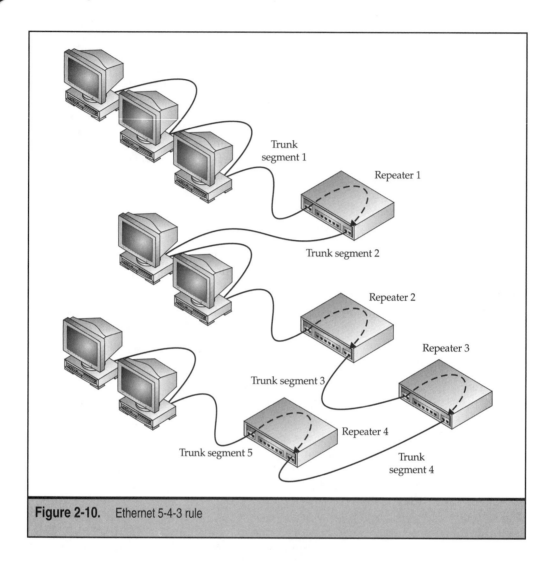

Figure 2-10. Ethernet 5-4-3 rule

10BaseT Currently, the most common use of Ethernet is across unshielded twisted-pair (UTP) cable. Its notation is 10BaseT, where *T* denotes "twisted pair." Keep in mind that Ethernet is a "logical" bus topology. The difference in this case is that instead of physically tapping into the bus or single wire segment, the network is wired in a "physical" star topology where each individual device is connected to a hub using a minimum of Category 3 UTP cabling. It's important to note that in practice, you should never use anything less than Category 5 UTP cabling.

Category numbers simply indicate how tightly the wires are twisted. The more twists in a given distance, the more magnetic waves (noise) will be cancelled out when the frequency of the wire increases. This becomes very important with faster Ethernet versions such as 100BaseT and higher, where each segment is wired from a hub to the node itself with a maximum cable length of 100 meters. Each segment can be populated with one device. The 10BaseT hub has multiple UTP ports making it the physical center of the "star." (The logical bus topology is maintained internally.) All of the devices are connected using RJ-45 connectors. The maximum number of hubs (also called repeaters) between any two nodes is four, and the maximum number of nodes on one broadcast domain is 1024.

10BaseF   The 10BaseF standard is the fiber-optic version of Ethernet. 10BaseFB allows up to 2 kilometers (km) per segment (on multimode fiber) and is designed for backbone applications such as cascading hubs (repeaters). (A *backbone* is a segment of the network that links several individual workgroup and department LANs or that links several building LANs in a campus environment.)

10BaseFL describes the standards for fiber-optic links between stations and repeaters, allowing up to 2 km per segment on multimode fiber. In addition, there are the 10BaseFP (passive components) standard and the Fiber-Optic Inter-Repeater Link (FOIRL) standard, which provides the specification for a fiber-optic Media Attachment Unit (MAU) as well as other interconnecting components. The 10BaseF standard allows a maximum of 1024 nodes on one broadcast domain.

Other Ethernet standards include 100BaseT, Gigabit Ethernet, and an obscure 10Broad36 version, but they are just variations of the standards we have already discussed. Now let's get into the real nuts and bolts of how Ethernet operates, starting with its media access method.

LAN Media Access Methods

Two methods are used to access the network medium itself: Carrier Sense Multiple Access with Collision Detection (CSMA/CD) and token passing methods such as Token Ring and FDDI. Let's start by talking about CSMA/CD, the media access method used by Ethernet and IEEE 802.3.

Carrier Sense Multiple Access with Collision Detection (CSMA/CD)

Ethernet is a shared medium. Nodes (any device attached to a network) need a way to determine when the network is available for sending packets. If two nodes at different locations attempt to send data at the same time, a *collision* results. To avoid conflicts and protect data integrity when sending data packets, Ethernet uses a media access method called Carrier Sense Multiple Access with Collision Detection (CSMA/CD).

"Multiple Access" means that every station is connected to the same medium, creating the necessity to "share the medium."

"Carrier Sense" means that before transmitting data, the station checks or "listens" to ensure that no other station is currently transmitting. If the station senses that there is no carrier, it can begin to transmit its data.

"Collision Detection" means that when a station transmits data, it listens to the wire to make sure that it does not detect a collision. Two different stations can listen to the wire, sense no carrier, and start transmitting data at the same time—causing their signals to collide a few nanoseconds later. What is happening is that each station sends data at 10Mbps Mbps. This allows 100 nanoseconds per bit. Light and electricity travel about one foot in a nanosecond. Therefore, after the electric signal for the first bit has traveled about 100 feet down the wire, the station will begin to send the second bit, and so on. These Ethernet cables can run for hundreds of feet. If you have two stations that are located, say, 250 feet apart, on the same cable, and they both start transmitting at the same time, they will be in the middle of the third bit before the signal from each reaches the other station.

Any system based on collision detection needs to make allowances for the time required for a "worst case" round-trip scenario through the entire network based upon the laws of physics. In Ethernet, this worst-case round trip is limited to 50 microseconds (millionths of a second). At a signaling speed of 10 million bits per second, this is enough time to transmit 500 bits. At 8 bits per byte, this is slightly less than 64 bytes. Therefore, to detect a collision accurately, Ethernet requires that a station continue transmitting a minimum of 64 bytes of data until the 50-microsecond period has ended. If the station has less than 64 bytes of data to send, then it simply pads the data by adding zeros at the end.

Collision Detection   While the sending computer is transmitting data, it monitors the channel, listening for collisions. If it detects a collision, it stops transmitting its data and immediately transmits a *jamming* signal. The jamming signal is designed to be long enough to make the round trip through the network to make sure that all other stations have detected the collision. The transmitter must send data for at least twice the propagation delay (distance divided by speed of the medium). This collision window or contention period allows the transmitter to always detect collisions, making it unnecessary for all receivers on the network to transmit jamming signals. After the collision window has passed, the transmitter has safely acquired the channel, and no more collisions should occur.

Binary Exponential Back-Off Algorithm   After the actual collision occurs, all of the stations run a *back-off algorithm*. This algorithm causes each station to wait for a randomly chosen number of *slot times* (the maximum amount of time it takes an Ethernet frame to travel the full length of the network) before trying to transmit the packet again. The slot time is usually a little longer than the collision window, allowing a station to access the network during the slot time it selected without causing a collision in the next slot time. Mathematically it is possible (though highly unlikely) that two or more stations involved in the collision could select the same number of slot times, causing another collision to occur. If this happens, the process doubles the interval range chosen to generate the random number and then repeats itself. The new collision may occur with a station in the original collision, or it may occur with a new station. After 16 attempts at retransmission, the controller will cease attempts and generate an error message.

In the beginning, when Ethernet was dominated by Thicknet coaxial cable, it was easy to translate the 50-millisecond limit and other electrical restrictions into rules about cable length, number of stations, and number of repeaters. But as Ethernet has evolved, by adding new media (such as fiber-optic cable) and smarter hubs (switches, routers, and so on), calculating the physical distance limits with precision becomes more complex. But in the end, those limits work out, as they are still based on the same concept of calculating the worst-case round-trip scenario of the network as a whole. You could just as easily create your own Ethernet-like collision system with a 40-microsecond or 60-microsecond period. Now let's discuss token passing methods for LAN media access.

Token Passing

Before we get into the specifics of the other two major LAN technologies, let's talk about the concept behind the methods used. The access method used on Token Ring and FDDI networks is called *token passing*. Token passing is a *deterministic* access method, as opposed to the *probabilistic* Ethernet method. In token passing, collisions are prevented by ensuring that only one station can transmit at any given time. This is accomplished by passing a special packet called a *token* from one station to another around a ring. A station can only send a packet when it gets the free token. When a station gets the free token and transmits a packet, the packet travels in one direction around the ring, passing all of the other stations along the way. As with Ethernet, the packet is usually addressed to a single station, and when it passes by that station, the packet is copied. The packet continues to travel around the ring until it returns to the sending station, which removes it and sends the free token to the next station around the ring. This is the basic idea behind any token passing scheme.

Token Ring and IEEE 802.5   Another major LAN technology is Token Ring. Token Ring rules are defined in the IEEE 802.5 specification. Like Ethernet, the Token Ring protocol provides services at the Physical and Data Link layers of the OSI model. Token Ring networks can be run at two different data rates, 4 MbpsMbps or 16 MbpsMbps.

The ring topology used in Token Ring networks is a collapsed ring that looks like a physical star. Each station is connected to a Token Ring wiring connector by a single twisted-pair cable with two wire pairs, as shown in Figure 2-11. One pair serves as the *inbound* portion of the ring (also known as the receive pair) and the other pair serves as the *outbound* (transmit) pair.

In Token Ring LANs, each station is connected to a Token Ring wiring concentrator, called a Multistation Access Unit (MAU), using an individual run of twisted-pair cable. The MAU is the equivalent of a hub in Ethernet technology.

Token Ring Frame Format   The Token Ring frame format breaks down as shown in Figure 2-12.

Three kinds of frames travel around the ring:

▼ Token frames

■ Data frames

▲ Command frames

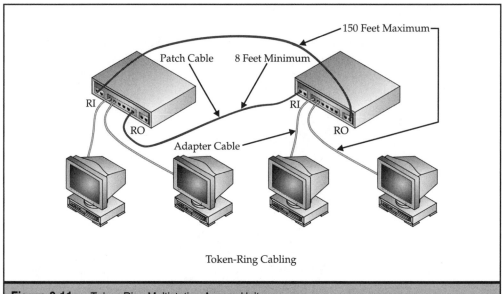

Figure 2-11. Token Ring Multistation Access Unit

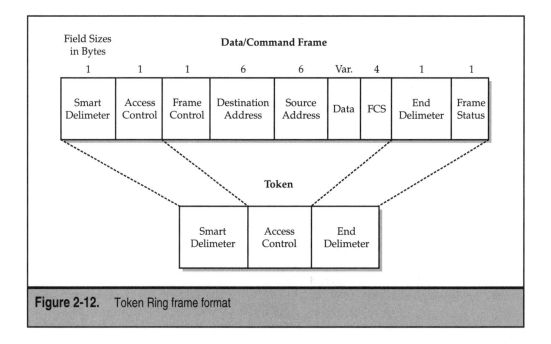

Figure 2-12. Token Ring frame format

The fields used in these frames are described in the following text.

Token Frame Fields

▼   **Start Delimiter**   This field alerts each station of the arrival of a token (or data/command frame). It includes signals that distinguish the byte from the rest of the frame by violating the encoding scheme used elsewhere in the frame (thereby differentiating it from the rest of the fields).

■   **Access Control**   This field contains the Priority field (the most significant 3 bits) and the Reservation field (the least significant 3 bits), as well as a token bit (used to differentiate a token from a data/command frame) and monitor bit (used by the active monitor to determine whether a frame is circling the ring endlessly).

▲   **End Delimiter**   This field signals the end of the token or data/command frame. It also contains bits to indicate a damaged frame and to identify the last frame in a logical sequence.

Data and Command Frame Fields   Data and Command frames have the same three fields as Token frames (Start Delimiter, Access Control, and End Delimiter), plus several others:

▼   **Start Delimiter**   This field alerts each station of the arrival of a token (or data/command frame). This field includes signals that distinguish the byte from the rest of the frame by violating the encoding scheme used elsewhere in the frame.

■   **Access Control**   This field contains the Priority field (the most significant 3 bits) and Reservation field (the least significant 3 bits), as well as a token bit (used to differentiate a token from a data/command frame) and a monitor bit (used by the active monitor to determine whether a frame is circling the ring endlessly).

■   **Frame Control**   This field indicates whether the frame contains data or control information. In control frames, this byte specifies the type of control information.

■   **Destination and Source Addresses**   Two 6-byte address fields identify the destination and source station addresses.

■   **Data**   Length of field is limited by the ring token holding time, which defines the maximum time a station can hold the token.

■   **Frame Check Sequence (FCS)**   The source station files a calculated value for this field that depends on the frame contents. The destination station recalculates the value to determine whether the frame was damaged in transit. If so, the frame is discarded.

■   **End Delimiter**   This field signals the end of the token or data/command frame. It also contains bits to indicate a damaged frame and to identify the last frame in a logical sequence.

▲   **Frame Status**   This 1-byte field terminates a command or data frame. The Frame Status field includes the address-recognized indicator and the frame-copied indicator.

IEEE 802.5 Token Passing Scheme   When a computer has data to send on to the ring, it must wait until it captures the token. Once the computer has possession of the token, it changes the frame format of the token so data can be sent. As the data travels around the ring, it passes through each computer until it reaches the destination computer. Each computer receives the frame, checks for errors, regenerates the frame, and then sends the frame back on to the ring. If an error is found, a special bit in the frame (called the Error Detection bit) is set so that other computers will not waste time reporting the same error.

Once the frame arrives at its intended destination, it is copied into the buffer memory of the destination's token ring card. Then the destination computer changes two series of bits on the frame and sends the frame back on to the ring. These two bits, called the Address Recognized Indicator (ARI) and the Frame Copied Indicator (FCI), let the rest of the computers on the ring know whether the frame has been received successfully. If the frame is not received successfully, these bits are not changed, alerting the sending station that it must retransmit the frame. If the two bits are changed, the frame continues around the ring until it arrives back at the source computer. Once this happens, the computer changes the frame format back into a token and sends it back on to the next computer downstream that has data to send.

Ring Monitors

Every station on a Token Ring network is either an Active Monitor or a Standby Monitor. The Active Monitor is chosen in an election process called *monitor contention*. The purpose of the Active Monitor is to:

▼   **Act as a master clock for the ring**   This places a master clock signal onto the wire, allowing all of the other stations to synchronize their internal clocks. This is so every computer on the ring determines when a bit time begins and ends.

■   **Provide a minimum 24-bit latency buffer**   The *latency buffer* ensures that no station other than the Active Monitor itself is ever stripping the initial bits of the token off the wire before it has finished sending the last bits.

■   **Initiate and monitor ring polling**   This causes an Active Monitor Present (AMP) frame to be sent every seven seconds. If the Active Monitor does not receive an AMP frame from its nearest upstream neighbor within seven seconds of initiating ring polling, it reports a Ring Poll Failure message to the ring's Error Monitor.

■   **Maintain token integrity**   The Active Monitor watches for problems in token passing processes and for lost tokens. To be considered "valid," a good token must be seen every 10 milliseconds.

▲   **Compensate for frequency jitter**   As stations resend data around the ring, they each introduce a small amount of phase shift in the signal. This phase shift is referred to as *jitter*. The Active Monitor compensates for jitter by generating a master clock signal based on its own internal clock instead of the clock signal being received from its upstream neighbor.

At any given time on a properly operating token ring network, there will be only one Active Monitor. Any station that is not acting as the Active Monitor is a Standby Monitor. If the Active Monitor fails, the Standby Monitors detect that an Active Monitor is missing and initiate the monitor contention process to elect a new Active Monitor. The purpose of the Standby Monitor is to:

▼ **Monitor token passing**   Standby Monitors watch token passing on the ring, but much more leniently. To be considered valid, a data frame or a good token must be seen at least once every 2.6 seconds. If not, the Standby Monitor initiates monitor contention.

■ **Monitor ring polling**   If a Standby Monitor does not see an Active Monitor Present (AMP) frame at least once every 15 seconds, it initiates monitor contention.

▲ **Monitor ring frequency**   If a Standby Monitor detects that the clock signal being produced by the Active Monitor differs significantly from its own internal clock, it initiates monitor contention.

Monitor Contention   Monitor contention refers to the process by which an Active Monitor is elected. Standby Monitors initiate monitor contention when they detect an event that suggests that the Active Monitor is not properly performing its duties. Because monitor contention sometimes implies a malfunction in the Active Monitor, a station that is acting as the Active Monitor does not participate in monitor contention. This prevents it from winning monitor contention and becoming the Active Monitor again.

Ring Polling   Ring polling (also called neighbor notification) is a process of "roll calling" that occurs every seven seconds. This allows every station on the ring to identify its nearest active upstream neighbor (NAUN). This helps isolate fault domains or "fence off" any errors to assist the network administrator in locating problems.

Ring Purge   A ring purge refers to an Active Monitor sending a ring purge frame around a Token Ring. It can also refer to the actual frame that is sent during a ring purge operation.

The purpose of a ring purge operation is to reset the ring to a known state. Any station receiving a ring purge frame immediately stops whatever it is doing, resets its timers, and enters *Bit Repeat* mode. Once the Active Monitor receives its own ring purge frame back, it knows that every station on the ring is now in Bit Repeat mode and waiting for a token.

Ring purges are generally sent out after a recovery operation has occurred and immediately before the Active Monitor generates a new token.

Beaconing   Beaconing is the process of isolating errors into a fault domain so that recovery actions can take place. The fault domain consists of:

▼ The station reporting the failure (the beaconing station)

■ The station upstream from the beaconing station

▲ The ring medium between them

The beacon process consists of transmitting beacon MAC frames every 20 milliseconds. Any computer that does not detect a receive signal generates beacon MAC frames and continues to do so until a receive signal is restored. A beacon MAC frame indicates the station's nearest active upstream neighbor (NAUN). Because each station knows the address of its nearest upstream neighbor, it can help identify the elements within the failure domain.

Fiber Distributed Data Interface (FDDI)   Fiber Distributed Data Interface (FDDI) has a data rate of 100 Mbps and 200 Mbps if both rings (which you will read about shortly) are used for data. Originally, FDDI networks used fiber-optic cable, but they can be run on UTP as well (referred to as "CDDI," where the *c* stands for copper). Fiber optic is still common in many FDDI networks because it can be used over much greater distances than UTP cable. FDDI, like Token Ring, uses a token passing media access method. Like Token Ring, FDDI is usually configured in a collapsed ring, or physical star topology, and is used primarily as a backbone.

The FDDI Standard   FDDI uses not only a counter-rotating dual ring topology that can act as built-in redundancy in the event of certain failures, but also uses sophisticated encoding techniques that help ensure data integrity. The FDDI standard specifies that the total length of the fiber-optic cabling used to connect the nodes may not exceed 200 km, or 100 km per ring. An unbroken FDDI network can run to 100 km with nodes up to 2 km apart on multimode fiber, and 20 km apart on single-mode fiber. Up to 500 nodes can exist on any single ring. If the primary ring fails and the FDDI is designed to circumvent failures, the total ring must not exceed 200 km, and the number of nodes must not exceed 1000. When the secondary ring exists purely for backup, the FDDI design should adhere to the 500-node and 100-km maximums. Figure 2-13 shows FDDI in relationship to the OSI model.
FDDI includes the following standards:

▼   **Physical Layer Medium Dependent (PMD)**   The PMD standard defines the physical characteristics of the media interface connectors and the cabling, as well as the services necessary for transmitting signals between nodes.

■   **Physical Layer Protocol (PHY)**   The PHY standard defines the rules for encoding and framing data for transmission, clocking requirements, and line states.

■   **Media Access Control (MAC)**   The MAC standard defines the FDDI timed-token protocol, frame and token construction and transmission on the FDDI ring, ring initialization, and fault isolation.

▲   **Station Management (SMT)**   The SMT standard defines the protocols for managing the PMD, the PHY, and the MAC components of FDDI. The SMT protocols monitor and control the activity of each node on the ring.

FDDI Frame Format   Figure 2-14 shows the FDDI frames.

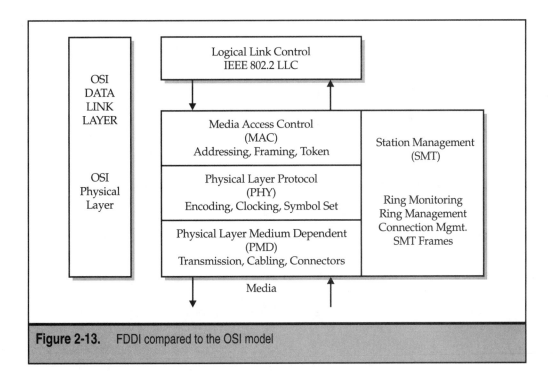

Figure 2-13.   FDDI compared to the OSI model

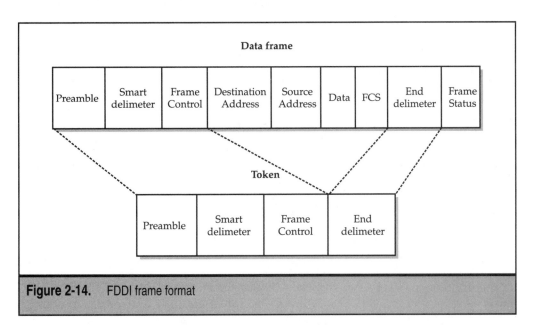

Figure 2-14.   FDDI frame format

FDDI frames contain the following fields:

▼ **Preamble**   This field is a unique sequence that prepares each station for an upcoming frame.

■ **Start Delimiter**   This field indicates the beginning of a frame by employing a signaling pattern that differentiates it from the rest of the frame.

■ **Frame Control**   This field indicates the size of the address fields, specifies whether the frame contains asynchronous or synchronous data, and provides other control information.

■ **Destination Address**   This field contains a unicast (singular), multicast (group), or broadcast (every station) address. As with Ethernet and Token Ring addresses, FDDI destination addresses are 6 bytes long.

■ **Source Address**   This field identifies the single station that sent the frame. As with Ethernet and Token Ring addresses, FDDI source addresses are 6 bytes long.

■ **Data**   This field contains either information destined for an upper-layer protocol or control information.

■ **Frame Check Sequence (FCS)**   This field is filled by the source station with a calculated *cyclic redundancy check* value, which depends on the frame contents, such as Token Ring or Ethernet. The destination address recalculates the value to determine whether the frame was damaged in transit. If so, the frame is discarded.

■ **End Delimiter**   This field contains unique symbols, which cannot be data symbols, that indicate the end of the frame.

▲ **Frame Status**   This field allows the source station to determine whether an error occurred and whether the frame was recognized and copied by a receiving station.

FDDI Dual Counter-Rotating Ring Architecture

FDDI architecture consists of two independent, counter-rotating rings: a *primary* ring and a *secondary* ring. Data flows on each of the rings but in opposite directions. The primary ring transmits data, and the secondary ring is used as a self-healing backup device. In some implementations, both rings are used to carry data, but when they do so, redundancy is lost because the secondary ring is utilized as well, leaving no backup ring.

The way the counter-rotating ring architecture prevents data loss in the event of a link failure, a node failure, or the failure of both the primary and secondary links between any two nodes is as follows:

▼ If a link on the primary ring fails, the secondary ring transmits the data.

▲ If a node or corresponding links on both the primary and secondary rings fail, one ring "wraps" to the other around the faulty components, forming a single ring.

As soon as the component is repaired or starts to function again, the architecture automatically reverts to the original dual-ring state.

FDDI Ring Operation

An FDDI ring consists of nodes arranged in a logical ring architecture, as shown in Figure 2-15. There are two classes of nodes: *stations* (nodes with no master ports) and *concentrators* (nodes with master ports).

The FDDI standards define two types of stations:

▼ **Single Attachment Stations (SAS)**   An SAS connects to only one ring. This means that it cannot wrap the ring if a fault occurs.

▲ **Dual Attachment Stations (DAS)**   A DAS connects to both the primary and secondary rings and does provide for fault tolerance.

The FDDI standards define three types of concentrators:

▼ **Dual Attached Concentrators (DAC)**   DACs provide a reliable connection between stations and the backbone.

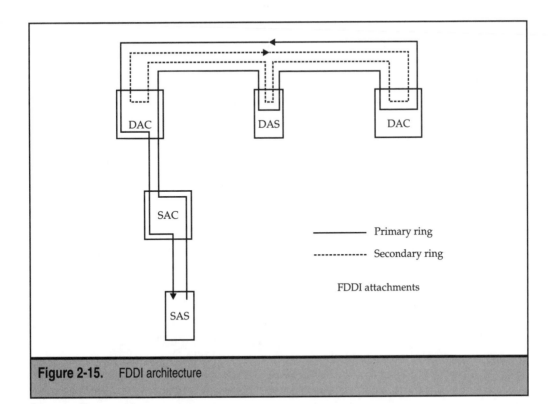

Figure 2-15.   FDDI architecture

- **Single Attached Concentrators (SAC)** SACs are less reliable because they provide only a single connection to the backbone.

▲ **Null Attached Concentrators (NAC)** NACs are not connected to any backbone.

FDDI initializes the ring and transmits data as follows:

1. The nodes on the ring establish connections with their neighbors. The Connection Management (CMT) portion of SMT (discussed in the section, "FDDI Station Management (SMT)") controls this process.

2. Using the claim token process, the nodes negotiate a Target Token Rotation Time (TTRT). The TTRT is the value that the MAC sublayer uses to time its operations. The claim token process determines which node initializes the ring (generates the token). The node with the lowest bid for the TTRT wins the right to generate the token.

3. After a node has initialized the ring, the ring begins to operate in steady state. In steady state, the nodes exchange frames using the Timed-Token Protocol (TTP). The TTP defines how the TTRT is set, the length of time a node can hold the token, and how a node initializes the ring. The ring remains in steady state until a new claim token process occurs, such as when a new node joins the ring.

4. The nodes pass the token from one node to another on the FDDI ring.

5. When a node on the ring wants to transmit data, it captures the token and then transmits the data to its downstream neighbors.

6. Each node reads and repeats frames as it receives them. If a node detects an error in a frame, the node sets an error indicator.

7. A frame circulates on the ring until it reaches the node that first transmitted it. That node removes the frame from the ring.

8. When the first node has sent all its frames, or exceeded the available transmission time, it releases the token back to the ring, and the process starts all over again.

Station Timers

Each node uses three timers to regulate its operation in the ring:

▼ **Token Rotation Timer (TRT)** The TRT measures the period between the receipt of tokens. TRT is set to varying values, depending on the state of the ring. During steady-state operation, the TRT expires when the token rotation time exceeds the TTRT.

- **Token Holding Timer (THT)** The THT controls the length of time that a node can hold the token to transmit frames. The value of the THT is the difference between the arrival time of the token and the TTRT.

▲   **Valid Transmission Timer (TVX)**   The TVX measures the period between valid transmissions on the ring. When the node receives a valid frame or token, the TVX is reset. If the TVX expires, the node starts a ring initialization sequence to restore the ring to proper operation.

FDDI Station Management (SMT)

SMT is a low-level protocol that manages the FDDI functions provided by the PMD, the PHY, and the MAC. SMT can run only on a single FDDI ring managing only the FDDI components and functions within a node.

SMT contains three components: Connection Management (CMT), Ring Management (RMT), and SMT frame services.

Connection Management (CMT) performs the following tasks:

▼   Inserts and removes stations at the PHY level.

■   Connects PHYs and MACs with a node.

■   Uses trace diagnostics to identify and isolate faulty components.

▲   Manages the physical connection between adjacent nodes. This involves testing the quality of the link before establishing a connection, establishing connections, and monitoring link errors continuously when the ring is operational.

Ring Management (RMT) receives status information from the MAC and CMT, and then reports this information to SMT and higher-level processes like SNMP. It detects stuck beacons and duplicate addresses (which prevent proper ring operation) and determines when the MAC is available for transmitting frames.

SMT frame services manage and control the FDDI network and the nodes on the network. Different SMT frame classes and types implement these services. *Frame class* identifies the function that the frame performs. *Frame type* specifies whether the frame is an announcement, a request, or a response to a request. FDDI SMT frames are limited to a single FDDI ring. This means that the frames cannot move across WANs or multiple FDDI rings. The frames do not manage functions outside FDDI.

FDDI Ring Maintenance

Each node is responsible for monitoring the integrity of the ring. By using the TVX, nodes can detect a break in ring activity. If the interval between token receptions exceeds the value of the TVX, the node reports an error condition and initiates the claim process to restore ring operation.

If it cannot generate a token, the node that detected the problem initiates *beacon* frames. Beacon frames indicate to the other nodes that the ring is broken. If the beacon transmission exceeds the value set in the stuck beacon timer—which is controlled by the Ring Management portion of SMT—then the RMT will attempt to restore the ring to normal operation. If the ring does not return to normal operation in a specified period, the RMT will initiate a *trace*. A trace is a diagnostic function that isolates a fault on the ring.

The trace is sent to the upstream neighbor on the secondary ring. The upstream neighbor acknowledges the trace, and both stations leave the ring to test the path between them. If the test passes, the stations rejoin the ring. If one station fails, it remains off the ring while the other station ring wraps.

WIDE AREA NETWORKS (WANS)

LANs are essentially infrastructures located on the premises of the organization whether it is a campus, building, or high-rise tower. That means that cabling, equipment, and infrastructure are owned and controlled by that organization's administrators. Those devices can be owned by either the organization or the public carrier. When it is owned by the organization, it is often referred to as "CPE," which stands for "customer premise equipment."

So basically, a WAN is a data communications network that connects other LANs and covers a relatively broad geographic area using public transmission facilities provided by one or more common carriers, such as telephone companies. LANs tend to see WANs as one big "cloud" between it and the other LAN in question. That is why WANs tend to be considered point-to-point topologies. Take a look at Figure 2-16 to see a WAN from the LANs perspective.

The fact is that inside that "cloud," the WAN involves the connection of multiple public WANs communicating with each other over many kinds of protocols common to the telecommunications industry.

WAN Technology Basics

Before going into the protocols and technologies used in WANs, let's go over some of the equipment used in WANs.

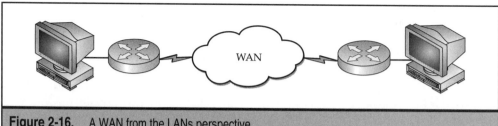

Figure 2-16. A WAN from the LANs perspective

WAN Equipment

WANs use numerous types of devices that are specific to WAN environments. WAN switches, modems, access servers, CSU/DSUs, DTE and DCE devices, and Integrated Services Digital Network (ISDN) terminal adapters are defined as follows:

▼ **WAN switch**   A WAN switch is a multiport internetworking device used in carrier networks. These devices typically switch traffic such as Frame Relay, X.25, and Switched Multi-megabit Data Service (SMDS) and operate at the Data Link layer of the OSI reference model.

■ **Modem**   A modem is a device that interprets digital and analog signals, enabling data to be transmitted over voice-grade telephone lines. At the source, digital signals are converted to a form suitable for transmission over analog communication facilities. At the destination, these analog signals are returned to their digital form.

■ **Access server**   An access server acts as a concentration point for dial-in and dial-out connections.

■ **Channel Service Unit/Digital Service Unit (CSU/DSU)**   This is a digital-interface device (sometimes two separate digital devices) that adapts the physical interface on a Data Terminal Equipment (DTE) device to the interface of a Data Circuit-terminating Equipment (DCE) device in a switched-carrier network. The CSU/DSU also provides signal timing for communication between these devices.

■ **ISDN terminal adapter**   This device is used to connect ISDN Basic Rate Interface (BRI) connections to other interfaces, such as the serial port of a computer. A terminal adapter is essentially an ISDN modem.

■ **Data Terminal Equipment (DTE)**   This equipment is designed to communicate with the transmission method used on the internal network. It is used in combination with DCE for conversion to the transmission method of the public carrier. Computers, printers, and terminals are examples of DTE.

▲ **Data Communication Equipment, or Data Circuit-terminating Equipment (DCE)**   This equipment is designed to communicate with the public-carrier transmission method. It is used in combination with a DTE to convert the transmission method used on the internal network. A modem is an example of DCE.

Connectionless and Connection-Oriented Communication Concepts

In a network, not every computer has to be directly connected to the others. Hosts should be able to forward data packets according to an address field that has been attached to the packet. The Network layer provides the service of routing and delivering data packets to any host in the network. Once the packets leave the LAN, they travel throughout the WAN through intermediate systems that will only process the packets based on information in the first three layers of the OSI model. More specifically, the Data Link layer (Layer 2) and the Network Layer (Layer 3). These two layers can operate in two different modes: connection-oriented services or connectionless services.

Connection-Oriented Services   A connection-oriented service is a service that guarantees delivery of information to that service for protocols like FTP or HTTP. A guaranteed service provides reliability, ensures segments are delivered and reassembled in sequence, and is error free. When data cannot be sent reliably or in sequence, an error is sent to the user's Application layer. Connection-oriented protocols will use TCP to establish a connection to a destination before any form of data is transferred.

A telephone service is a good example of a connection-oriented service. Before the caller can start a conversation, the call setup and data transfer phases must be completed. After those stages have been completed, the caller can begin talking. When the caller finishes the conversation, the call termination phase takes place. Each of these phases in a telephone call is characteristic of connection-oriented services. Connection-oriented services consist of the following phases:

▼ Connection establishment or setup phase

■ Transfer of data

▲ Connection release or teardown phase

These connections generally take one of the following forms. (These connections will be made clearer when we discuss the protocols and technologies used in WANs later in the chapter.)

▼ Physical or actual circuit connections

■ Virtual circuit connections

▲ Logical circuit connections

Almost all WAN protocols and technologies tend to use connection-oriented services. You do have a lot of overhead and management of data when compared to connectionless services. I mean what costs more, phone calls or stamps?

Connectionless Services   A connectionless service, on the other hand, packages data and sends it off without specifying the path the data will take on the way or knowing whether it will arrive. In connectionless services, the units of transfer are usually called *packets* or *datagrams*. A good analogy for a connectionless service would be a postal service. To mail a letter, the sender takes an envelope (packet), writes the name and address of the person for whom it is intended, adds postage, and sticks it in the mailbox. The postal service does not guarantee what path the letter will take to get to its destination or that the letters will arrive in the same order in which they were sent. Also, nobody from the post office will call you and let you know that your letter arrived safe and sound, let alone was read.

In connectionless mode systems, higher-layer protocols are relied upon to handle resequencing and retransmission of lost packets, and to guarantee reliable data transfer. That's the job of the Transport layer. Connection-oriented transport protocols hide the complexities of several networking concepts such as quality of service, error correction, flow control, and transferring packets that are larger than the lower layer protocols being used on the network can handle. (This is known as *segmenting,* where the data is separated into packages that will fit into the lower-layer protocol's specifications carrying the data, and numbered so that they can be reassembled in proper order upon arrival.)

Almost all LAN technologies are based upon connectionless protocols, which creates advantages and disadvantages.

Connectionless services provide the following advantages:

▼ Packet-switched networks provide high performance.

■ Call establishment is not required.

▲ Management and overhead is lower.

Connectionless protocols have the following disadvantages:

▼ Delivery of packets is not guaranteed.

■ Higher-level protocols must be relied upon to perform end-to-end circuit data management.

■ Flow control is minimal.

■ Error detection or correction is frequently not provided.

■ Bandwidth efficiency is lower.

▲ Message sequencing is required.

WAN Switching Technologies

When data is sent over long distances, the information must traverse several different communications infrastructures. No single entity owns or controls these infrastructures, which are based on different kinds of systems with varying speeds and capacities. To send this data, a series of links must be established from the local station to the destination station. To create these links, a process called *switching* must occur.

Switching, in this case, means selecting a method and a pathway that will get information to its destination. This may involve *message* switching (selecting the best paths between machines that store messages on the way). It could also mean *circuit* switching (creating circuits from one point to the other, like the phone company does). Or, it might involve *packet* switching (breaking up data messages into small units, such as cells or packets, sending each one as its own entity, and directing each packet down the best path). I discuss these methods in the following sections.

Message Switching

In message switching (also called store-and-forward), the message is sent in its entirety and routed through multiple switches, where it can either be stored until a better time presents itself (such as when there is less traffic) or it can be sent. It can even be forwarded ahead of other messages depending on its priority level. This process, shown in Figure 2-17, continues until the complete message arrives at its intended destination fully intact.

Mail service is a variation of message switching. A letter follows several paths that are either most efficient or most economical depending on the postage (message priority). When mailing a letter, the sender addresses the envelope, adds postage, and places it into a mailbox. The mail carrier transports the letter from the mailbox to the post office. The letter is kept (stored) at the local post office until it is processed. Once the letter is processed locally, decisions are made as to the most efficient or economical way of getting the letter to the post office nearest its destination. The letter may go by air or truck to a receiving area that holds (stores) the letter until a postal service representative picks it up and takes (forwards) it to the remote post office. Once there, it is held until it is processed and the proper delivery route is selected. Then the mail carrier places the letter in a mailbag and delivers it to the address written on the envelope, placing the letter in the mailbox until the recipient checks it. In message switching, the complete message (just like regular mail) is sent in its entirety, routed through multiple switches where it can either be stored until a better time presents itself (such as less traffic). It could even be forwarded ahead of other message received at these intermediate switches based on information on the message address indicating a higher priority. This continues until the complete message arrives at its intended destination fully intact. Take a look at Figure 2-17.

One of the best examples of message switching technology is e-mail. In most e-mail networks, users send an e-mail from their computer to the local mail server, where it is stored on the hard disk. Because many e-mail messages are not urgent, many e-mail serv-

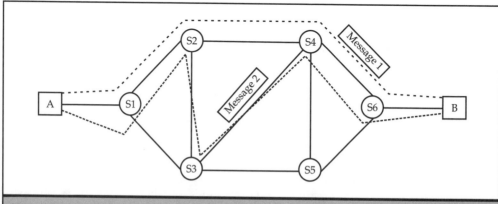

Figure 2-17.   Message switching or "store-and-forward" data path

ers only connect intermittently. E-mail servers examine the message and decide when to send it based on various priority codes embedded in the message.

The message itself contains all the information as to where it is going. No dedicated pathway is required to get the message to its destination. The intermediate message switching equipment along the way can temporarily store the message and select the next best route to send it on, based on any number of criteria such as cost, congestion, and time, and it can maintain a record of the pathways used. The message is stored as many times as is necessary to forward it to the right place.

These message switching devices are not much different from other WAN switches with the exception that they generally require very large hard drives, which can be expensive and slow, to accommodate message storage. Message switches also have the capability of sending multiple copies of a message so that it can be duplicated for broadcast to other nodes.

The downside of message switching, with regard to WANs, is that the storing and forwarding process is slow: to store a whole message, select a route, and then retransmit the message takes time. Because this mechanism is unsuitable for real-time applications such as voice, video, and other multimedia, it is not very common these days.

Circuit Switching

Circuit switching involves the creation of an actual physical communications channel. This physical path for data flow between the sender and receiver is set up through the network all the way from the sending device to the receiving device. This physical path, shown in Figure 2-18, is called a *circuit*.

During this time, the connection is used exclusively by both entities for the duration of the communication, providing a fixed data rate channel, where both users operate at the same rate. The same method is used when making a telephone call to someone using the phone system.

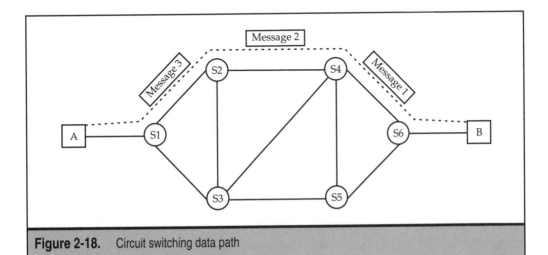

Figure 2-18.   Circuit switching data path

The advantages associated with the physical pathway created by circuit switching offer reliability of transfer, consistent bandwidth, security, and create an environment in which there is no chance of other devices contending for the path.

The disadvantage of circuit switching is the massive overhead required to create the physical pathway. It takes time to put all the links in place to complete the circuit. And though the circuit offers dedicated bandwidth to the devices while they are actively sending and receiving data, it wastes tremendous amounts of bandwidth when the channel becomes idle.

Integrated Services Digital Network (ISDN) and leased-line services such as T1 and T3 are examples of WAN technologies that use circuit switching.

Packet Switching

Packet switching involves the breaking up of messages into smaller components called *packets*. Packets range in size from about 600 bytes to over 4000 bytes depending on the system involved. Each packet contains source and destination information and is treated as an individual message. These mini-messages are received and routed through optimal pathways by various switches on a WAN, as shown in Figure 2-19.

There are two major types of packet switching: *datagram* packet switching and *virtual circuit* packet switching. We'll start by examining datagram packet switching.

Datagram Packet Switching   The first type of packet is called a datagram. Datagrams are sent onto the network individually on the way to its destination. There is no guarantee that all of the packets will arrive or that the message will remain intact. This is because when packet switches receive these datagrams, there may be more than one route to choose from when the forwarding decision is made. Each individual packet is directed down what seems to be the optimal path at that point in time, based on the criteria of whatever routing protocol is being run on the switch. Over time the quality of these pathways can become better or worse depending on current congestion levels or whether the

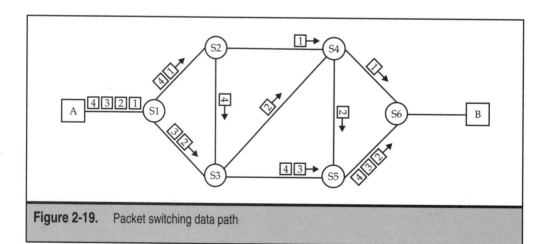

Figure 2-19.   Packet switching data path

links are operating at all. So a datagram from a message may end up taking any number of different pathways on the way to its final destination. It doesn't really matter as long the packets are kept in the right order when the message is reassembled. If the any of the packets are out of order, the message will be altered.

To ensure that packets are placed in the right order as they are received, packet-switched networks incorporate a special device known as a Packet Assembler Disassembler (PAD). How can the PAD determine what the right order is? It is really very simple. Every datagram header in the message has a sequence number that designates its position. The PAD simply looks at that number to reassemble the packets in the correct order. In fact, the PAD on the sending station is responsible for taking the messages that are being sent, breaking them up into packets, and then assigning sequence numbers to each packet.

Datagram packet switching does not allow the sender and receiver to establish data transmission details such as the packet size that will be used during the transmission or to ensure that all packets are acknowledged once they are received. However, virtual circuit packet switching does provide these capabilities.

Virtual Circuit Packet Switching   As a datagram makes its way toward its destination, the network makes decisions on the fly concerning the pathway each packet will take. To improve reliability, a decision concerning the best pathway to a destination could be made prior to any data being sent. In this manner, a single, static path could be set up between two communicating parties and used exclusively for communicating with one another. This pathway is known as a *virtual circuit*.

The idea behind virtual circuits is to increase the reliability of packet transmission. When creating a virtual circuit, the sender and receiver agree on which path will be used and on packet size. Then, during the process of communicating, acknowledgments are sent from receiver to sender to verify receipt of the packets. Typically, the two communicating entities trade information about errors and transfer speed. These two factors are known as error control and flow control. Virtual circuits can be set up for use during several communication sessions or on a session-by-session basis.

The whole purpose behind virtual circuits is reliability. Though virtual circuits create overhead for communication, they are necessary to ensure that data travels safely from senders to receivers. This is especially important for critical applications.

In comparing datagram and virtual circuit packet switching with other switching technologies, you must consider several factors. First, packet switching is faster because messages are not stored in their entirety for later retrieval. Each packet is small enough to be stored in a routing machine's memory until it can be routed an instant later. Second, packet switching allows the avoidance of pathway failure due to excessive traffic loads or mechanical problems. This is accomplished by routing the packets along pathways based on congestion and integrity. Finally, packet switching allows networks to use pathways that may not ordinarily get much traffic. Instead of concentrating use on a few paths that are always busy, packet switching spreads the load of communication across several paths.

Packet switching requires sophisticated machinery capable of making intelligent pathway choices. Virtual circuit packet switching methods require that packets be held while pathway choices are being made; this takes time and adds overhead. Given that fact, the very nature of temporary pathways will be less reliable than transmitting data along a fixed physical link.

Examples of protocols that use packet switching technologies are Asynchronous Transfer Mode (ATM), Frame Relay, SMDS, and X.25.

WAN Protocols and Technologies

LAN protocols operate at the lowest two layers of the OSI model, whereas WAN protocols tend to function at the lowest three layers of the OSI reference model, as shown in Figure 2-20.

In this section, we will discuss some of these protocols and technologies.

X.25

X.25 is a connection-oriented packet switching protocol that uses the telephone network for data communications. It was designed in 1964 to allow computers to send data over analog telephone services, which at the time were unreliable because they were not designed for that purpose. X.25 provides extensive error checking and recovery techniques with buffering so that end stations can ensure that communication over the network is error free.

X.25 uses the telephone network for data communications. To begin communication, one computer calls another to request a communication session. The called computer can accept or refuse the connection. If the call is accepted, the two systems begin full-duplex

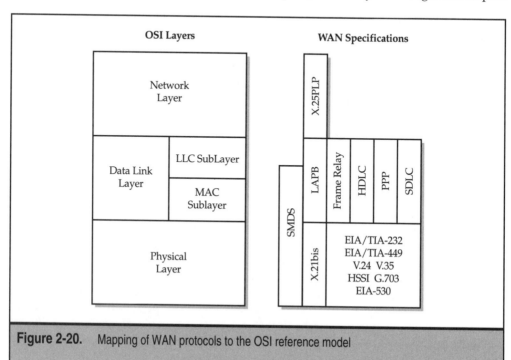

Figure 2-20.   Mapping of WAN protocols to the OSI reference model

information transfer. Either side can terminate the connection at any time. End-to-end communication between DTEs is accomplished through a bidirectional link called a *virtual circuit*. Virtual circuits permit communication between distinct network elements through any number of intermediate nodes—without the dedication of all portions of the physical medium that characterizes circuit switching technologies.

Virtual circuits can be either permanent or switched (temporary). Permanent virtual circuits (PVCs) are typically used for data transfers. Switched virtual circuits (SVCs) are used for sporadic data transfers. When a virtual circuit is established, the DTE sends a packet to the other end of the connection by sending it to the DCE using the proper virtual circuit. The DCE reads the virtual circuit number to determine how to route the packet through the X.25 network. The Layer 3 X.25 protocol multiplexes between all the DTEs served by the DCE on the destination side of the network, and the packet is delivered to the destination DTE.

X.25 operates at the lowest three layers of the OSI model:

▼ **Layer 1 protocols (Physical layer)**   X.25 uses X.21/X.21bis protocols to provide the handshaking and electrical requirements at the Physical layer. EIA/TIA-232/449/530 and G.703 can also be used. X.21bis is the equivalent of RS232C and RS449/442. Most routers support X.21, V.35, RS232C, and RS449/442.

■ **Layer 2 protocols (Data Link layer)**   At the Data Link layer, X.25 uses the Link Access Protocol Balanced (LAPB). This protocol provides link setup, control, sequencing, and error recovery. It also provides a windowing mechanism. LAPB is very robust; it can be applied directly to serial links and is often used on satellite links. Private X.25 networks can use LAPB, but when connecting to public X.25 networks, only X.25 encapsulation can be used.

▲ **Layer 3 protocols (Network layer)**   Layer 3 is concerned with end-to-end communication involving both PVCs and SVCs. X.25 uses the Packet Layer Protocol (PLP) at the Network layer. This allows up to 4095 virtual circuits over one physical interface. Regular Network layer protocols are tunneled inside of PLP packets. Network layer addresses are mapped onto the X.121 addresses. This is connection-oriented service and forces sequencing. X.75 is used to link different X.25 networks from other countries.

Frame Relay

Frame Relay is an example of a packet-switched connection-oriented (usually) technology. It is a high-performance WAN protocol that operates at the Data Link layer of the OSI reference model.

Frame Relay is essentially a streamlined version of X.25. The main difference is that it doesn't have the overhead associated with X.25's error checking properties. Frame Relay includes a means of detecting corrupted transmissions through a cyclic redundancy check (CRC), which can detect whether any bits in the transmission have changed between the source and destination. But it does not include error correction capabilities. Error correction and retransmission is left up to the higher-layer protocols running on the end systems. This is because Frame Relay typically operates over WAN facilities that offer more reliable connection services and a higher degree of reliability than the platforms for X.25 WANs during the late 1970s and early 1980s.

Frame Relay is strictly a Layer 2 (Data Link layer) protocol suite, whereas X.25 provides services at Layer 3 (the Network layer) as well. Frame Relay was designed for use on modern digital and fiber technologies, so it has little need for accounting and error checking. Frame Relay is a newer, faster, and less cumbersome form of packet switching than X.25.

Devices on a Frame Relay network do not have to repackage or reassemble frames during transmission. This enables Frame Relay to offer higher performance and greater transmission efficiency than X.25 and makes Frame Relay suitable for current WAN applications. Packet-switched networks enable end stations to dynamically share the network medium and the available bandwidth. In Frame Relay, variable-length packets, called frames, are used for more efficient and flexible transfers. Frames can be as large as 4KB in size. Frame Relay can operate at speeds from 56 Kbps to 1.544 Mbps.

Frame Relay uses a statistical method of multiplexing the packets onto the network, making it even more efficient.

Statistical Time Division Multiplexing   Frame Relay uses Statistical Time Division Multiplexing. (Do not confuse this with STDM, which stands for "Synchronous Time Division Multiplexing" and is also called TDM. If anything, Statistical Time Division Multiplexing would be referred to as "Asynchronous TDM.") This makes Frame Relay technology much more efficient than T1 and T3 carriers. The best way to explain Statistical Time Division Multiplexing is to compare it to a long freight train. The boxcars of this freight train are painted in groups of four all the way to the end in the colors red, green, blue, and purple. Each boxcar can carry a maximum of three sacks of grain at a time, and four people ship sacks of grain on this railroad.

In normal TDM, a different color would be assigned to each person. One shipper has red designated for his shipments, another shipper has green assigned, and so on. Assume that the red shipper has three sacks of grain to ship, and the other three shippers have 30 sacks. This scenario is inefficient: after the first set of four boxcars goes by with three sacks loaded on each of them, the red car in the following groups of boxcars would be empty.

In Statistical TDM, the other shippers can take turns using the red boxcar after the red shipper is finished. They can even do clever slants on this arrangement where the purple shipper convinces the others that his shipment is "really" important and that after the red shipper has finished, the red boxcars can be used for the purple shipper's remaining sacks of grain. Then, once the purple shipper has shipped all of his sacks, the remaining two shippers can alternate turns. Consider the description of that last slant an idea of what Quality of Service (QoS) is all about.

Frame Relay Devices   Two types of devices are used in a Frame Relay network: Data Terminal Equipment (DTE) and Data Circuit-terminating Equipment (DCE). DTEs are usually reserved for a specific network and are typically owned by and located on the premises of the customer. Examples of DTE devices are terminals, personal computers, routers, and bridges.

DCEs are usually carrier-owned internetworking devices—usually packet switches. DCEs provide clocking and switching services in a network to transmit data through the WAN, as shown in Figure 2-21.

Frame Relay SVCs, PVCs, and DLCIs   As explained earlier, Frame Relay provides connection-oriented Data Link layer communication. This means that a defined communication exists between each pair of devices and that these connections are associated with a connection identifier. This service is implemented by using a *Frame Relay virtual circuit*, which is a logical connection created between two DTE devices across a Frame Relay Packet-Switched Network (PSN).

Virtual circuits provide a two-way communications path from one DTE device to another. A number of virtual circuits can be multiplexed into a single physical circuit for transmission across the network. This capability can reduce the equipment and network complexity required to connect multiple DTE devices.

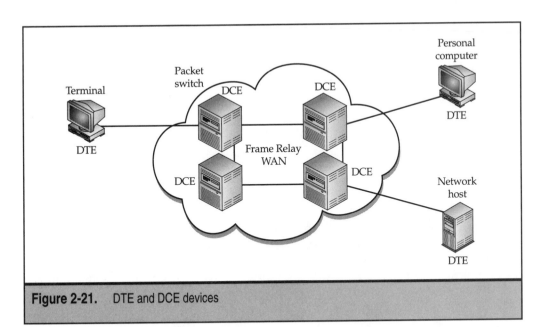

Figure 2-21.   DTE and DCE devices

A virtual circuit can pass through any number of intermediate DCE devices (switches) located within the Frame Relay PSN. As with virtual circuits in an X.25 network, there are two kinds of virtual circuits: switched virtual circuits and permanent virtual circuits.

- **Switched virtual circuits (SVCs)**   SVCs are temporary connections used in situations requiring only sporadic data transfer between DTE devices across the Frame Relay network. A communication session across an SVC consists of four operational states:

  - **Call Setup**   The virtual circuit between two Frame Relay DTE devices is established.

  - **Data Transfer**   Data is transmitted between the DTE devices over the virtual circuit.

  - **Idle**   The connection between DTE devices is still active, but no data is transferred. If an SVC remains in an idle state for a defined period, the call can be terminated.

  - **Call Termination**   The virtual circuit between DTE devices is terminated.

After the virtual circuit is terminated, the DTE devices must establish a new SVC if there is additional data to be exchanged. It is expected that SVCs will be established, maintained, and terminated using the same signaling protocols used in ISDN. Few manufacturers of Frame Relay DCE equipment, however, support switched virtual connections. Therefore, their actual deployment is minimal in today's Frame Relay networks.

- **Permanent virtual circuits (PVCs)**   Permanent virtual circuits (PVCs) are permanently established connections used for frequent and consistent data transfers between DTE devices across the Frame Relay network. Communication across a PVC does not require the call setup and termination states that are used with SVCs. PVCs always operate in one of the following two operational states:

  - **Data Transfer**   Data is transmitted between the DTE devices over the virtual circuit.

  - **Idle**   The connection between DTE devices is active, but no data is transferred.

Unlike SVCs, PVCs will not be terminated under any circumstances due to being in an idle state. DTE devices can begin transferring data whenever they are ready because the circuit is permanently established.

Frame Relay virtual circuits are identified by Data Link Connection Identifiers (DLCIs). DLCI values are usually assigned by the Frame Relay service provider. Frame Relay DLCIs have local significance, which means that the values themselves are not unique in the Frame Relay WAN. Two DTE devices connected by a virtual circuit, for example, may use a different DLCI value to refer to the same connection. Figure 2-22 shows how a single virtual circuit may be assigned a different DLCI value on each end of the connection.

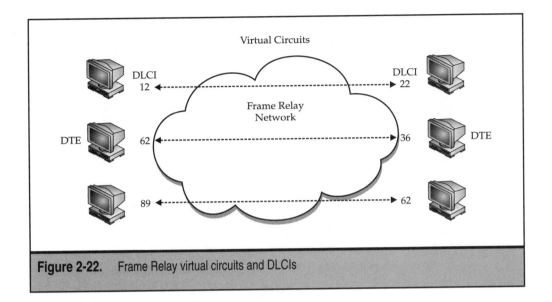

Figure 2-22.   Frame Relay virtual circuits and DLCIs

Congestion Notification Mechanisms   Frame Relay reduces network overhead by implementing simple congestion notification mechanisms rather than explicit, per-virtual-circuit flow control. Because Frame Relay typically is implemented on reliable network media and flow control can be left to high-layer protocols, data integrity is not sacrificed. Frame Relay uses two congestion notification mechanisms that are controlled by a single bit contained in the Frame Relay header:

▼   Forward—explicit congestion notification (FECN)

▲   Backward—explicit congestion notification (BECN)

The FECN bit is part of the Address field in the Frame Relay frame header. The FECN mechanism is initiated when a DTE device sends Frame Relay frames into the network. If the network is congested, DCE devices (packet switches) set the value of the frame's FECN bit to 1. When the frames reach their destination, the receiving DTE device will know that the frames experienced congestion in the path from source to destination. The DTE device can relay this information to a higher-layer protocol for processing. Depending on the implementation, flow control may be initiated, or the indication may be ignored.

The BECN bit is also part of the Address field in the Frame Relay frame header. DCE devices set the value of the BECN bit to 1 in frames traveling in the opposite direction of frames with their FECN bit set. This informs the receiving DTE device that a particular path through the network is congested. The DTE device can then relay this information to a higher-layer protocol for processing. Depending on the implementation, flow control may be initiated, or the indication may be ignored.

Frame Relay Discard Eligibility   The Address field in the Frame Relay frame header also contains a *Discard Eligibility* (DE) bit, which is used to identify less important traffic that can be dropped during periods of congestion. DTE devices can set the value of the DE bit of a frame to 1 to indicate that the frame has lower importance than other frames. When the network becomes congested, DCE devices will discard frames with the DE bit set to 1 before discarding those that are not set. This reduces the likelihood of critical data being dropped by Frame Relay DCE devices during periods of congestion.

Frame Relay Frame Format   The following descriptions summarize the basic Frame Relay frame fields illustrated in Figure 2-23.

▼ **Flags**   Flags delimit the beginning and end of the frame. The value of the Flag field is always the same and is represented either as the hexadecimal number 7E or the binary number 01111110.

■ **Address**   The Address field contains the following information:

   ■ *DLCI*   The 10-bit DLCI is the essence of the Frame Relay header. This value represents the virtual connection between the DTE device and the switch. Each virtual connection that is multiplexed onto the physical channel will be represented by a unique DLCI. The DLCI values have local significance only, which means that they are unique only to the physical channel on which they reside. Therefore, devices at opposite ends of a connection can use different DLCI values to refer to the same virtual connection.

   ■ *C/R*   The Command/Response is the bit that follows the most significant DLCI byte in the Address field. The C/R bit is not currently defined.

   ■ *Extended Address (EA)*   The EA field is used to indicate whether the byte in which the EA value is 1 is the last addressing field. If the value is 1, the current byte is determined to be the last DLCI octet. (Bytes each have eight bits and are sometimes called octets.) Although all current Frame Relay implementations use a two-octet DLCI, the EA capability allows longer DLCIs to be used in the future. The eighth bit of each byte of the Address field is used to indicate the EA.

   ■ *Congestion Control*   These last three bits in the Address Field control the Frame Relay congestion notification mechanisms. They are the forward-explicit congestion notification (FECN), backward-explicit congestion notification (BECN), and discard eligibility (DE) bits, which were discussed in the preceding sections, "Congestion Notification Mechanisms" and "Frame Relay Discard Eligibility."

■ **Information**   This field contains encapsulated upper-layer data. Each frame in this variable-length field includes a user data or payload field of up to 16,000 octets. This field serves to transport the higher-layer protocol packet (PDU) through a Frame Relay network.

▲ **Frame Check Sequence (FCS)**   This value is computed by the source device and verified by the receiver to ensure the integrity of transmitted data.

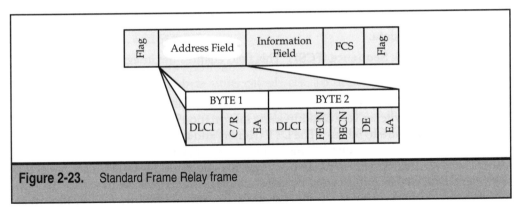

Figure 2-23. Standard Frame Relay frame

Local Management Interface (LMI) A major development in Frame Relay technology occurred in 1990 when Cisco Systems, Digital Equipment, Northern Telecom, and StrataCom formed a consortium to focus on Frame Relay technology development. This consortium developed a specification that conformed to the basic Frame Relay protocol, but extended the protocol to provide additional capabilities for complex internetworking environments. These Frame Relay extensions are referred to collectively as the Local Management Interface (LMI).

Since publication of the consortium's specification, many vendors have announced their support of this extended Frame Relay definition. American National Standards Institute (ANSI) and Consultative Committee for International Telegraph and Telephone (CCITT) have subsequently standardized their own variations of the original LMI specification, and these standardized specifications now are more commonly used than the original version.

LMI Frame Format The following descriptions summarize the LMI frame fields illustrated in Figure 2-24.

▼ **Flag** Flags delimit the beginning and end of the frame.

■ **LMI DLCI** This field identifies the frame as an LMI frame instead of a basic Frame Relay frame. In the LMI consortium specification, DLCI = 1023.

■ **Unnumbered Information Indicator** This field sets the poll/final bit to zero.

■ **Protocol Discriminator** This field always contains a value indicating that the frame is an LMI frame.

■ **Call Reference** This field always contains zeros. It is not currently used.

■ **Message Type** This field labels the frame either as a *status inquiry message*, which allows a user device to inquire about the status of the network, or as a *status message*, which responds to status inquiry messages. Status messages include keep-alives and permanent virtual circuit (PVC) status messages.

■ **Information Elements (IEs)** This field contains a variable number of individual information elements. IEs consist of the following fields: *IE Identifier* fields uniquely identify the IE, and *IE Length* fields indicate the length of the IE.

An ATM endpoint (also called an end system) is the cell's final destination and contains the ATM network interface adapter. Computers, workstations, routers, and LAN switches are examples of ATM endpoints.

ATM Network Interfaces   ATM networks consist of a set of ATM switches interconnected by point-to-point ATM links, which are called *interfaces*. ATM switches support two primary types of interfaces:

▼   User-to-network interfaces (UNIs)

▲   Network-to-network interfaces (NNI)

The UNI connects ATM end systems, like computers and routers, to an ATM switch. The NNI connects ATM switches only.

Depending on whether the switch is owned and located at the customer's premises or publicly owned and operated by the telephone company, UNI and NNI can be further subdivided into public and private UNIs and NNIs. A private UNI connects an ATM endpoint and a private ATM switch. Its public counterpart connects an ATM endpoint or private switch to a public switch. A private NNI connects two ATM switches within the same private organization. A public NNI connects two ATM switches within the same public network managed by one company or organization. This can be seen in Figure 2-26.

ATM Cell Header Format   An ATM cell header can be in either UNI format or NNI format. The UNI header is used for communication between ATM endpoints and ATM switches in private ATM networks. The NNI header is used for communication between ATM switches. Figure 2-27 shows the basic ATM cell format, the ATM UNI cell header format, and the ATM NNI cell header format.

Unlike the UNI header, the NNI header does not include the Generic Flow Control (GFC) field. In addition, the NNI header has a longer Virtual Path Identifier (VPI) field

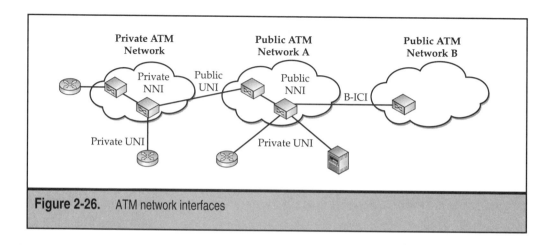

Figure 2-26.   ATM network interfaces

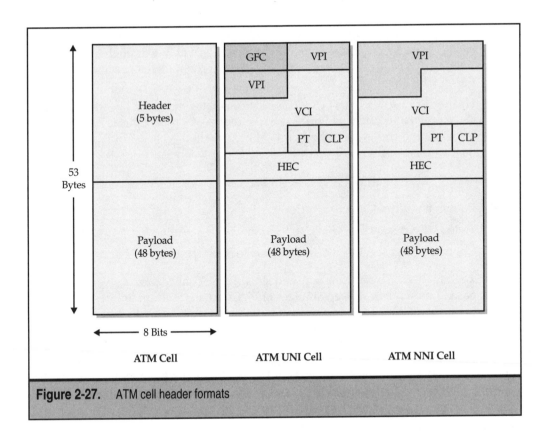

Figure 2-27. ATM cell header formats

that occupies the first 12 bits, allowing for larger trunks between public ATM switches. The following text describes the ATM cell header fields:

▼ **Generic Flow Control** (GFC)   This field provides local functions, such as identifying multiple stations that share a single ATM interface. This field is typically not used and is set to its default value.

■ **Virtual Path Identifier (VPI)**   In conjunction with the VCI, this field identifies the next destination of a cell as it passes through a series of ATM switches on the way to its final destination.

■ **Virtual Channel Identifier (VCI)**   In conjunction with the VPI, this field identifies the next destination of a cell as it passes through a series of ATM switches on the way to its final destination.

■ **Payload Type (PT)**   This field indicates in the first bit whether the cell contains user data or control data. If the cell contains user data, the second bit indicates congestion level, and the third bit indicates whether the cell is the last in a series of cells that represent a single AAL5 frame (AAL is the ATM Adaptation Layer and is defined in the ATM reference model section).

■ **Congestion Loss Priority (CLP)** This field indicates whether the cell should be discarded if it encounters extreme congestion as it moves through the network. If the CLP bit equals 1, the cell should be discarded in deference to cells where the CLP bit equals 0.

▲ **Header Error Control (HEC)** This field calculates checksum (an error checking algorithm that is calculated prior to transmission of a packet that is run again on the receiving end to see if the "sum" of the equation "checks" out) only on the header itself.

ATM Services There are three types of ATM services:

▼ Permanent virtual circuits (PVCs)

■ Switched virtual circuits (SVCs)

▲ Connectionless service (which is similar to SMDS)

PVCs allow direct connections between sites. These connections are very similar to typical leased line services. The advantages of PVCs are that they guarantee availability of a connection, and they do not require call setup procedures between switches. The disadvantages of PVCs include static connectivity and manual setup.

SVCs are created and released dynamically and remain in use only as long as data is being transferred. In this sense, they are similar to a telephone call. Dynamic call control requires a signaling protocol between the ATM endpoint and the ATM switch. The advantages of SVCs include connection flexibility and call setup that can be handled automatically by a networking device. Disadvantages include the extra time and overhead required to set up the connection.

ATM Virtual Connections ATM networks are fundamentally connection-oriented, which means that a *virtual channel* must be set up across the ATM network prior to any data transfer. A virtual channel is roughly equivalent to a virtual circuit.

Two types of ATM connections exist: *virtual paths* and *virtual channels*.

Virtual paths are identified by *virtual path identifiers* (VPIs). Virtual channels are identified by *virtual channel identifiers* (VCIs) in combination with the VPI.

A virtual path is a bundle of virtual channels, all of which are switched transparently across the ATM network on the basis of the common VPI. All VCIs and VPIs, however, have only local significance across a particular link and are remapped, as appropriate, at each switch.

ATM Reference Model The ATM architecture uses a logical model to describe the functionality it supports. ATM functionality corresponds to the Physical layer and part of the Data Link layer of the OSI reference model, as shown in Figure 2-28.

Multimedia includes various types of information that have different characteristics and are handled differently, both by the devices that work with them and by higher-level networking protocols. To use ATM, something must interface with the different devices and must package their different types of data into ATM cells for transport. That something is an ATM-capable node that handles the conversions specified in the ATM reference

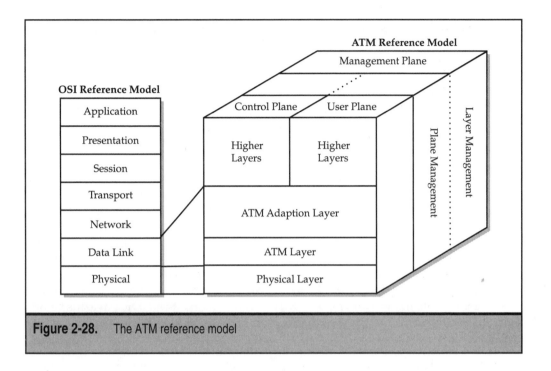

Figure 2-28.   The ATM reference model

model. The ATM reference model is composed of the following planes, which span all the layers of the ATM model:

▼ **Control**   This plane is responsible for generating and managing signaling requests.

■ **User**   This plane is responsible for managing the transfer of data.

▲ **Management**   This plane contains two components: *layer management*, which manages layer-specific functions, such as the detection of failures and protocol problems, and *plane management*, which manages and coordinates functions related to the complete system.

The ATM reference model is composed of the following ATM layers:

▼ **Physical layer**   This layer corresponds to the Physical layer of the OSI reference model. Like the OSI model, it is concerned with moving information. In this case, it moves the 53-byte ATM cells on to the communications medium. As already mentioned, this medium can be any of a number of different physical transports, including fiber-optic based SONET (Synchronous Optical Network), T carriers, and even modems. The medium and the message are clearly separable because ATM is a transport method that is independent of the transmission medium over which the messages travel.

- ■ **ATM layer** The ATM layer is responsible for establishing connections and passing cells through the ATM network. Combined with the ATM Adaptation layer, the ATM layer is roughly analogous to the Data Link layer of the OSI reference model. This layer attaches headers to the ATM payloads. To do this, it uses information in the header of each ATM cell, which includes information that identifies the paths and circuits over which the cell will travel. This allows the ATM switches to deliver cells accurately to their intended destinations. The ATM layer also multiplexes the cells for transmission before passing them down to the Physical layer.

- ▲ **ATM Adaptation layer (AAL)** The AAL is responsible for isolating higher-layer protocols from the details of the ATM processes. Combined with the ATM layer, the AAL is also roughly analogous to the Data Link layer of the OSI model. This layer sits between the ATM devices and the higher-level network devices and protocols that send and receive the different types of information over the ATM network. AAL, as the "adaptation" in its name suggests, mediates between the ATM layer and higher-level protocols, altering the services of one so that they fit the services of the other. The AAL takes in all the different forms of data (audio, video, data frames) and repackages the information into 48-byte payloads before sending them down to the ATM layer for further processing.

After ATM filters all this information down through the AAL, ATM, and Physical layers, the Physical layer sends the cells to their destinations over connections that might switch them from one circuit to another. Along the way, the ATM switches maintain connections that provide the network with the bandwidth necessary to provide the data with the Quality of Service (QoS) negotiated during transfer.

When the cells arrive at their destinations, they are sent up the Physical layer and then passed up to the ATM layer. The ATM layer forwards the cells to the appropriate services (voice, data, video, and so on) in the AAL, where the cell contents are converted back to their original form. After everything is checked to be sure that it arrived correctly, the original information is delivered to the destination device.

Switched Multi-Megabit Data Service (SMDS)

Switched Multi-megabit Data Service (SMDS) is a broadband public networking service offered by communications carriers as a means for businesses to connect LANs in separate locations. Unlike most WAN technologies, SMDS is a connectionless packet-switching technology. This means that the data can run in a mode that is "native" to most LAN technologies and doesn't require the added steps of conversion. This was designed to provide a less expensive means of linking networks than through the use of dedicated leased lines commonly used in WANs. SMDS is well suited to the traffic bursts characteristic of LANs. This makes it ideal for LAN-to-LAN communications.

Because SMDS is connectionless, it is available when and as needed, rather than being on at all times. It is a fast technology operating at speeds of 1 Mbps to 45 Mbps. The network address design of an SMDS connection is identical to a telephone number that includes country code and area code, as well as the local number. This address is assigned by the carrier and is used to connect LANs with LANs. A group address can also be used to broadcast information to a number of different LANs at the same time.

Users who need to transfer information to one or more LANs simply select the appropriate addresses to indicate where the information is to be delivered. SMDS takes it from there and makes a best effort to deliver the packets to their destinations. It does not check for errors in transmission, nor does it make an attempt at flow control. Those tasks are left to the communicating LANs.

The packets transferred through SMDS vary in length. They consist of the source address, the destination address, and up to 9188 bytes of payload data. The packets are routed individually and can contain data from any protocol the sending LAN uses, whether it is an Ethernet packet, a Token Ring packet, or another protocol. SMDS essentially passes the information from one place to the other and doesn't deal with the form or format of the data. In other words, SMDS acts somewhat like a courier service. It picks up and delivers, but does not concern itself with the contents of its packages.

ISDN

Integrated Services Digital Network (ISDN) technology was developed in the mid-1980s. What makes ISDN different from the standard phone service in use today is that ISDN allows data to be transmitted digitally. Normal analog service requires a modem to convert the digital signal from the DTE into an analog signal for transmission over the Public Switched Telephone Network (PSTN). Once the signal reaches its destination, the modem then has to convert the analog signal back into a digital signal. An ISDN line can simultaneously transmit voice, data, and video over the same physical line.

At the customer premises, the ISDN line terminates on a device called a *terminal adapter* (TA), which is sometimes referred to as an ISDN modem. One side of the TA connects to the incoming ISDN line from the public carrier, and the other side connects to the serial interface on the DTE.

ISDN supports many different protocols, including TCP/IP, IPX, and AppleTalk. Available encapsulations that can be used with ISDN are protocols like PPP, HDLC, X.25, and V.120. Most internetworking designs use PPP as the encapsulation.

The distance limitation of ISDN is 18,000 feet from the carrier's central office (CO). This distance can be increased using repeaters, although using repeaters may increase the cost of service.

The ISDN carrier provides a service profile identifier (SPID) number to identify the line configuration of the BRI service. It is the equivalent of a phone number. There is no standard format for SPID numbers, so SPID numbers vary depending on the switch type and the carrier.

If the SPIDs are not configured or are configured incorrectly on the device, the Layer 3 initialization fails, and the ISDN services cannot be used. In North America, the most common types of ISDN switching devices are 5ESS, DMS100, and NT-1. For proper ISDN operation, it is imperative that the correct switch type is configured on the ISDN device.

ISDN Services   Two kinds of services are available with ISDN: Primary Rate Interface (PRI) and Basic Rate Interface (BRI).

▼ **Primary Rate Interface (PRI)**   In the United States, a PRI consists of 23 64-Kbps B channels and one 64-Kbps D channel for a total bandwidth of 1536 Kbps. In Europe, a PRI consists of 30 B channels and one 64-Kbps D channel for a total bandwidth of 1984 Kbps.

▲ **Basic Rate Interface (BRI)**   A BRI consists of two 64-Kbps B channels and one 16-Kbps D channel. The channels can be used separately or bound together.

PRI and BRI ISDN services uses two kinds of transmission channels:

▼ **Bearer channels (B channels)**   B channels occupy 64-Kbps bandwidth and can transmit voice and data simultaneously. B channels are aggregated into a single logical high-speed connection using a channel aggregation protocol, such as B channel bonding or Multi-link PPP.

▲ **Data channels (D channels)**   D channels are used for signaling to pass dial-string information. The D channel occupies 16 Kbps on a BRI service, and 64 Kbps on a PRI service. The D channel can also be used for datagram delivery over X.25.

ISDN and the OSI Layers   ISDN occupies the Physical, Data Link, and Network layers of the OSI model and uses the following protocols:

▼ **Physical layer**   The ISDN Basic Rate Interface (BRI) Physical layer specification is defined in International Telecommunication Union Telecommunication (ITU-T) Standardization Sector I.430. The ISDN Primary Rate Interface (PRI) Physical layer specification is defined in ITU-T I.431.

■ **Data Link layer**   The ISDN Data Link layer specification is based on Link Access Procedure on the D channel (LAPD) and is formally specified in ITU-T Q.920 and ITU-T Q.921.

▲ **Network layer**   The ISDN Network layer specification is defined in ITU-T I.450 (also known as ITU-T Q.930) and ITU-T I.451 (also known as ITU-T Q.931). Together these two standards specify user-to-user, circuit-switched, and packet-switched connections.

ISDN Equipment Types   ISDN specifies two basic terminal equipment types:

▼ **Terminal Equipment Type 1 (TE1)**   A TE1 is a specialized ISDN terminal, for example, a piece of computer equipment or a telephone. It is used to connect to ISDN through a four-wire, twisted-pair digital link.

▲ **Terminal Equipment Type 2 (TE2)**   A TE2 is a non-ISDN terminal such as Data Terminal Equipment (DTE) that predates the ISDN standards. A TE2 connects to ISDN through a terminal adapter (TA). An ISDN TA can be either a stand-alone device or a board inside the TE2.

ISDN specifies a type of intermediate equipment called a *network termination* (NT) device. NTs connect the four-wire subscriber wiring to two-wire local loops. There are three supported NT types:

▼ **NT Type 1 (NT1) device**   An NT1 device is treated as customer premises equipment (CPE) in North America, but is provided by carriers elsewhere.

■ **NT Type 2 (NT2) device**   An NT2 device is typically found in digital private branch exchanges (PBX). An NT2 performs Layer 2 and Layer 3 protocol functions and concentration services.

▲ **NT Type 1/2 (NT1/2) device**   An NT1/2 device combines functions of separate NT1 and NT2 devices. An NT1/2 is compatible with NT1 and NT2 devices and is used to replace separate NT1 and NT2 devices.

ISDN Reference Points   ISDN reference points define logical interfaces. Four reference points are defined in ISDN. Figure 2-29 shows some typical ISDN configurations:

▼ **R reference point**   The R reference point defines the reference point between non-ISDN equipment and a TA.

■ **S reference point**   The S reference point defines the reference point between user terminals and an NT2.

■ **T reference point**   The T reference point defines the reference point between NT1 and NT2 devices.

▲ **U reference point**   The U reference point defines the reference point between NT1 devices and line-termination equipment in a carrier network. (This is only in North America, where the NT1 function is not provided by the carrier network.)

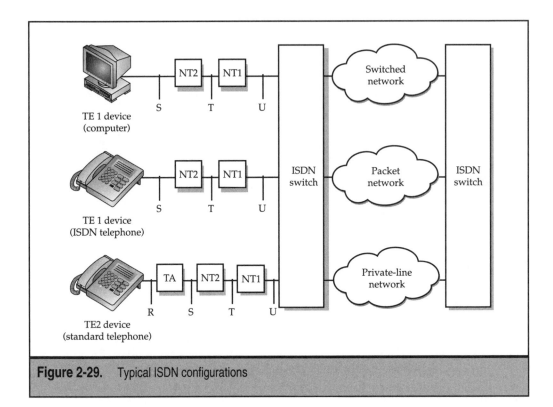

Figure 2-29. Typical ISDN configurations

CHAPTER 3

The TCP/IP Protocol Stack

In 1969, the U.S. Department of Defense Advanced Research Projects Agency (DARPA) created a reference model named, oddly enough, the Department of Defense (DOD) model. The DOD project started as the result of a resource sharing experiment called Advanced Research Projects Agency Network (ARPANET). This project was started to design a network that could withstand destruction of multiple geographically dispersed locations without causing any disruption in the flow of data throughout the network itself. This project became the global network you know as the Internet. It was this DOD project and reference model for which the TCP/IP protocol suite was designed. This chapter will discuss TCP/IP, the protocols, and their characteristics.

The DOD model consists of four layers instead of the seven layers that make up the OSI reference model. (See Figure 3-1.)

Why doesn't the Department of Defense model emulate the OSI specifications? As you will recall from the first chapter, the DOD warned all computer companies that they would not purchase any products that didn't comply with the OSI model by 1993. This is because the DOD model was designed more than a decade before the OSI. Besides, you must have been given the "do as I say, not as I do" line at some point in your life. Now that TCP/IP is the standard protocol for internetworking, everyone still refers to the OSI model as a common or educational reference.

TCP/IP has proven to be a very impressive protocol suite. It has scaled far beyond the imaginations of anyone involved with its design and proven to be robust. The protocol suffers from one weakness, however: security. The original designers didn't include any security measures because they had no way of knowing just how large and publicly accessible the Internet would become. This is why you are reading a book on VPNs in the

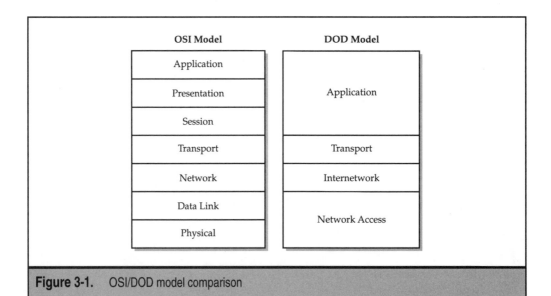

Figure 3-1. OSI/DOD model comparison

first place. VPNs provide an opaque "tunnel" for data that is normally transparent when traveling over the Internet.

THE DOD REFERENCE MODEL LAYERS

The DOD model is based on four distinct layers. Two naming conventions exist for these layers, as shown in the following list. In this book, we'll refer to the layers by their more common names. The less common names are shown here in parentheses:

▼ Application layer (Process layer)

■ Transport layer (Host-to-Host layer)

■ Internetwork layer (Internet layer)

▲ Network Access layer (Network Interface layer)

Application Layer

The Application layer consists of all applications that use the network to transfer data. The Application layer does not care about how the data gets from point to point, and it knows very little about the status of the network. Applications pass data to the next layer in the TCP/IP stack and then continue to perform other functions until a reply is received. Some of the more common protocols occupying this layer include:

▼ HyperText Transfer Protocol (HTTP)

■ Simple Mail Transfer Protocol (SMTP)

■ Domain Name Services (DNS)

■ File Transfer Protocol (FTP)

■ Dynamic Host Configuration Protocol (DHCP)

■ Simple Network Management Protocol (SNMP)

■ Telnet

▲ Network File System (NFS)

The main purpose of the Application layer is to maintain a common command structure or syntax between the applications being used during the communication process. This is what makes data communication between any two systems possible regardless of the operating system or network architecture the two systems might be using at the time.

Transport Layer

The Transport layer is responsible for error detection and correction in the DOD model. Transport protocols provide for reliable, connection-oriented communications and keep track of the state of sessions between connected computers. The desired method of data delivery determines the transport protocol. The two protocols that are most often associated with the Transport layer are

▼ Transmission Control Protocol (TCP)

▲ User Datagram Protocol (UDP)

TCP provides the virtual circuit service necessary for applications requiring reliable end-to-end connections. Data transfer is made reliable using packet sequencing and acknowledgement of receipt.

UDP is the connectionless protocol used at this layer and does not use connections or acknowledgements, so it is less reliable but faster.

Internetwork Layer

The Internetwork layer is mainly responsible for data addressing, packaging, and routing functions. The main protocols occupying this layer are

▼ Internet Protocol (IP)

■ Address Resolution Protocol (ARP)

■ Reverse Address Resolution Protocol (RARP)

■ Internet Control Messaging Protocol (ICMP)

▲ Internet Group Management Protocol (IGMP)

We will be covering these in some detail later in this chapter. The Internetwork layer does not take advantage of sequencing and acknowledgment services that may be present in the Data Link layer. TCP/IP always assumes an unreliable Network Access layer. Connection-oriented communications requiring any session establishment packet sequencing and acknowledgment are left up to the Transport layer.

Network Access Layer

The Network Access layer is responsible for placing TCP/IP packets on the network medium and receiving TCP/IP packets off the network medium. TCP/IP does not concern itself with network access methods such as frame formats or type of medium being used. This was intentional so that it could be used to connect different network types. This includes LAN technologies such as Ethernet or Token Ring and WAN technologies such as X.25 or Frame Relay. Due to its independence from any specific network technology, TCP/IP can be easily adapted to virtually any new network transport technologies.

Where the TCP/IP Protocols Fit

Now let's look at the TCP/IP protocols and how they fit in relation to the DOD and OSI models. As shown in Figure 3-2, the TCP/IP protocols exist in only the three layers of the four-layer DOD model. If you were to overlay the TCP/IP protocols onto the OSI model, you would see that they do not correspond one-to-one with the layers of the OSI model.

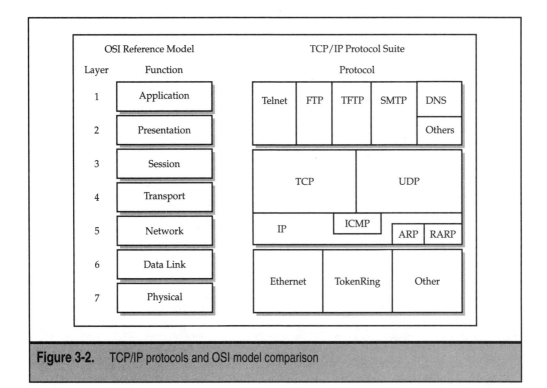

Figure 3-2.   TCP/IP protocols and OSI model comparison

As we mentioned at the beginning of the chapter, TCP/IP protocols do not concern themselves with the bottom two layers of the OSI model (Data Link and Physical layer) by design. The protocols populate the top three layers with respect to the DOD reference model and the top five layers with respect to the OSI model. They begin in the Network layer, where protocols like the Internet Protocol (IP) and Internet Control Messaging Protocol (ICMP) reside. The Transport layer contains the Transmission Control Protocol (TCP) and User Datagram Protocol (UDP). The Application layer houses the utilities and protocols that make up the rest of the TCP/IP suite. These protocols use the services of the two lower layers to provide the communications and networking functions required by the applications being used by computers on the network whether they are local or remote. You can see which services are used for some of these protocols in Figure 3-3.

PROTOCOL DESCRIPTIONS

We will now discuss the individual protocols in more detail. To protect your network, it is important to learn the protocols that are subject to attack and where their weaknesses lie. (We will talk about those weaknesses and attacks in later chapters and throughout the book.) Let's discuss the Network layer protocols first.

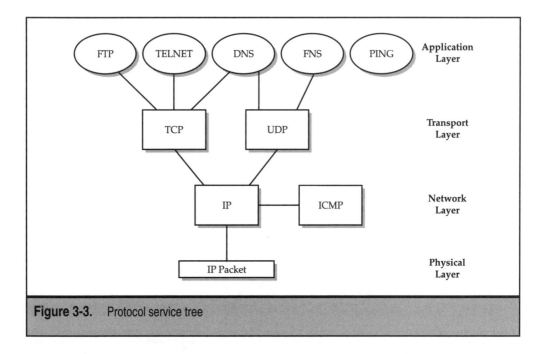

Figure 3-3.   Protocol service tree

Network Layer Protocols

We will start with the ARP protocol first. ARP and RARP are almost identical, as you will soon see.

Address Resolution Protocol (ARP)

Upper-layer protocols communicate using logical addresses, which in TCP/IP's case is the IP address. Whenever one of these upper-layer protocols wants to transmit data to a destination IP address such as 10.20.30.40, it needs a way to resolve the physical address of the destination device being used by the underlying network. The underlying network does not understand logical addresses. Instead, it communicates using Data Link layer addresses or the physical address of the device's NIC attached to the network. This is accomplished using the ARP protocol.

When a transmitting device wants to find the hardware or physical address of another device on the same network, it uses a lookup table, known as the *ARP cache,* to do this translation. If the address is not found in the sending devices ARP cache, it broadcasts a message called an *ARP request* onto the network. This ARP request includes the IP address of the device whose physical address is being requested. Whenever a message is broadcast, it causes all of the devices on the network to process the ARP packet. Only the device whose IP address is in the ARP packet will respond by sending a directed transmission, or unicast, to the device requesting the physical address. This unicast message is called an *ARP reply* and contains the physical hardware address of the device. The reply can also include source route information. Now the physical address and source route information will be stored in the ARP cache of the sending device for a specified period. All subsequent data being sent to this destination IP address can now be translated to a physical address without having to broadcast an ARP request.

Reverse Address Resolution Protocol (RARP)

Some network hosts, such as diskless workstations, do not know their own IP address when they are booted. To determine their IP address, they use a mechanism similar to ARP called RARP. This is because now it is the physical or hardware address of the host that is known and the IP address for the host that is being requested. It differs more fundamentally from ARP in the fact that a "RARP server" must exist on the network. This RARP server maintains a database of mappings from hardware address to protocol address. In case there are multiple RARP servers on the network, the RARP requester only uses the first RARP reply received on its broadcast RARP request and discards the others. The reverse address resolution procedure is performed the same way as the ARP address resolution. ARP and RARP use the same header structure. The header structure can be seen in Figure 3-4.

Hardware Type Field: This is an 16-bit field that identifies the network medium being used.

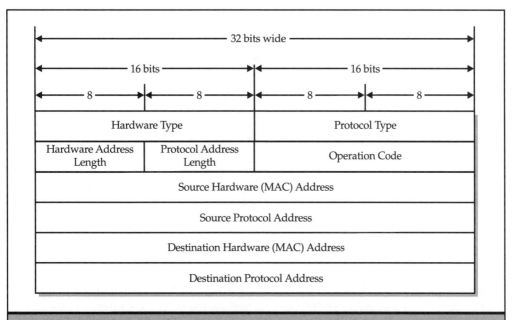

Figure 3-4.   ARP/RARP header

Value	Description
1	Ethernet
2	Experimental Ethernet
3	Amateur Radio AX25
4	Proteon ProNET Token Ring
5	Chaos
6	IEEE 802
7	ARCNET
8	Hyperchannel
9	Lanstar
10	Autonet Short Address
11	LocalTalk
12	LocalNet (IBM PCNet or SYTEK LocalNET)
13	Ultra link
14	SMDS
15	Frame Relay
16	Asynchronous Transmission Mode (ATM)
17	HDLC
18	Fiber Channel
19	Asynchronous Transmission Mode (ATM)
20	Serial Line
21	Asynchronous Transmission Mode (ATM)
22	MIL-STD-188-220
23	Metricom
24	IEEE 13941995
25	MAPOS
26	Twixaxial
27	EUI-64
28	HIPARP

Protocol Type field: 16 bits.
Hardware Address Length field: 8 bits. Length of the hardware address in bytes.
Protocol Address Length field: 8 bits. Length of the protocol address in bytes.
Operation Code field: 16 bits.

Value	Description	References
1	Request	RFC 826
2	Reply	RFC 826, RFC 1868
3	Request Reverse	RFC 903
4	Reply Reverse	RFC 903
5	DRARP Request	RFC 1931
6	DRARP Reply	RFC 1931
7	DRARP Error	RFC 1931
8	InARP Request	RFC 1293
9	InARP Reply	RFC 1293
10	ARP NAK	RFC 1577
11	MARS Request	
12	MARS Multi	
13	MARS MServ	
14	MARS Join	
15	MARS Leave	
16	MARS NAK	
17	MARS Unserv	
18	MARS SJoin	
19	MARS SLeave	
20	MARS Grouplist Request	
21	MARS Grouplist Reply	
22	MARS Redirect Map	
23	MAPOS UNARP	

Source Hardware Address field: Variable length.
Source Protocol field: Variable length.
Destination Hardware Address field: Variable length.
Destination Protocol field: Variable length.

Internet Protocol (IP)

The Internet Protocol (IP) occupies the Network layer of the OSI model as well as the TCP/IP model. IP is a *connectionless* service, which means that no virtual circuit is nailed up during the transmission. The packets are simply sent on a best-effort basis. IP provides no guarantee that the packets being sent will ever reach their destination. IP packets are also referred to as *datagrams*.

If the packet or datagram being sent is one whose destination lies on a network other than its own, then the datagrams have to go through a series of routers before they reach their destination. At each router along the way, the packet is sent up the protocol stack only as high as the Network layer, where the router examines the logical address and determines the physical address of the next hop (see Figure 3-5).

IP Packet

Now we are going to dissect the IP packet in depth. We'll do this because this is what routers do, as well as the people who want to break into your network. You can see what an IP datagram looks like in Figure 3-6.

A typical IP header is usually 20 bytes long if it has no options set. If there are any options, the IP header can be as large as 24 bytes. Let's look at what comprises the IP header portion of the IP datagram. Use Figure 3-7 as a reference for the field descriptions that follow.

Let's start with the mandatory fields in the header:

▼ **Version field** This is a 4-bit field that dictates which IP version number of the protocol is being used. The version number tells the receiving device how to decode the rest of the header. This field is almost always set to version 4; however, several systems are now testing IPv6 or IPng (next generation). The Internet and most LANs do not support IP version 6 yet, but the change is inevitable.

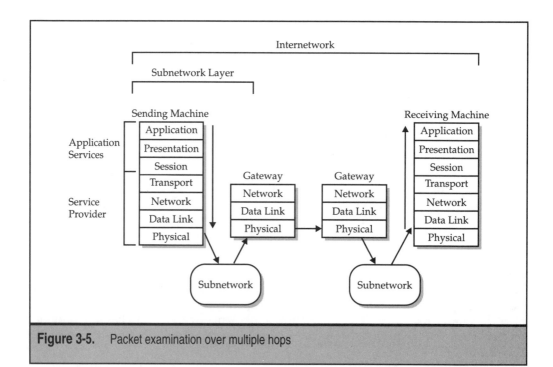

Figure 3-5. Packet examination over multiple hops

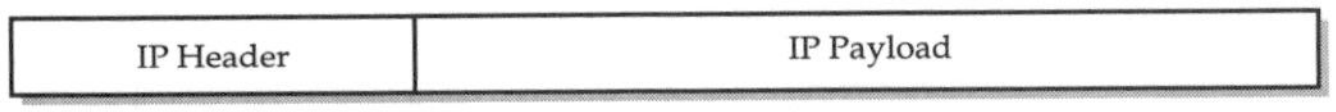

IP Header | IP Payload

Figure 3-6.   IP datagram

■ **Header Length field**   This 4-bit field specifies the length of the IP packet header in 32-bit words. The minimum value for a valid header is 5 (20 bytes). With use of optional fields, you can increase the header size to its maximum of 6 words (24 bytes). The reason for the field is to properly decode the header—IP must know where the IP header ends and the data begins.

▲ **Type of Service (TOS) field**   The 8-bit Type of Service field tells IP how to process the datagram properly. The precedence portion of the field was an early attempt at Quality of Service (QoS). The first 3 bits specify the datagram's precedence, indicating an importance that increases proportionally to the value. That binary value can range from 0 (normal) through 7 (network control).

Precedence is a measure of the nature and priority of this datagram:

▼ *Routine:* 000

■ *Priority:* 001

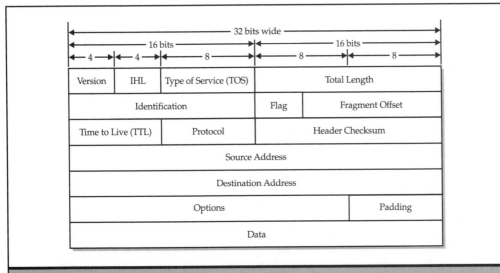

Figure 3-7.   IP header

- *Immediate:* 010
- *Flash:* 011
- *Flash Override:* 100
- *Critical:* 101
- *Internetwork Control:* 110
- ▲ *Network Control:* 111

However, most current implementations of TCP/IP and practically all hardware that uses TCP/IP ignore this field, treating all datagrams with the same priority. The next 4 bits are 1-bit flags that control:

- ▼ *4—Delay:* If the bit is set to 0, the setting is normal. A bit set to 1 implies low (D)elay.
- *5—Throughput:* If the bit is set to 0, the setting is normal. A bit set to 1 implies high (T)hroughput.
- *6—Reliability:* If the bit is set to 0, the setting is normal. A bit set to 1 implies high (R)eliability.
- ▲ *7—Monetary cost:* If the bit is set to 0, the setting is normal. A bit set to 1 implies low (R)eliability.

The last bit of the field is not used and is set to 0. For most purposes, the values of all the bits in the Type of Service field are set to 0 because differences in precedence, delay, throughput, reliability, and monetary cost between machines are virtually nonexistent unless a special network has been installed. The following illustration shows how the TOS field in the header breaks down into bits.

- ▼ **Total Length field**   This 16-bit field indicates the total length of the datagram, including the header, in bytes. The length of the data area itself can be computed by subtracting the header length from this value. The size of the Total Length field is 16 bits, which obviously means that the entire packet cannot exceed 65,535 bytes, including the header. This field is used to determine the length value to be passed to the transport protocol in order to set the total frame length.

- **Identification field**   This 16-bit field contains a value that is a unique identifier created by the sending device. This number is necessary for the reassembly of fragmented messages to make sure that the fragments of one message are not mixed in with others. Each unit of data received by the IP layer from a higher protocol layer is assigned one of these unique identifiers when the data arrives. If a datagram is fragmented, each fragment has the same unique identifier. The sending device numbers each packet, incrementing the value after each packet is sent. This value doesn't necessarily have to start at zero and also doesn't necessarily increment by one on all IP implementations.

■ **Flags field**   The Flags field is a 3-bit field. The first bit is reserved and should always be set to 0. The other two bits are called the DF (Don't Fragment) and the MF (More Fragments). When the DF flag is set to 1, the datagram cannot be fragmented under any circumstances. If this bit is set to 1, then the datagram will be discarded and an error message sent back to the sending device. When the MF flag is set to 1, it means that the current datagram is to be followed by more packets or subpackets. Then these packets need to be reassembled sequentially to re-create the original message. When a series of packets has its DF bits set for fragmentation, then the last fragment sent will have its MF flag set to 0 so that the receiving device knows the final fragment has been received and can then reassemble the rest of the fragments. Since the fragments may not arrive in the order in which they were sent, the MF flag is used in conjunction with the Fragment Offset field (the next field in the IP header) to tell the receiving machine the full extent of the message. The next illustration shows how the Flags field in the header breaks down into bits.

0	1	2	3	4	5	6	7
Precedence			D	T	R	M	O

■ **Fragment Offset field**   If the MF (More Fragments) flag bit is set to 1, meaning that these are the fragments of a larger datagram, the 13-bit Fragment Offset field identifies the position in the chain of fragments that is occupied within the current datagram. This allows IP to reassemble the fragments in the right sequence. These offsets are relative to the beginning of the message. These offsets are calculated in units of 8 bytes, corresponding to the maximum packet length of 65,535 bytes. The IP layer on the receiving machine uses the identification number, which indicates the message the transmitted datagram belongs with, to reassemble the complete message.

■ **Time To Live field**   This is an 8-bit field that indicates the amount of time, in seconds, that a datagram can remain on the network before being discarded. This value is set by the sending device. The TTL field is generally set at 32 or 64. The value of the TTL field is decreased by no less than one second for each device that the packet travels through. When the packet is processed by a router, the arrival time is noted. If the datagram has to wait in the queue, the TTL can be decremented accordingly. If that router is overloaded significantly, the packet can actually die in the queue. Once the TTL value reaches 0, it is immediately discarded. When this happens, an ICMP (Internet Control Message Protocol) message is sent to notify the original sending device that the packet was discarded so that the device can retransmit the datagram. The purpose of the TTL field is to prevent IP packets from circulating through networks forever.

■ **Protocols field**   This is an 8-bit field that contains the identification number of the transport protocol in the IP payload. Obviously, this allows for a maximum of 256. Here is a list of the more common numbers you are likely to encounter:

Value	Protocol
0	IPv6 Hop-by-Hop Option
1	Internet Control Message Protocol (ICMP)
2	Internet Group Multicast Protocol (IGMP)
3	Gateway to Gateway Protocol (GGP)
4	IP in IP Encapsulation
6	Transmission Control Protocol (TCP)
8	Exterior Gateway Protocol (EGP)
9	IGRP
17	User Datagram Protocol (UDP)
18	Multiplexing
27	Reliable Data Protocol (RDP)
28	Internet Reliable Transaction Protocol (IRTP)
33	Sequential Exchange Protocol
35	Inter-Domain Policy Routing Protocol (IDPR)
37	Datagram Delivery Protocol
42	Source Demand Routing Protocol (SDRP)
43	IPv6 Routing header
44	Ipv6 Fragment header
45	Inter-Domain Routing Protocol (IDRP)
46	Reservation Protocol (RSVP)
47	General Routing Encapsulation (GRE)
50	Encapsulating Security Payload (ESP)
51	Authentication Header (AH)
53	IP with Encryption
57	SKIP
83	VINES
88	EIGRP
89	Open Shortest Path First Routing Protocol (OSPF)
92	Multicast Transport Protocol (MTP)

Value	Protocol
94	IP-within-IP Encapsulation Protocol
108	IP Payload Compression Protocol (IPPCP)
111	IPX in IP
112	Virtual Router Redundancy Protocol (VRRP)
115	Level 2 Tunneling Protocol (L2TP)

■ **Header Checksum field**   This 16-bit field is a checksum for the protocol header field only in order to enable faster processing. Because the TTL field is decremented at each device, the checksum, too, changes with every device that the datagram passes through. The checksum algorithm takes the one's complement of the 16-bit sum of all 16-bit words. This is a fast algorithm, but it can miss unusual errors like the loss of an entire 16-bit word containing only 0's. Since the data checksums used by both TCP and UDP cover the entire packet, these types of errors usually can be caught as the frame is being assembled for the network transport.

■ **Source IP Address field**   This 32-bit field contains the logical source IP addresses of the sending device and is not altered at any time along the way to its destination.

■ **Destination IP Address field**   This 32-bit field contains the logical destination IP addresses of the receiving device and is not altered at any time along the way to its destination.

▲ **Options field**   This is a variable-length field that includes several codes of variable length. More than one option can be used in the datagram. If more than one option is used, then they are listed consecutively in the IP header. Take a look at the next illustration and the list of options within that comprise the byte.

0	1	2	3	4	5	6	7
C	Class		Options				

This byte is usually divided into three fields:

1. *Copy Flag*   The 1-bit Copy field of this byte is used to stipulate how the option is handled when fragmentation is necessary in a gateway. When the bit is set to 0, the option should be copied to the first datagram but not subsequent ones. If the bit is set to 1, the option is copied to all the datagrams.

2. *Option Class field*   The 2-bit Option Class field in this byte indicates the type of option. When the value is 0, the option applies to datagram or network control.

A value of 2 means the option is for debugging or measurement purposes. The values 1 and 3 are currently reserved for future use.

3. *Option Number*   The Options Number field in this byte includes options that enable the routing and timestamps to be recorded. These are used to provide a record of a datagram's passage across the internetwork, which can be useful for diagnostic purposes. Both these options add information to a list contained within the datagram. There are two kinds of source routing indicated within the Options field:

   a. *Loose Source Routing*   Loose source routing provides a series of IP addresses that the machine must pass through, but it enables any route to be used to get to each of these addresses, usually routers.

   b. *Strict Source Routing*   Strict source routing enables no deviations from the specified route. If the route can't be followed, the datagram is abandoned. Strict routing is frequently used for testing routes but rarely for transmission of user datagrams because of the higher chances of the datagram being lost or abandoned.

Both settings are highly suspect in today's security-conscious networks and are almost always denied at any router or firewall because of the obvious damage that can be caused when anyone can control the paths of packets. Shown next is a list of the options available in the Options field and the RFCs (IETF Request for Comments) that refer to them.

Option	Copy	Class	Value	Length	Description	References
0	0	0	0	1	End of Options List	RFC 791
1	0	0	1	1	No Option (used for padding)	RFC 791
2	1	0	130	11	Security (military purposes only)	RFC 791, RFC 1108
3	1	0	131	Variable	Loose Source Route	RFC 791
4	0	2	68	Variable	Timestamp (adds fields)	RFC 781, RFC 791
5	1	0	133	3 .. ?	Extended Security	RFC 1108
6	1	0	134		Commercial Security	

Option	Copy	Class	Value	Length	Description	References
7	0	0	7	Variable	Record Route (adds fields)	RFC 791
8	1	0	136	4	Stream ID	RFC 791, RFC 1122
9	1	0	137	Variable	Strict Source Route	RFC 791
10	0	0	10		Experimental Measurement	
11	0	0	11		Maximum Transmission Unit (MTU) Probe	RFC 1063
12	0	0	12		Maximum Transmission Unit (MTU) Reply	RFC 1063
13	1	2	205		Experimental Flow Control	
14	1	0	142		Experimental Access Control	
15	0	0	15			
16	1	0	144		IMI Traffic Descriptor	
17	1	0	145		Extended Internet Protocol	
18	0	2	82		TraceRoute	RFC1393
19	1	0	147		Address Extension	RFC 1475
20	1	0	148	4	Router Alert	RFC 2113
21	1	0	149	Variable	Selective Directed Broadcast Mode	RFC 1770
22	1	0	150		NSAP Addresses	

Option	Copy	Class	Value	Length	Description	References
23	1	0	151		Dynamic Packet State	
24	1	0	152		Upstream Multicast Packet	

▼ **Padding field**   The value of the padding area depends on the options selected. The padding is used to make sure that the datagram header ends on and that the data starts on a 32-bit boundary.

Internet Control Message Protocol (ICMP)

This protocol is exploited often when malicious hackers attack a network. The reason is evident in the name of the protocol itself. The Internet Control Message Protocol carries a lot of weight in the IP stack because it controls how computers and routers behave when messages are received. Nothing in the protocol allows for checking the credibility of the message. It is kind of like you trusting me unconditionally when I tell you that Highway 101 is closed and that you should take a street through a bad part of town to get to work. While you are on that street, you are subject to many dangers, just as are packets being re-directed from their proper path through a malicious network where they can be examined, copied or altered even before being sent on their way.

ICMP messages are generated for several reasons:

▼   When a datagram cannot reach its destination

■   When the router does not have the buffering capacity to forward a datagram

▲   When the router can direct the host to send traffic on a shorter route

IP was not designed to be reliable. The purpose of ICMP is to provide information about problems in the communication path, not to make IP reliable. There are still no guarantees that a datagram will be delivered or, for that matter, that a control message will be returned. Some datagrams may still be undelivered without any report of their loss. It is up to the higher-layer protocols to implement their own reliability procedures. No ICMP messages are sent about the fates of other ICMP messages.

You can see the fields that make up the ICMP header in Figure 3-8, followed by the descriptions of each.

▼ **Type field**   The 8-bit Message Type field in the ICMP header contains the type of message being sent. Here is a list of the most common instances of this value.

Value	Description
0	Echo Reply
3	Destination Not Reachable

Value	Description
4	Source Quench
5	Redirection Required
8	Echo Request
11	Time to Live Exceeded
12	Parameter Problem
13	Timestamp Request
14	Timestamp Reply
15	Information Request (now obsolete)
16	Information Reply (now obsolete)
17	Address Mask Request
18	Address Mask Reply

- **Code field**   The 8-bit Code field expands on the message type, providing a little more information for the receiving machine.

- **Checksum field**   This 16-bit field in the ICMP header is calculated in the same manner as the normal IP header checksum. (The checksum is an error checking algorithm.)

- **ICMP Message field**   Many different types of ICMP messages can be sent, as shown in Figure 3-9:

  - *Source Quench ICMP*   These messages are used for flow control. When a given device's buffer is full, it sends a Source Quench message for each datagram that it discards. When a device receives a Source Quench message, it slowly reduces the transmission until it stops receiving Source Quench messages.

  - *Destination Unreachable and Time Exceeded*   These messages are not only used for obvious reasons but also for other instances, such as when a datagram must be fragmented but the DF flag is set.

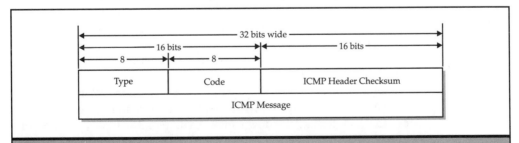

Figure 3-8.   ICMP header

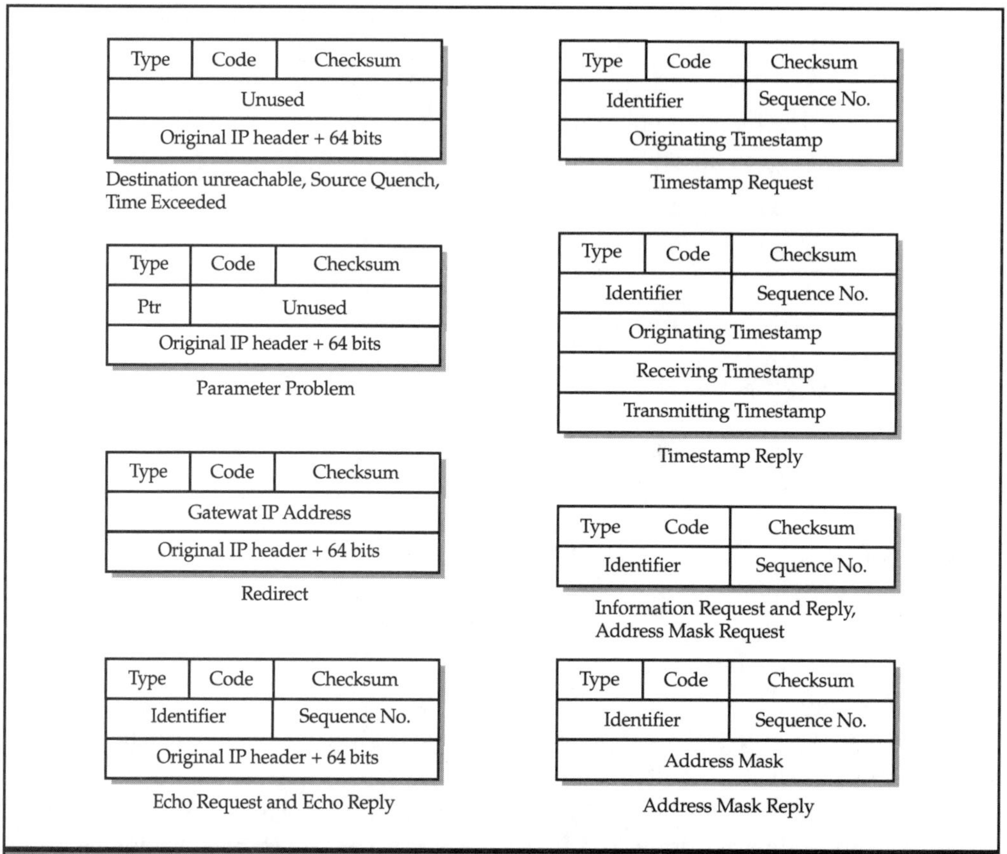

Type	Code	Checksum
Unused		
Original IP header + 64 bits		

Destination unreachable, Source Quench,
Time Exceeded

Type	Code	Checksum
Ptr	Unused	
Original IP header + 64 bits		

Parameter Problem

Type	Code	Checksum
Gatewat IP Address		
Original IP header + 64 bits		

Redirect

Type	Code	Checksum
Identifier	Sequence No.	
Original IP header + 64 bits		

Echo Request and Echo Reply

Type	Code	Checksum
Identifier	Sequence No.	
Originating Timestamp		

Timestamp Request

Type	Code	Checksum
Identifier	Sequence No.	
Originating Timestamp		
Receiving Timestamp		
Transmitting Timestamp		

Timestamp Reply

Type	Code	Checksum
Identifier	Sequence No.	

Information Request and Reply,
Address Mask Request

Type	Code	Checksum
Identifier	Sequence No.	
Address Mask		

Address Mask Reply

Figure 3-9. ICMP messages

■ *ICMP Redirect* These messages are sent to a device when a better route is known. When a router receives a datagram and sees that it has a "better" route, it sends an ICMP Redirect message to the source of the datagram with the IP address of what it considers to be the better route. When the router sends a Redirection message, a value is placed in the Code field of the header to indicate the conditions under which the new route applies. A value of 0 means that datagrams for any device on the destination network should be redirected. A value of 1 indicates that only datagrams for the specific device should take the new route. A value of 2 implies that only datagrams for the network with the same TOS, read from one of the IP header fields, should take the new route. A value of 3 will only reroute for the host with the same TOS.

- *Parameter Problem*   These messages are used whenever a syntax is detected in the IP header. When a Parameter Problem message is sent back to the sending device, the Parameter field in the ICMP error message contains a pointer to the byte in the IP header where the error was detected.

- *ICMP Echo Request and Echo Reply*   These messages are easily the most common forms of IP troubleshooting used today and probably into the future. These are also referred to as PING (Packet InterNet Groper) packets. The PING command sends a series of Echo Requests and waits for Echo Replies. Whenever an Echo Request is sent, a device or router down the path sends an Echo Reply back to the requesting device.

- *Timestamp Requests and Replies*   These messages can be used to measure the amount of time a packet takes to cross a given distance in a network. When combined with strict source routing, this procedure can be used to identify bottlenecks.

- *Address Mask Requests and Replies*   These messages are used for testing within the parameters of a specific network or subnet.

TCP/IP Transport Layer Protocols

Now we move into the Transport layer, where upper-layer protocols are spared the complexities of the Network layer, and where the protocols tell packets which applications will use their contents. The two players in this field are the User Datagram Protocol (UDP) and the Transmission Control Protocol (TCP). Here you will be introduced to the concept of ports and sockets. Before we talk about UDP and TCP, it is important that you understand the concepts behind ports and sockets.

Ports

When one process wishes to communicate with another process, it first identifies itself to the upper-layer protocols by what are known as *ports.* A port is a 16-bit number, so mathematically it must be in a range from 1 to 65535. The Transport layer protocols such as UDP and TCP use ports to identify which upper-layer protocol or application program (process) the incoming messages are to be delivered to. Many protocols such as Telnet, FTP, and SMTP to name a few, use the same port number in all TCP/IP implementations. These ports have been assigned to these protocols by the Internet Assigned Numbers Authority (IANA) and are referred to as *well-known ports.* The standard applications are referred to as *well-known services.* The assigned well-known ports occupy the port numbers in the range of 0 to 1023.

The ports with numbers in the range 1024—65535 are not controlled by the IANA and on most systems can be used by ordinary user-developed programs. However, these days it is usually agreed that the range from 1024 to 49151 is considered better left to standard uses, and these ports are generally referred to as *registered ports.* Ports with numbers in the range from 49152 to 65535 are considered to be public, and anybody can use them.

These are generally referred to as *ephemeral ports*. As a side note, until recently the registered ports were only up to 5000, and above that they were ephemeral. However, with the explosive growth of Internet applications, this number was soon expanded. It is important to keep aware of these port numbers when setting up filters on your routers and firewalls. Usually port 0 is used as a wildcard and for requesting a kernel to choose a port for us—we do not care which.

Sockets

Every network connection going into and out of a host's port is uniquely identified by a combination of two numbers: the IP address of the machine and the port number in use. These two numbers make up the *socket*. Since at least two computers are involved in any connection, there are sockets on both the sending and receiving ends of the connection. Both the IP addresses and the ports are unique to each machine, so the sockets are also unique. Applications talk to each other across the network based entirely on the socket number. Each machine on the connection maintains port lists of all active ports, and the two machines have reversed entries for each session between them. This process is called *binding*. For example, if one machine has a source port at 23 and the destination at 25, the other machine has a source port at 25 and a destination at 23.

There are two kinds of sockets:

1. *Stream Socket*   This is connection oriented in nature and relates to TCP.
2. *Datagram Socket*   This is connectionless in nature and relates to UDP.

User Datagram Protocol (UDP)

UDP does very little as transport protocols go. Apart from multiplexing and simple error checking, it adds nothing to IP. If the application developer chooses UDP over TCP, the application ends up communicating almost directly with the IP protocol. UDP takes messages from application processes, attaches source and destination port number fields for the multiplexing service, adds length and checksum fields, and then passes the resulting "segment" to the Network layer. The Network layer encapsulates the segment into an IP datagram and then makes its best effort to deliver the segment to the receiving host. If the segment arrives at the receiving host, UDP uses the port numbers and the IP source and destination addresses to deliver the data in the segment to the correct application process. With UDP there is no handshaking between sending and receiving Transport-layer entities before sending a segment. This is why UDP is considered to be a *connectionless* protocol.

See Figure 3-10 to see what fields exist in the UDP segment.

▼   **Source Port**   This 16-bit field is an optional field. If it is not used, it is set to zero. Otherwise, it specifies the port of the sender.

■   **Destination Port**   This 16-bit field indicates the port on the destination machine.

■   **Length**   This 16-bit field specifies the length in bytes of the UDP header and the encapsulated data. The minimum value for this field is 8.

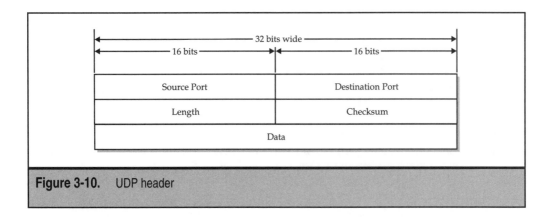

Figure 3-10. UDP header

▲ **Checksum**   This is computed as the 16-bit one's complement of the one's complement sum of a pseudoheader of information from the IP header, the UDP header, and the data, padded as needed with zero bytes at the end to make a multiple of two bytes. Wow! That's a mouthful. (But not a very important one; go ahead and spit that out if you want. OK, you can rinse now. Essentially it is defining the error checking procedure that we alluded to earlier in the chapter.)

Transmission Control Protocol (TCP)

TCP, the Transmission Control Protocol, provides connection-oriented and acknowledged services to the IP layer and the upper-layer protocols. These services allow an application to be sure that datagrams sent out over the network are received in the order sent and in their entirety. Unlike UDP, TCP provides for reliable communications. If a datagram is damaged or lost, TCP handles the retransmission, rather than the applications in the higher layers. It does this by sequencing the segments with a forward acknowledgment value that tells the destination which segment the source expects to receive next. If an acknowledgement is not received within a given period, those segments are retransmitted. The reliability mechanism of TCP makes it possible to handle missing, delayed, duplicate, or damaged packets. A timeout mechanism allows devices to detect lost packets and request retransmission.

Many people think that since TCP frequently is mentioned in the same breath as TCP/IP, it somehow relies on or is tied to the IP protocol. This is not the case. TCP is a completely independent protocol capable of being used with almost any underlying protocol. TCP is used in the File Transfer Protocol (FTP) and the Simple Mail Transfer Protocol (SMTP), neither of which uses or relies on the IP protocol. This protocol is typically used by applications that require guaranteed delivery.

Sliding Window   TCP uses a sliding-window algorithm to provide a method for efficient segment transfers as well as for handling both timeouts and retransmissions. The byte stream is transferred in segments. The window size determines the number of bytes of data that can be sent before an acknowledgement from the destination device is received. This can be seen in Figure 3-11.

TCP establishes a full-duplex virtual connection between the two endpoints. Each endpoint is defined by an IP address and a TCP port number, which defines its state. Since the connection state resides entirely in the two end systems and not in the intermediate network elements, such as routers, the intermediate network elements do not maintain the TCP connection states. Instead, the intermediate routers are completely oblivious to the TCP connections; they see only datagrams, not connections.

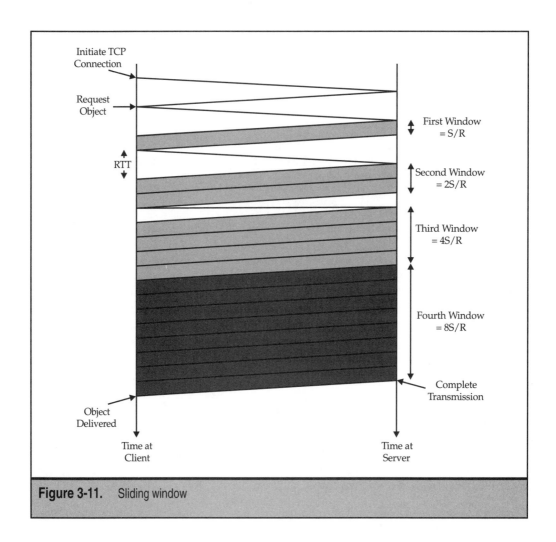

Figure 3-11.   Sliding window

Three-Way Handshake   Communications begin with what is known as a *three-way handshake.* The three-way handshake is a connection establishment procedure. This procedure synchronizes both ends of a connection by allowing both sides to agree upon initial sequence numbers. This procedure also guarantees that both sides are ready to transmit and receive data. This is necessary so that this data is not transmitted or retransmitted during session establishment or after the session is finished.

When host A wants to open a connection to host B, A sends an initial segment to B. This initial segment has the Initial Sequence Number (ISN) that B needs to use to send data to A. This initial segment is identified by the SYN bit set to 1 in the TCP header. If the SYN bit is set, the 32-bit sequence number in the header is interpreted as the ISN. In all other cases (when the SYN bit is not set), the 32-bit sequence number identifies the sequence number of the first data byte contained in that segment. B, on receiving the SYN from A, has to respond with another SYN, as well as acknowledge the SYN sent by A. This is indicated by "SYN+ACK" in the state machine diagram. Take a look at Figure 3-12.

When closing a connection, the three-way process is repeated. Connection release uses the FIN in place of the SYN. See Figure 3-13.

TCP Segment Header   Now we will discuss the elements of the TCP segment header. Take a look at these fields in Figure 3-14.

▼ **Source Port field**   This 16-bit field identifies the local TCP user (usually an upper-layer application program).

■ **Destination Port field**   This 16-bit field identifies the remote machine's TCP user.

■ **Sequence Number field**   This 32-bit field represents the sequence number of the first data byte in this segment. If the SYN bit is set, the sequence number is the initial sequence number, and the first data byte is the initial sequence number + 1.

■ **Acknowledgment Number field**   If the ACK bit is set, this 32-bit field contains the value of the next sequence number the sender of the segment is expecting to receive. Once a connection is established, this is always sent.

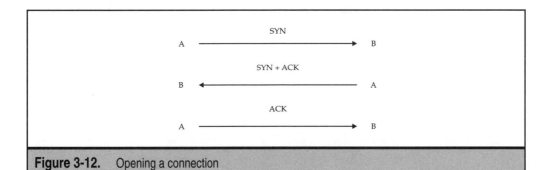

Figure 3-12.   Opening a connection

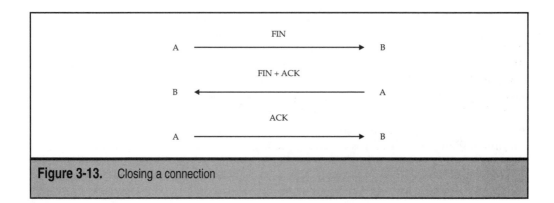

Figure 3-13.   Closing a connection

- **Data Offset field**   This 4-bit field represents the number of 32-bit words in the TCP header. This indicates where the data begins. The length of the TCP header is always a multiple of 32 bits.

- **Reserved field**   This 6-bit field is reserved for future use. The six bits must be set to 0.

- **ECN, Explicit Congestion Notification** 2 bits.

- **Control Bits field**   This field is illustrated next.

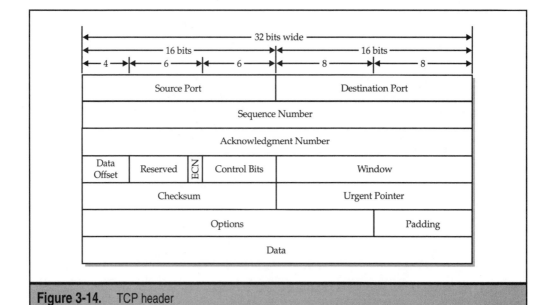

Figure 3-14.   TCP header

0	1	2	3	4	5
U	A	P	R	S	F

- *U, URG*   This is the Significant Urgent Pointer field.

- *A, ACK*   This is the Significant Acknowledgment field.

- *P, PSH*   This is the Push Function field.

- *R, RST*   This is the Reset Connection field.

- *S, SYN*   This is the Synchronization Sequence Number field. This indicates that the sequence numbers are to be synchronized. This flag is used when a connection is being established.

- *F, FIN*   This is the End of Data field. This indicates that the sender has no more data to send. This is the equivalent of an end-of-transmission marker.

- **Window field**   This 16-bit field represents the number of data bytes beginning with the one indicated in the acknowledgment field that the sender of this segment is willing to accept.

- **Checksum field**   This is computed as the 16-bit one's complement of the one's complement sum of a pseudoheader of information from the IP header, the TCP header, and the data, padded as needed with zero bytes at the end to make a multiple of two bytes. The pseudoheader contains the following fields:

- **Urgent Pointer field**   This 16-bit field is used if the URG flag was set; it indicates the portion of the data message that is urgent by specifying the offset from the sequence number in the header. No specific action is taken by TCP with respect to urgent data; the action is determined by the application.

- **Options field**   Similar to the IP Header Option field, this is used for specifying TCP options. Each option consists of an option number (one byte), the number of bytes in the option, and the option values. Only three options are currently defined for TCP:

  - *0*   End of option list

  - *1*   No operation

  - *2*   Maximum segment size

- ▲ **Padding field**   This is used to make sure that the header is a 32-bit multiple.

TCP/IP APPLICATION LAYER PROTOCOLS

In this section, we will talk about some of the Application layer protocols. We will mainly cover the most likely protocols to be attacked. This is where you can apply what you learned in the "Ports" and "Sockets" sections. We will start with DNS.

Domain Name System (DNS)

Domain Name System is a distributed database system used to map hard-to-remember IP addresses into easy-to-remember common names. A good example would be a phone directory. When you want to call someone, you know their name, and look up their number. When you want to connect to a given web site, all you need to remember is the URL, but the computer needs the numerical IP address in order to communicate with this web site. That's where DNS comes in.

The domain name server keeps a table of hosts and their IP addresses. When a computer needs to resolve a name to an IP address, it does what is known as a *lookup*—it sends the name to the domain name server (DNS), and the DNS server responds with the IP address. It does this by sending either UDP or TCP (depending on whether it is a request or a zone transfer), but most of the time it will send UDP packets to port 53. The message has a fixed 12-byte header and four variable-length fields, as shown in Figure 3-15. We won't discuss DNS in depth because you don't need to know it as intimately as other protocols for purposes of security. However, you should be aware that DNS is one of the most vulnerable and attack-prone protocols.

The reason DNS servers are so attractive to malicious hackers is that once a DNS server has been compromised, it can be fooled into redirecting virtually any clients that make use of the server directly to and through the hackers' computer. Because of this, impersonation and "man-in-the-middle" attacks are easy to perform. If you don't harden your DNS servers against false zone-transfer packets, false information can be fed to it, and then the hacker will be in charge of your name space. That is not what you want to happen, trust me.

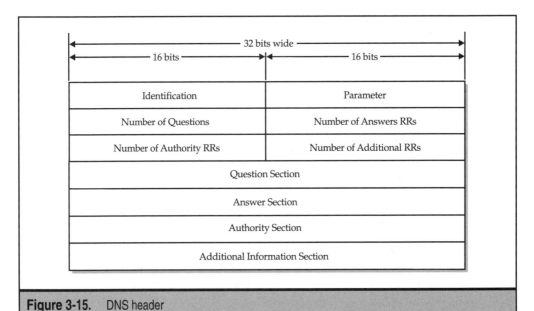

Figure 3-15.   DNS header

Chargen

Chargen is generally used as a troubleshooting tool and is useful for testing network applications. It sends out a continuous stream of ASCII characters. It is not a "danger" to your system per se, but it can be used to annoy or overload systems that are vulnerable to a continuous flood of ASCII characters in a weak kind of denial of service attack to use up system resources. It uses UDP and TCP port 19. It is a data-only packet that uses no headers.

Echo

This is another tool that is useful in testing networks. Echo essentially repeats to a client whatever the client sends to it. Again it is not a threat to your system in terms of being able to take control, but it is a popular protocol to use in denial of service attacks. It uses UDP port 7. It is a data-only packet that uses no headers.

Finger

This service is a tool that is used to help people discover things such as the last time users checked their e-mail, as well as things like office hours, telephone numbers, projects, and other personal data. This is essential information to probing by malicious hackers trying to glean usernames and password possibilities. Finger doesn't represent much danger to the network system itself, but should be blocked. It is rarely used anymore for obvious reasons. It used TCP port 79. It is a data-only packet that uses no headers.

File Transfer Protocol (FTP)

This is a protocol used for just what it says: file transfers. It is very popular and can also be used to remotely update web content on web servers. Its usefulness makes it tough to block completely, and it usually must have a way through the firewall. The way it works is that a client will open a connection to the FTP control port (TCP port 21) of an FTP server. In order for the server to be able to send data back to the client machine later, a second FTP data port (TCP port 20) connection must be opened between the server and the client. To make this second connection, the client sends a PORT command to the server machine. This command includes parameters that tell the server which IP address to connect to and which port to open at that address, which in most cases is intended to be an arbitrary higher-numbered port on the client machine. The server then opens that connection, with the source of the connection being port 20 on the server and the destination being the arbitrary higher-numbered port specified in the PORT command parameters.

If you are not careful, a malicious hacker could use the nature of this process to bypass a firewall or filtering router in certain network configurations. For example, let's assume that a site has its FTP server behind the firewall. Using any number of port scanning techniques, our malicious hacker can determine that an internal web server at that site is available on port 8080, a port normally blocked by a firewall. By connecting to the public FTP server at the site, the malicious hacker can initiate a further connection between the FTP server and an arbitrary port on a nonpublic machine at that site (for instance, the internal web server at port 8080). As a result, the attacker establishes a connection to a machine that would otherwise be protected by the firewall. This is just one of the ways hackers can use ports and sockets to circumvent lax security.

HyperText Transfer Protocol (HTTP)

HTTP is a connectionless protocol running over a TCP communications stream. An HTTP client sends a URL (Uniform Resource Locator), which contains a hostname and path portion (although the hostname can also include a port number for web servers that do not necessarily have to use the standard of TCP port 80), to an HTTP server to establish a TCP stream connection. Even though the client establishes a connection with a server, the protocol is called *connectionless* because once the server has responded to the client's request, the connection between client and server is dropped and forgotten. There is no "memory" between client connections. HTTP server implementations treat every request as if it were brand-new. Other protocols typically keep the connection open, such as in an FTP session where you can move around in remote directories, and the server can keep track of who you are and where you are.

An HTTP connection generally consists of a header followed by an empty line and some data. The header will specify things such as the action expected of the server, or the type of data being returned, or a status code. The use of header fields sent in HTTP transactions gives the protocol a lot of flexibility. These fields allow descriptive information to be sent in the transaction, enabling authentication, encryption, and/or user identification. The header is a block of data preceding the actual data and is often referred to as *meta-information*. "Meta" means simply "something about something." So, meta-information is simply information about more information. HTTP is a complex protocol, which is why it is a favorite of malicious hackers. The more complex something is, the harder it is to protect. Since HTTP is not dynamic, many add-ons such as ActiveX and Java make it a very big security risk, indeed, and in both directions (coming into and going out of your network). That means there are sites just as malicious as individuals trying to get into your site. Once an ActiveX control has been downloaded and run on a client computer in your network, it can assume the same degree of privilege as any program running on that machine. All of this takes place behind your firewall and without your knowledge. You must take great care when setting up a firewall to screen for all of these things.

NetBIOS

Meet the malicious hackers' best friend. Windows is easily the most ubiquitous operating system in the world and is a favorite of the least sophisticated users. That is a bad combination. Don't misunderstand, NetBIOS by itself is not the danger. It becomes dangerous when you leave all of its ports and features open. Unsophisticated users don't know that they should, or how to, turn off the services for them. The first rule of network security is to block incoming access to TCP ports 135, 137, and 138, as well as UDP port 139 on your router. These ports are used by the NetBIOS over the TCP/IP protocol stack.

Network File System (NFS)

Meet the UNIX equivalent of NetBIOS. NFS is not secure because the protocol was not designed with security in mind. NFS sends all file data over the network in clear text. Even if a malicious hacker is unable to get your server to send him a sensitive file, he can just wait until a legitimate user accesses it, and then can pick it up as it goes by on the network. NFS uses UDP and TCP port 2049. You should always block this port as well.

Remote Procedure Call (RPC)

The Remote Procedure Call protocol is one that allows two devices to coordinate the execution of software routines. A program on one device will use RPC to transfer the execution of a subroutine to another computer and have the result returned via RPC back to this same device. RPC can be dangerous, and you should not let RPC traffic through any firewall.

Simple Mail Transport Protocol (SMTP)

Simple Mail Transfer Protocol is the protocol by which mail clients, and other user agents, send mail messages over the Internet. SMTP transactions take place over a TCP communications stream, usually on TCP port 25. SMTP is used solely for sending mail. Obviously, you need to allow this protocol, but it is important that you know which port it uses.

Simple Network Management Protocol (SNMP)

The Simple Network Management Protocol (SNMP) is a protocol that allows for the exchange of management information between network devices. SNMP enables network administrators to manage network performance, find and solve network problems, and plan for network growth. This protocol can be dangerous because it can do just what the name implies—it can manage devices. Computers by their nature are very trusting and can't wait to do what you tell them. Many versions of this protocol have no way of authenticating the source of the command. You should always block SNMP at your firewall. It uses UDP port 161.

Telnet

Telnet is simply a remote login protocol that allows a user on one host to establish a connection with a remote host and interact with the user's terminal as though he or she were physically present. The problem from a security standpoint is that none of the traffic is encrypted, and when you are logging on, both your username and password can be read by any device listening on the network. Telnet uses TCP port 23, and you should always consider blocking this port on your firewall.

CHAPTER 4

Security

n this chapter, we will be discussing the concepts and components of information security. This subject is very broad and requires the careful consideration of many different factors, methods, and products. We will talk about how to recognize and quantify vulnerabilities and then plan and design security procedures. We will define the five security characteristics essential to creating and maintaining a secure environment for the network upon which the very existence of a company relies.

WHAT IS SECURITY?

To answer that question, we will start by taking a look at security as it applies to people. Everyone can identify with the concept of security at home. The saying "A man's home is his castle" provides a very good example, because castles and homes rely on layers of security. With castles you had moats, then drawbridges, then gates with huge locks, then stairways to different floors, then doors to rooms, and finally chastity belts to…well, you get the idea. All of these measures are, to outward appearances, good ways of implementing security at different levels. But everything has its vulnerabilities. A well-placed fire arrow into the ropes holding the drawbridge up could quickly eliminate that barrier. The gates with huge locks might have small hinge pins that could be knocked out, and so forth.

The term "security," as it applies to computers, networks, and information, is so broad and so vague that there really is no single definition. I think it would best be described as a "state of being." This section explores some of the methodologies and tools that can help in making your data infrastructure more secure. VPNs are tools that can be used to secure information as it travels over public infrastructures. *Firewalls* are devices, or more commonly, a group of devices used to protect you from hostile or malicious data that originates from sources outside your network. Policies, procedures, assessments, and training are some examples of the methodologies used to create and maintain this state of security. These tools and methodologies must be monitored and updated continuously. This must all be carefully orchestrated to achieve these goals. You will be way ahead of the game if you can keep the following points in mind:

▼ *Security is inversely proportional to complexity.* This holds true in every facet of security. Every factor, whether it's vulnerability, risk assessment, policy, procedure development, deployment of mechanisms, or even training, should be as uncomplicated and simple as possible. The more instructions, equipment, or procedures you have, the greater the chance of failure.

■ *Security is inversely proportional to usability and productivity.* Strong security policies are more inconvenient than lax or no security policies. People don't like using hard-to-remember passwords nor do they like to change them frequently. Firewalls, VPNs, and intrusion detection systems are very literal in terms of what they allow in and out of your network. That means that when new technologies become available on the Internet they need to be explicitly allowed in. This can make certain valuable services (along with all those that are most entertaining like music and graphic content) on the Internet unavailable.

■ *Your security is only as strong as your weakest link.* A prime example includes the subject of this book. When you implement VPNs, you open yourself up to a whole new angle of entry into the system. The purpose of VPNs is to make connections from home and remote locations private to the internal corporate network. You need to find a way to make sure that the notebook PCs and home computers of your mobile users maintain the same policies as those in the central office. If one of these employees has children at home who use the computer for games or Internet access, holes can be opened up inadvertently by downloading what seems to be harmless material. Once that wall has been breached while that remote computer is attached to the Internet, it becomes a nice, fully encrypted tunnel into the corporate network.

■ *Concentrate on known threats and probable threats.* Perceived threats that are created by sales hype, news reports, and e-mail "warnings" can waste your limited budget faster than almost all other real and probable threats combined.

■ *Security should be considered an investment, not an expense.* Human nature makes paying for something intangible unpleasant. It's like insurance—you never really feel like you are getting anything while you are paying. Yes, you are doing it for peace of mind, but nobody proudly displays their insurance policy in a beautifully mounted frame over their fireplace. This trait doesn't make convincing upper management any easier. Investing in computer and network security that can adapt to the constantly changing business requirements and risks makes it possible to satisfy the constantly changing applications, operating systems, protocols, computers, and more can be very challenging from a technical standpoint. Everyone knows what that means from a financial standpoint, and, unfortunately, people whose professions involve a thorough understanding of finances and technology are very rare indeed.

▲ *Make sure that the only people who figure out ways to break into your system either work for you or are extremely forgiving and charitable.* The people that fall into the latter category number so few that I wouldn't even make a column in my spreadsheet to keep track of them. This means that you need to constantly test the integrity and robustness of your system. This also means that you need to monitor any activity of those individuals whose intentions are not in your best interest that might also be testing that same integrity and robustness without your permission.

Making sure that corporate information systems run correctly and consistently is extremely important—even essential—in most businesses. Let's hope that the people in charge of the security budget and monitoring of the Air Traffic Control network consider themselves in the "absolutely essential" category. To put it simply, the objective of information security is to put measures in place that can eliminate or reduce significant threats to an acceptable level.

INFORMATION SECURITY DOMAINS

In today's world, information and the systems that contain it are the key assets of most companies. You need to keep that information protected at every point of entry in every boundary, but those boundaries are not always clear and are often overlooked. Remember, all it takes is one hole into your organization to breach security. I could have a state-of-the-art alarm system in my car, but I would be left with a false sense of security if I failed to consider that the alarm system requires 12 volts of continuous power and overlook the fact that the power cable can be cut simply by reaching under the car. A false sense of security is just as bad as no security and in some cases worse.

This goes to the very heart of this subject. You must consider every possible point of entry, and to do that, you need to understand the following. Information can take many forms—the most obvious forms tend to be overlooked—and the methods necessary to secure that information take many forms as well. Instead of dividing information into categories based on content, you need to consider where and how it is kept. Security as it applies to information falls under three general domains.

The Three Information Domains

Those three information domains can be seen in Figure 4-1.

▼ **Logical/Network** Information is also stored on computers and accessed via networks. Information can be stored offsite that users can access through URLs, UNCs, or other abstract methods of storage. The actual location of the data is often unknown to the user. Along with this abstraction comes a certain loss of accountability and responsibility should they accidentally or purposely alter or destroy that data.

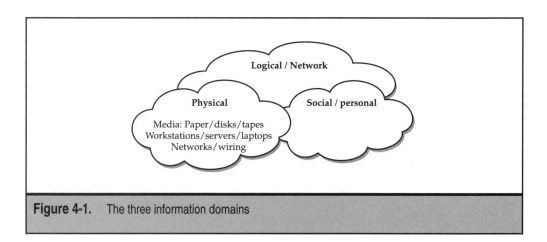

Figure 4-1. The three information domains

- ■ **Social/Personal** Successful organizations recognize the value of their personnel, the knowledge they hold in their heads, and their ability to use that knowledge for or against the organization's best interests.

- ▲ **Physical** Information can be in printed or written form, stored somewhere using a box, safe, floppy disk, server, or other media. Classical security concentrates on physical protection of buildings, server rooms, access controls, and so on.

Possible Access Points into the Domain

In each of these domains there exist possible points of access to the outside. These can be seen in Figure 4-2.

The numbers in parentheses in the following list correspond to the numbered interfaces in Figure 4-2.

- ▼ **Logical/Network** Computers and networks provide storage and retrieval of corporate information and processes. The increased complexity, speed of technical evolution, market movements, and organizational changes of the '90s have made securing this domain a real challenge.

  1. Telephone and voicemail security is often overlooked. Telephone and voicemail systems are computer based and increasingly complex and may interface to the corporate intranet.

  2. Dial-up modems remain the predominant access method for telecommuting and remote access to corporate data. Dial-up networks can be an easy entry point for attackers. Dial-up networks are countless, difficult to monitor, and rarely exhibit much in the way of security. Analog connections are easier to compromise than ISDN connections; unfortunately, there are infinitely more analog connections.

  3. The Internet is fast becoming the preferred method for information exchange whether via the Web, e-mail, or extranets using encrypted tunnels or VPNs. The Internet connection offers a way to communicate with millions of people globally, but it's a two-way street. You can only control security to the edges of your network infrastructure. Transmission over this medium makes it open season for eavesdropping, identity spoofing, and denial of service by anyone.

  4. Large corporations have many links to partners and vendors, often using many different technologies that were implemented before the company had Internet access. These partners and vendors will now most likely also have their own Internet access, and connections to them are often not as secure as they should be. The problem that is most common arises because each company has its own IT staff. Should any one of those partners have a lax, or worse yet, no security policy, your whole security system could unwittingly be as weak as theirs. Partners don't always stay partners, and

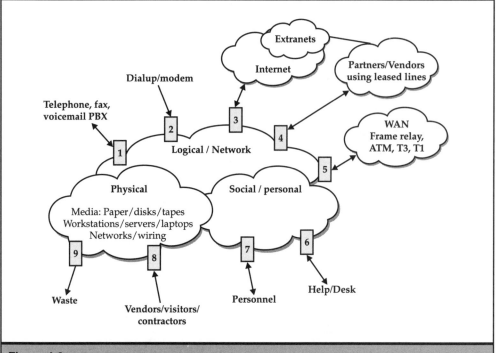

Figure 4-2.   Possible access points into the domain

in many cases, they are also your competitors. Being connected to those partner networks makes it easy for attackers within that partner organization as well as any attackers who may have already penetrated the partner's network leaving your network with a "back door" situation.

5. Wide area networks (WANs) are used to extend the corporate intranet to remote areas. The corporate intranet may extend across several cities and countries. Multiple technologies can be involved using different Data Link layer protocols in conjunction with TCP/IP. Though the complexity of WANs makes penetration difficult, it doesn't protect you from those individuals employed within the service provider itself. The main point here is that dedicated leased-line WAN facilities are administered by telephone companies, and depending on your specific situation, you might not want security policies that are outside of your control.

■ **Social/Personal**   Employees have relationships with others within and outside the company. Employees can be reached by methods such as telephone, fax, e-mail, and personal meetings.

1. *Social engineering* (taking advantage of people's naiveté and trusting nature) can be used to trick personnel into divulging information or providing access.

2. Helpdesks may also be subject to social engineering, providing modem numbers, passwords, and user identification to be disclosed unwittingly to unauthorized persons.

▲ **Physical** Most of us live in the physical world of people, buildings, equipment, wiring, computers, disks, paper documents, and so on. Many physical access points can allow unauthorized people to enter. These access points can be things like doors, windows, wiring closets, electrical rooms, and the like. However, physical security extends beyond the property and buildings under corporate control. What many companies overlook is that by leasing space, they become subject to landlords, maintenance personnel, custodians, and so on.

1. Many people who are not employees will have access to buildings in one way or another. Obvious threats include theft, damage, and copying.

2. Sensitive information, if not carefully disposed of, can yield a smorgasbord of valuable information to anyone willing to sift through it.

Other physical threats include theft or loss of company laptops, backup tapes, and so on. Theft or duplication of media during transport also needs consideration.

Ultimately, any one of these threats can result in critical information being lost, copied, deleted, accessed, or modified. These activities can also provide enough information to deny service. All of these possibilities open the door for loss of confidentiality, integrity, or availability.

GENERAL SECURITY CONSIDERATIONS

The security requirements for one organization are never the same as for another. Before you start implementing a security system, you need to ask yourself several questions. Once you've answered those questions, you need to put a plan in place to quantify and evaluate the answers. This is the hardest part of implementing any kind of security system. Some of the questions you should ask are

▼ *What are we trying to protect?* This is the question that counts and, when you think about it, it's the hardest to quantify. You must think not only about what those assets are, but also just exactly how the value of those assets can be measured. You need to decide their values and locate these assets. The process isn't as easy as you might think because not all assets are tangible or clearly quantifiable. You will quickly discover that responses to this question will include:

■ Equipment

■ Client information

■ Product information

- Employees
- Marketing strategies
- Public image
- ▲ Reputation

Once your asset list is complete, you may find the following questions even more disturbing and difficult to answer.

▼ *How do we quantify the value of these assets?* Assets are easy to quantify when they have a "market value," such as equipment and property. But what is the going price for reputation or corporate image? It is different for everyone. (The obvious exception is public utilities. Their motto tends to be: "We don't care—we don't have to.")

■ *What are our vulnerabilities?* This can be a tougher question than it seems. From a physical standpoint, the location of server farms, networking equipment, computers, and such will seem obvious. Unfortunately, the more subtle ones tend to get by without notice—for example, a vulnerability related to an individual's personality or character traits. You can purchase and implement the most elaborate and latest combination of firewalls, VPNs, guard dogs, laser beam and motion sensing equipment, and even electric fences coated with cyanide, surrounded by poison oak bushes that shoot trained bees at you, to try to protect those systems. All of that will immediately become useless if just one of these people, possessing or knowing who possesses access and/or authorization capabilities, has a personality trait that makes the discovery of a key password possible for anyone clever enough to exploit those personality traits.

▲ *Who is most likely to attack us?* This question goes back to the beginning of time. Julius Caesar could have benefited from knowing who his attacker was likely to be. Attackers are not always obvious and are not always outside the walls of your information "castle." In fact, some surveys estimate that most damage and theft breaks down as follows:

- Individuals within the organization: 81 percent
- Individuals outside of the organization: 13 percent
- Former employees: 6 percent

The best course of action is to do some profiling to determine the likelihood of these percentages based on the nature of your business. There is one very important thing that you must never forget in this category that is unexpected and often overlooked. Most of today's protocols allow access to services based on permissions. The natural assumption is that permissions are for people, but many machines and individual services running on those machines are given permissions, in many cases by default. A clever person can simply disguise himself as one of these machines or services and have full access to all of the services on the network.

▼ *What methods can be used to attack us?* Remember when you were little and some other kid would come up and threaten you? The good old "Oh yah, what are you gonna do about it" canned phrase could always be counted on. (If you don't remember that, then obviously you were never a kid.) The methods available for an attack differ in many ways depending on the type of business involved. For the most part, however, those attacks will generally fall into one of these four categories:

- Unauthorized entities accessing computing resources and/or information

- The removal, destruction, modification, disclosure, or misuse of information

- Monitoring, viewing, or copying information as it travels over the network

- Preventing or altering the availability of the services provided on the network (denial of service)

We will go over these in more detail later.

■ *What would the consequences of a successful attack be?* This obviously relates to the value of the item or items compromised. This is where the main concern comes into play—cost. The one basic rule of security is that you need to evaluate the amount you are willing to invest into securing your data as it relates to what the cost would be of losing that data. To simplify that, it is similar to asking whether you are willing to hire 24-hour security guards to guard your mailbox. That decision would obviously depend on the type of mail you receive. If it involves multi-million-dollar royalty payments (not unlike those that I am anticipating from the sales of this book), then the answer might be yes. If your mail consists of flyers, monthly invoices, and checks from Ed McMahon informing you of the possibility that they may be worth $10,000,000, then I would opt for no.

▲ *What can we do to protect ourselves from attack?* You need to consider what means of protection are available to you. This would include people, equipment, policies, laws, and so on. Protecting your computers would also require you to evaluate whether these factors can be controlled. If your company is international, it is likely that local laws or logistics involving environmental or geographical factors do not make allowances for some or any forms of data protection in those areas. This is also where you decide what kinds of software and hardware products are available and compatible with your existing requirements.

You may have other things to consider depending on your situation. Once you feel you have answers to these questions, you will need to find an approach to implementing this security.

PLANNING YOUR SECURITY APPROACH

A number of organizations provide templates that spell out very specific guidelines for creating a security plan or policy. The template you need to follow depends on the nature and size of your organization. In this section, I look at two approaches to security: the reactionary method and a more deliberate plan.

The Reactionary Method

Let's look at a hypothetical situation. Yours is a medium-size company. You have a medium or large network installed and currently operating. You have even developed some ad hoc policies as your company has grown. (This ad hoc policy structure is representative of probably 90 percent of all companies out there.) Now you suspect that you have recently been victimized by an attack or feel one is imminent. This is probably the most common reason that companies ever implement a security policy, hence the name "reactionary."

This company might approach the situation with a primary interest in putting out the perceived existing fires, but with an eye toward comprehensive protection further down the road. This approach favors speed at the expense of precision. It is also a good way to force yourself into creating an actual legitimate security policy...intentionally, this time!

So we have pretty much established that you know what it is you want to protect. You know from whom you want to protect it and to what degree:

▼ You need to examine what your existing policies are, such as network topology, operational procedures, and your average user practices and characteristics in use.

■ You then need to create an attack profile, determining the nature of the attack you just suffered. Approach this as though you were the attacker; how did the attacker compromise your network, and what other means might be tried next? What network access points, modems, phone lines, computers, or server services are visible to those outside the network? Where are key systems kept, and how accessible are they either physically or logically? In other words, try to locate as many weaknesses as you can.

■ Examine any weaknesses discovered earlier, make a list of those current weaknesses, and, given their nature and locations, try to determine what future threats are likely to be encountered.

■ Define some kind of formal security policy, and classify the information in the policy. Try to make clear what information is most important and ensure that everyone knows it. Include whatever procedures or training might be required of your users.

■ Define and create a specific user policy, making sure that it is signed and understood by each user and that you include whatever instruction might be necessary to educate the users.

■ Create technical guidelines covering the secure installation, maintenance, and production of your networks, servers, and any other perceived weak points.

▲ Audit and test the integrity of those newly installed systems and procedures on a regular basis.

This pretty much represents the reactionary method—one that will be implemented "when trouble strikes."

A Deliberate Plan

Now let's take an example of another medium to large corporation. Their approach will be more deliberate and precise. It will take much more initial investment in time, money, and planning. This type of approach is usually in response to witnessing the dilemma of a company that was forced to use the reactionary method. Keep in mind that the empirical numbers that you will see in the following approach imply a previously tested formula to apply. This will just be an example of the possible conclusions derived from this company's risk assessment.

This company's approach might be the following:

▼ *Start by assembling a detailed list of your assets.* Decide what needs protection. Classify your information and your processes. Identify and itemize the importance of these assets. Find out what kinds of information you have. Know where this information is stored. Know how and in what form this information is stored. Quantify the possible cost or damage that will be suffered should these assets be lost. Once you have done this, ensure that the measures taken to protect those assets correspond proportionally to the value of assets.

■ *Examine existing measures.* List the rules that currently exist. Take a look at your security policies, if any. List standards or practices that have already been established.

■ *Define what your overall security objectives will be.* Decide what level of availability you will need. Decide on the amount of redundancy that will be required. List the methods you will use to maintain the confidentiality of your information. Know how you intend to maintain and monitor the integrity of your information.

■ *List what threats you think you might be vulnerable to.* Make sure you have some way of knowing what the system needs to be protected from. List what kinds of threats you are most likely to encounter. Classify those threats into categories such as disgruntled employees, terminated employees, vengeful hackers, curious hackers, corporate espionage, and technical failures.

■ *Anticipate and list what the impact of a successful attack will be.* Have a way to quantify the impact or consequence should one or more of these threats succeed. This will always vary and is always specific to your situation. Classify what these losses will be regarding loss of company secrets, modification of accounting data, falsified records, money transfers, and any others you can think of. It is important that these losses be judged by business experts, not technical experts. When classifying these impacts, separate short-term effects and long-term effects. When quantifying, you assign a value to the impact. Let's say you choose to break it down into some number between one and six with one being the lowest impact. For example:

1. The impact is negligible.

2. The effect is minor, major business operations are not affected.

3. Business operations are unavailable for a certain amount of time, revenue is lost, but customer confidence is affected minimally, and the company is unlikely to lose customers.

4. Significant loss to business operations, customer confidence, or market share. Customers will be lost.

5. The effect is disastrous, but the company can survive, at a significant cost.

6. The effect is catastrophic to the point where the company cannot survive.

▲ *Quantify your risks.* Define a method to determine the likelihood of a threat occurring. In this case, it is important that this be judged by technical experts as opposed to business experts. We will also assign a value to those likelihoods. Again, let's break it down into a number between one and six, with one being the lowest likelihood. For example:

1. The threat is highly unlikely to occur.

2. The threat is likely to occur less than once a year.

3. The threat is likely to occur once a year.

4. The threat is likely to occur once a month.

5. The event is likely to occur once a week.

6. The event is likely to occur daily.

Then create a formula to quantify the actual risk. You could multiply the impact by the likelihood to determine the risk. The risk would then have a minimum value of 1 (no risk) and a maximum of 36 (extremely dangerous risk). You then just need to decide on a number between the two extremes that you consider acceptable. All risks having a value higher than this number will be considered unacceptable risks requiring the necessary measures to avoid them. For our purposes, we will set the acceptable risk at 15 and use this value to determine the severity of the measure used to fight the risk.

▼ *Examine requirements outside of your control.* This would also require you to evaluate whether these factors can be controlled. Any large organization needs to consider national laws, international laws, corporate requirements, corporate culture, contractual requirements, budgets, and so on. You must also take into account any logistics involving environmental or geographical factors. This is also where you decide what kinds of products to consider, such as software, hardware, or a combination that is compatible with your existing requirements.

■ *Define your overall security approach or policy.* Revisit your original security objectives, and then define what countermeasures, policies, roles, processes, responsibility, mechanisms, and other pertinent actions or standards apply to your unique situation. You need to see if, when these policies are applied, your risks can be reduced to an acceptable level. If not, and the cost seems too high, see if the remaining risks can be economically insured, or revisit your quantifications and measures until a happy medium can be found.

■ *Develop an implementation strategy.* Consider your strategy carefully, because you must coordinate changes, notices, training, and all other factors over a period that is achievable while still maintaining productivity. Define a security group that includes representatives from all departments related to its existence. Users, administrators, managers, and executives should have clearly defined roles, responsibilities, and understanding of the consequences. You then need to run pilot tests and fine-tune your policies, processes, and standards until you obtain satisfactory results. Once you have done this, you are ready to switch over completely and then start the hard part (listed next).

▲ *Monitor, evaluate, and test the integrity of the system*   Reevaluate risks and security strategy regularly, and keep abreast of any changes that can affect the system in any way. That is no small undertaking. If you don't find the weaknesses, someone out there will be happy to discover them for free. (Naturally, that just covers weakness discovery charges; everything after that first free service costs…well, more.)

COMPONENTS OF A SOUND SECURITY ARCHITECTURE

You should now understand that to protect something, you must first identify your assets as well as the risks associated with those assets. Once you have gathered these facts, you can start building your security architecture. Security architecture is a combination of conceptual objects and tangible objects. Information security can best be understood by breaking it down into two categories.

1. The five security characteristics of information:

   ■ Identity
   ■ Integrity
   ■ Confidentiality
   ■ Availability
   ■ Auditing

2. The three states of information:

   ■ Processing
   ■ At rest
   ■ In transit

These are not absolute definitions of informational states. You may run across some definitions that are more granular, where you will see terms like access control, auditing, and others, but all of those or any others you will see can be found in our earlier definitions. In this section, I discuss these two categories in more detail.

THE FIVE SECURITY CHARACTERISTICS OF INFORMATION

The control and maintenance of these five characteristics tend to define the policies or the conceptual objects of your security architecture.

Identity

Identity, as it applies to computers and security, involves three processes. Those processes are

▼ Identification

■ Authentication

▲ Authorization

Identification Definition

Identification is just a way of finding out what or who is being presented to you. To allow access to a computer network, finding out who wants it isn't enough. Suppose your mother-in-law comes over to your house, knocks on the door, and you ask, "Who is it?" You then hear through the door: "It's me! Ezmerelda, your mother-in-law." What do you do now? You know the identity. Now you need to decide whether you will let her in. (That decision, of course, probably depends on whether your spouse is home at the time.)

Authentication Definition

You are also faced with another dilemma—you have heard a response through the door and need some way to make sure that it is, in fact, your mother-in-law. This is the process equivalent to *authentication*. You can use several methods to authenticate your mother-in-law's identity. You could recognize her voice (although the "you good for nothing so-and-so" would probably be all you need), or you could look through the peephole and ask that she show you two forms of identification. (That's always fun; mothers-in-law just seem to love it when you do that.)

Authorization Definition

Now for the "real" dilemma. Do you let her in? This is where *authorization* comes in. This can also be referred to as *access control*. Authorization is dependent on the other two factors and doesn't stop at the door, so to speak. Even if you do let your mother-in-law in, you need to determine what you will allow her to do (with the default option of "leaving" already active). Can she go into the bedroom, the kitchen, your medicine cabinet, your wallet, the bathroom? The authorization can even be hierarchical. For example: yes, she can go into the bathroom and use the toilet, but she can't open the medicine cabinet. There are some other subtleties with this subject involving two added concepts.

▼ Access

▲ Permission

Allowing your mother-in-law into the bathroom and into your house was controlling access. Once you allowed *access,* you needed to decide what *permission* you would give. Allowing her to use the facilities was the permission or right.

Identification Methods

There are many different ways of identifying an individual. Listed next are some of the more common methods used today.

User IDs   Being identified to the network resources requires a user ID. User IDs and associated passwords for authentication are inexpensive and widely integrated into today's systems.

Proprietary Tokens   Tokens are physical cards similar to credit cards that work in conjunction with a user ID to identify a user to the system. They combine something a person knows, such as a password or PIN, with something they possess, a token card. Token cards commonly generate either dynamic passwords or a response in a challenge-response communication between the user and the system. Tokens are commonly used for secure remote access where high levels of security are required.

Biometrics   A biometric is a unique, measurable, physical or behavioral characteristic of a human being. Biometrics are handy to use because you never have to remember to bring anything. Biometric characteristics can include fingerprints, hand and face geometry, signature, and voice. Each of these methods has different degrees of accuracy, cost, social acceptability, and intrusiveness. An extreme example of an intrusive technique would be requiring a DNA sample. Voice identification would be an example of a nonintrusive and socially acceptable technique.

All biometric products operate in a similar way. First, a system captures a sample of the biometric characteristic during an enrollment process. Unique features are then extracted and converted by the system into a mathematical code. This code is then stored as the biometric template for that person. The template may be stored in the biometric system itself, or in any other form of memory storage, such as a database, a smart card, or a barcode. When a user needs to be identified, a real-time sample is taken and matched against stored templates. If the match is within predefined tolerances, the individual's identity is established.

There is no perfect biometric technique for all uses. Different biometric techniques are more suitable for particular situations. You wouldn't want to install an eloborate retinal scan system for a cook to gain access to your meat locker. You also don't want nuclear weapon access based on a telephone voiceprint system. Factors can include the desired security level and the number of users. For instance, identifying a user for their known biometric template or one-to-one match. Biometric systems are not 100 percent accurate. Accuracy in biometrics is measured by false acceptances and false rejections. *False acceptances* occur when an unauthorized user is allowed access. A *false rejection* is when an authorized user is denied access. Thresholds can be adjusted to reduce one type of error at the expense of increasing the other. The choice of threshold depends on the level of security required, the acceptability of the type of error, and user acceptability. The accuracy of biometrics can also be improved by combining two techniques such as fingerprint identification and face recognition.

▼ **Fingerprint biometrics** Fingerprints have traditionally been used as an identification tool in law enforcement. Fingerprint recognition systems convert a scanned image of a fingerprint into a mathematical representation of the features. The main strengths of fingerprint recognition are its long history, the uniqueness of fingerprints, ease of use, cost, and accuracy. Additionally, it has the potential to be integrated into inexpensive devices such as smart cards and keyboards. A disadvantage may be its lack of social acceptability due to its association with those who get arrested and its invasion of privacy.

■ **Hand geometry** Hand geometry has features similar to fingerprints, though perhaps a higher social acceptability. Similar devices are used in both cases. Hand geometry is less accurate than fingerprints because of fewer features and less variability in these features. It may be acceptable when a user is matched against a known template. It would be less acceptable when trying to match against a large set of templates.

■ **Face geometry** Face geometry uses a standard video camera to capture facial images. The system extracts from the images features that don't easily change, such as the geometry of the eyes and nose. The template created is matched against real-time images. People change, and facial hair, positioning, and glasses can affect accuracy. Face geometry is less accurate than fingerprint biometrics.

■ **Voice biometrics** Voice biometrics is based on distinguishing the sound of a human voice based on the resonance of the human vocal tract. It is different from voice recognition, which recognizes spoken commands or words. The system is trained by repeating a phrase that will be used as an access code. One shortcoming of voice biometrics is false rejects that deny a legitimate user access. This is due to medium to low accuracy rates and dependence on the type of equipment used. It is mainly used for outdoor situations and telephone access.

■ **Signature recognition** Signature verification depends on the rhythm, relative trajectories, speed, and number of pen touches. It measures the method of signing rather than the finished signature and so is different from the comparison of a signature. A pen-based computer or digitizing pad is required for signature capture during enrollment and during verification. It has a relatively low level of accuracy. A good application for this might be where a history of signature use exists, such as retail transactions and document authentication.

▲ **Smart cards and card readers**   A smart card is a tamper-resistant computer embedded in a credit card–sized card. The cards have embedded integrated circuits that implement a CPU, application data storage, and RAM used by the CPU. A smart card and associated host software are used both as an application platform and an identification and authentication device.

Identification security for smart cards is based on:

▼ The user having physical possession of the smart card

■ The user knowing the password or PIN to activate the card's function

■ The security functions available on the cards

▲ The tamper-resistant qualities of the card

The host software supports the attachment of the smart card reader to the host platform (for example, workstation), and the identification functions require interacting with the smart card. Neither the reader nor the smart card trusts each other without a successful completion of the security process. A smart card together with a user password or PIN forms the basis of identification. The user must enter the correct password or PIN before the card will allow access to its resources. If the attempts to access the card exceed a user-specified number of attempts, the card disables or even destroys itself and its contents. Like a password, the card can be reenabled after failed attempts, unless it has destroyed itself. Smart cards are designed to resist tampering and external access to information on or used by the card. The ultimate decision on whether to carry out the identification transaction is made by the card, strengthening the security of smart card–based identification. Smart card authentication is based on cryptography and is used in the authentication process. This authentication is based on both public-key and secret-key cryptography. Cryptography and authentication are discussed in more detail in later chapters.

Authentication Methods

Authentication methods involve technologies and standards such as cryptography and various algorithms. We will be discussing those subjects and algorithms more fully in later chapters. For now, I will provide a brief overview.

Authentication methods involve the following:

▼ **Cryptography**   This is a technology used to scramble data in order to prevent unauthorized individuals from reading the data. A cryptographic key is a sequence of characters used in scrambling and unscrambling the data.

■ **Public-key/private-key cryptography**   This is a cryptographic technique that gives a user a "public" key for others to use to communicate with the user and a "private" key that is used as a digital signature.

■ **Public-key certificate**   This is an electronic document that contains a user's public key. It is made available to anyone wanting to verify the digital signature or communicate confidentially with a certified user.

■ **Message digest**   This is a method to ensure that information cannot be modified without detection. It is used in the digital signature process.

■ **Digital signature**   This is a process that uses a private key to scramble the information. Since only the signer's public key can be used to unscramble the information, this is considered sufficient proof of the signer's identity.

▲ **Public-key infrastructure**   These are the functions required to issue and manage the public-key certificates needed for authentication.

Authorization Methods

Authorization is the permission to use a computer resource. Access is the ability to do something with whatever system is being used, such as edit, add, or delete. Access control is the means by which the ability is explicitly enabled or restricted in some way, usually through physical and system-based controls.

Access control often requires that the system be able to identify and differentiate users. Access control is most often based on *least privilege*—the system grants just enough permission to do the specific part of the job the user is responsible for. User accountability requires the linking of activities on a computer system to specific individuals and, therefore, requires the system to identify users.

Access controls provide a technical means of controlling permissions such as what information users can utilize, the programs they can run, and the modifications they can make.

Computer-based access controls are called *logical access controls.* Logical access controls can be used to decide not only who or what is to have access to a specific system resource, but also the type of access that is permitted. These controls may be built into the operating system, applications programs, or specialized security packages. Logical access controls may be implemented internally to the computer system being protected or may be implemented in external devices.

The concept of *access modes* or *rights* is fundamental to access control. Common access rights, which can be used in both operating systems and applications, include the following:

▼ Read

■ Write

■ Delete

■ Create

▲ Execute

In deciding whether to permit someone to use a system resource, logical access controls examine whether the user is authorized for the type of access requested based on access criteria such as:

▼ **Identity**   It is probably fair to say that the majority of access controls are based upon the identity of the user or object, which is usually established through identification and authentication.

■ **Job or department**   Access to information may also be controlled by the job assignment of the user who is seeking access, like those you would find in network administration, human resources, or accounting.

■ **Location**   Access to particular system resources may also be based upon physical or logical location. For example, users can be restricted based upon network addresses.

■ **Time**   Time of day restrictions are common limitations on access. For example, use of confidential personnel files may be allowed only during normal working hours and denied at all other times.

▲ **Service constraints**   Service constraints refer to those restrictions that depend upon the parameters that may arise during use of the application or that are preestablished by the resource owner. For example, a particular software package may only be licensed by the organization for five users at a time. Access would be denied for a sixth user, even if the user were otherwise authorized to use the application. Access may also be selectively permitted based on the type of service requested. For example, users of computers on a network may be permitted to exchange electronic mail, but may not be allowed to log into each other's computers.

Integrity

Integrity control ensures that all data is kept in its original state and that it can't be corrupted, damaged, altered, appended, or removed. Digital signatures and hash algorithms can be used to guarantee the integrity of files and applications. One of the most common means of gaining unauthorized access to a computer system is to install altered copies of operating system programs that provide access for the intruder when they are executed. It is important, therefore, that the integrity of operating system components can be verified. Attackers understand this all too well.

Antivirus software is designed to record the state of the original system files and applications so that it can verify and monitor those files regularly to make sure they haven't been tampered with. Most packages on the market perform a self-verification of their own integrity as well.

Data integrity serves as a control mechanism for other security services such as access control for data stored on a hard drive and data that is in transit or being transmitted over the network infrastructure. Integrity services make sure that problems are avoided by verifying that the message content has not been altered. It also records the sequence of message components as they are being transmitted and ensures that the sequence has been preserved. This is especially important when data is in transit over the network, because it cannot be protected by any of the security services in the operating system or application software.

Confidentiality

Information must not be read or copied by anyone who has not been explicitly authorized to do so. Confidentiality involves protecting information at rest or stored on the server, and in transit across the network infrastructure. Encryption is the method used to ensure confidentiality by making the data meaningless to anyone who might be monitoring the flow of traffic, but who is not authorized to read it.

The objective of confidentiality is similar to that of access control. Confidentiality in its broadest definition extends the philosophy of access control to policing the data itself, but also the nature of the operation being performed at the time. Maintaining confidentiality is analogous to a customer's session with a bank. The security mechanisms should guarantee that no one can tell whether the transaction being completed is that of depositing money into an account, withdrawing money, or transferring money.

Availability

People have come to depend upon and demand high availability of service without interruption. This includes, in many cases, a low tolerance for degradation of performance. All users, employees, customers, partners, and suppliers rely on this predictability and availability. Much of this is due to the success of the telephone companies. They have managed to accomplish this availability with surprising consistency over many years. Availability is also an important aspect of security. Each piece of equipment usually provides some services that fulfill this requirement.

The first step in analyzing the risks associated with availability is to identify Single Points of Failure (SPOF). You also need to evaluate the impact on your system should the SPOF fail. The impact of the SPOF determines the grade of availability that has to be reached.

Availability is usually graded as a percentage. Systems that host mission-critical data should usually require availability in the range of 99.0 to 99.5 percent. Some systems require 99.95 percent and above. These systems are classified as High Availability.

You will learn more about this later, but availability means protecting the network from what are known as denial of service (DoS) attacks. A DoS attack makes the intended network unable to carry legitimate users' data. Another more common and much more severe version of this attack is the distributed denial of service (DDoS) attack. In a DDoS attack, one or more malicious hackers break into multiple poorly protected networks across the Internet, hijack their systems, and, without the owners' knowledge, use them to flood the intended target network with millions of bogus packets simultaneously. This kind of attack overloads the target's servers, connections, and routers. A DDoS attack is easier than it may seem because anyone can download easy-to-use hacker utilities from the Web.

To maintain constant availability, an organization must have redundant WAN and LAN connections, routers, and servers and, if possible, multiple geographic locations. The types of equipment used to maintain this availability usually include firewalls, VPNs, and intrusion detection systems. These systems examine every packet coming in and leaving the network. Should one of these systems be compromised for whatever reason, communications could be brought to a halt.

To maintain availability, you should focus on:

▼ Keeping the contents of your data private

■ Maintaining the integrity of your data

▲ Making sure that your data and network remain accessible

Combine and integrate these three objectives so that the characteristics for one of these do not interfere with the others. For example, if you prevent others from accessing your data, you help both integrity and privacy. If the network remains accessible, you also help ensure integrity. One aspect of maintaining accessibility is preventing access by hackers.

Auditing

Constant auditing of resources and activities is essential to maintaining the robustness of your security architecture. Several auditing systems and methods are available, and many operating systems and devices have built-in auditing capabilities. Firewall and intrusion detection systems, for example, work well together.

It is very important that as soon as you can, you find some way to "baseline" your data use activity. That is the data that will be essential to knowing whether attacks are imminent or about nonmalicious activity outside the scope of the policy you have set forth. Once you have this information, you will have some frame of reference to identify users suspected of improper modification of data. An audit "trail" may record before-and-after versions of records.

Auditing allows you to maintain individual accountability. Letting users know that they are personally accountable for their actions, and that those actions are constantly being monitored and logged, will go a long way in promoting proper user behavior. It is like a convenience store clerk who is standing under a cash register with two cameras facing him and the register. Users are less likely to attempt to circumvent security policy if they know that their actions will be recorded in an audit log. It will help keep the honest people honest, and the idiots will provide "entertainment" for the IT staff. There is nothing you want less than an unhappy IT staff.

Audit trails work in concert with logical access controls, which restrict use of system resources. Granting users access to particular resources usually means that they need that access to accomplish their job. Authorized access, of course, can be misused, which is where audit trail analysis is useful. While users cannot be prevented from using resources to which they have legitimate access authorization, audit trail analysis is used to examine their actions. Someone in Human Resources might have access to those personnel records for which they are responsible. Audit trails can reveal that an individual is printing more records than the average user, duplicating records and the like, possibly indicating that this person might have a side job working for a local "headhunter."

Access to online audit logs should be strictly controlled. Computer security managers and system administrators or managers should have access for review purposes. However, security and administration personnel who maintain logical access functions may have no need for access to audit logs.

It is particularly important to secure your audit. You need to be sure to maintain the integrity of audit trail data against modification. Most intruders will try to cover their tracks by modifying audit trail records. Audit trail records should be protected by strong access controls to help prevent unauthorized access. The integrity of audit trail information may be particularly important when legal issues arise, such as when audit trails are used as legal evidence. The confidentiality of audit trail information should also be protected. Strong access controls and encryption can be particularly effective in preserving confidentiality.

Audit trails can be used to review what occurred after an event, for periodic reviews, and for real-time analysis. Reviewers should know what to look for to be effective in spotting unusual activity. They need to understand what normal activity looks like. Audit trail review can be easier if the audit trail function can be queried by user name, computer name, application name, date and time, or some other set of parameters to run reports of selected information.

▼ **Audit trail review after an event** Following a known system or application software problem, a known violation of existing requirements by a user, or some unexplained system or user problem, the appropriate system-level or application-level administrator should review the audit trails. Review by the application or data owner would normally involve a separate report, based upon audit trail data, to determine if their resources are being misused.

■ **Periodic review of audit trail data** Application owners, data owners, system administrators, data processing function managers, and computer security managers should determine how much review of audit trail records should be done based on the importance of identifying unauthorized activities. This determination should have a direct correlation to the frequency of periodic reviews of audit trail data.

▲ **Real-time audit analysis** Audit trails are usually analyzed in a batch mode at regular predetermined intervals. Audit records are archived during that interval for later analysis. Audit analysis tools can also be used in a real-time or near real-time fashion. Such intrusion detection tools are based on audit reduction, attack signature, and variance techniques. Manual review of audit records in real time is nearly impossible on large multiuser systems due to the volume of records generated. However, it might be possible to view all records associated with a particular user or application and to view them in real time.

THE THREE STATES OF INFORMATION

For the most part, the control and maintenance of the three states of information determine the tools—or tangible objects—your security architecture requires. These tools can be actual equipment, software, or a combination of both. Some of the tools or systems necessary for controlling and maintaining these states are firewalls, intrusion detection systems (IDSs), storage devices, and Virtual Private Networks (VPNs). (I can recommend a really good book for beginning VPNs; we will list that title at the end of this book.) These systems may be separate or combined into one product. We will be going into more detail on some of these systems in later chapters.

The three possible states of information are

▼ **Processing**   This is when the data is being processed in the computer's central processing unit (CPU). Should the computer processing that particular data be compromised during the processing of this data, it could be damaged or altered.

■ **At rest**   This refers to storage of the data in a medium such as hard disks, floppy disks, backup tapes, or CD-ROMs.

▲ **In transit**   This is when the data is being transmitted over the network and between transmitting stations. This is a vulnerable state because many different and undetectable methods can be used to view, reroute, copy, and alter the data.

CHAPTER 5

Threats and Attack Methods

In this chapter, we will be discussing some of the more common threats to a network. We will also go over the attack methods, tools, and techniques used by malicious hackers to break into those networks.

PACKET SNIFFING

One of the most common ways to glean information from your network is by using what is called a *packet sniffer* or *network analyzer*. If a computer on your network is packet sniffing or in promiscuous mode, it is viewing all network traffic on its local network.

Promiscuous Mode

When computers communicate across a network, they listen to the wire for any information intended for them, based on information in the frames passing by. Those frames contain a destination address. By default, a network interface card (NIC) is configured to recognize only data intended for it specifically or for data expressly intended for all computers (a *broadcast*). Normally, when a computer's network interface card sees a frame with anything other than its specific Layer 2 (Data Link layer), hardware, or MAC address (or all 1's indicating a broadcast), it discards the frame. It does this to keep the frame from traveling further up the protocol stack and interrupting the computer's CPU needlessly. Figure 5-1 shows the decision process used by the network interface card.

When a network interface card is configured to listen and pass up every frame that goes by, it is said to be in *promiscuous mode.* This mode is intended primarily for diagnostic purposes and is rarely used. It places a terrible strain on the resources of the card and the computer it is attached to. It also uses a lot of disk space in a short period, because it is logging and processing every packet that goes by. It's like being in nightclub and trying to listen to every conversation simultaneously while trying to remember and understand everything being said. I'm sure you have a good memory; I just bet that remembering everything would use a lot of it—or that you have a pen with plenty of ink and a big writing tablet. When attackers successfully compromise a computer, they can install what is known as a *packet sniffer.* A packet sniffer is a software tool that puts the computer into promiscuous mode. Many of these programs are widely distributed over the Internet, completely free, and are easy enough for the most novice hacker to use.

Monitoring network traffic can be done by attackers from outside the network using viruses, and more commonly, Trojan horses. People inside the company can easily load the software themselves and do the same thing. (Remember, most malicious activity on computer networks occurs from inside the organization.) Once any computer has been compromised, the attackers can monitor and record all network communications. Monitoring

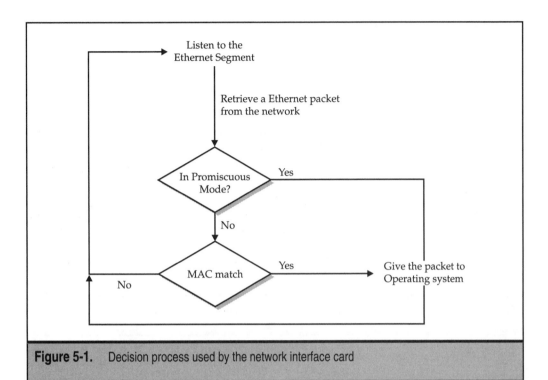

Figure 5-1.   Decision process used by the network interface card

network traffic can be very valuable because a lot of sensitive information can be gathered. Account names, passwords, and e-mail can be examined at the attackers' leisure. The attackers can then look for key words like "password," "login," "user," and so on, in the data stream that follows. This way, attackers can learn passwords for systems they are not even attempting to break into. Attackers can then use this information to compromise other computers and other networks. This is exactly why weak security in just one computer, network, or person can lead to a complete compromise of the entire network. You can get a better idea of this by looking at Figure 5-2. This might help visualize just how passwords and other information can be gleaned by using packet sniffing techniques.

From a network administrator's point of view, detecting a computer that is in promiscuous mode can be difficult if not impossible. The reason is that the computer is monitoring the network passively. Picture yourself in a conference room giving a speech and trying to determine who is listening and who is not. Sure, everyone can "hear" what is being said, but whether they are "listening" is an entirely different story.

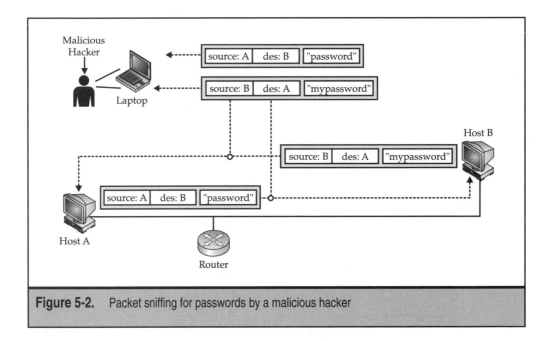

Figure 5-2.   Packet sniffing for passwords by a malicious hacker

Packet Sniffing Countermeasures

There are a number of ways to prevent or minimize packet sniffing and monitoring of network activity. These are some of the more common methods.

LAN Switches

One of the best ways you can reduce the threat of packet sniffing either remotely or at your location is to install LAN switches. Remember, the most common LAN technology by far is Ethernet, which, as we discussed in Chapter 2, uses a broadcast media access method, making it a "shared" network. All of the machines on a given Ethernet segment are attached logically (and possibly physically) to the same "wire" and can see every bit that goes by. The way switches can make promiscuous sniffing less of a threat is by making it appear to the computer attached to the network that it is the sole device transmitting on the wire. It does this logically by having all of the devices individually attached to its ports and controls all data transmission intelligently using the Application Specific Integrated Circuits (ASIC) chips in its internal switch fabric.

Antisniffing Application Methods

Quite a few good antisniffing applications are available. Most of these applications monitor the behavior of all the devices on the network. These applications do that by running tests that trick or fool various operating systems, DNS (Domain Name System) tests, and tests on network latency.

Tricking the Operating System Remember when we said that normally a network interface card is set to look only for packets with its specific hardware or MAC address? We also said that it will discard any packet before sending it up the stack for processing unless the packet is determined to be for that same address. That hardware or MAC address is a Data Link layer address. Some operating systems will only look at the IP address in the packets to determine whether they should be handed to the stack for processing. The IP address is a Network layer address that can be logically assigned, as opposed to being burned-in like MAC Data Link layer addresses are. Knowing this, some clever people decided to exploit this phenomenon by creating a packet with a MAC address that does not belong to any NIC in particular. Instead, a valid IP packet is created with the destination host's correct IP address. The IP Network layer address is encapsulated or wrapped inside the Data Link layer frame that contains the MAC address. This is how you can catch a promiscuous mode computer with its hand in the cookie jar. The computers whose network interface card is set to promiscuous mode will only check for a correct IP address before handing the packet up the appropriate stack. Should an attacker create an ICMP echo request or (PING packet) and place it inside the bogus Data Link layer frame, the vulnerable computer will respond to that PING packet when its NIC is in promiscuous mode. Had the NIC not been in promiscuous mode, monitoring every packet that goes by, then it would do just as it was supposed to do, ignore and discard the packet. It's analogous to e-mail viruses in that it doesn't matter how harmful contents of the package (packet) is if you discard it without every opening it.

DNS Tests One of the first items of business for someone attacking a network is to gain as much information as they can about the computers, users, and addresses inside the network. Some of the most valuable information that can be learned is the naming scheme used inside the network. This is because people don't remember computers by IP addresses—they remember them by names. That is the very service that DNS provides, hence the name Domain Name *Service.*

DNS tests operate on the premise that many network data-gathering tools used by an attacker perform IP-to-inverse-name resolution to learn DNS names in place of IP addresses. Having these names can be useful to an attacker, because many people use obvious naming schemes inside their network. (A corollary hint: Do not use obvious naming schemes.) Attackers are far more likely to show interest in a machine that has a name like "HRServer.abc.com" and "PayrollServer.abc.com" than something like "JMairsPC.abc.com."

When a computer that is in promiscuous mode starts performing DNS reverse lookups, the computer changes from being a passive network device to an active network device. A computer that sees that the packets are not destined for it will not attempt to resolve the IP addresses in the packets. If it does find that the information in the packets might pay off, the computer will place itself in promiscuous mode and start sending packets out on the network destined to fictitious hosts. The minute the antisniffing application starts seeing any reverse lookup requests going by on the wire involving those fictitious hosts, the address of the computer sending the requests will be recorded and flagged as a computer in promiscuous mode.

Network Latency Tests   The nice thing about network latency tests is that they can detect machines on the local network that are in promiscuous mode regardless of the operating system they might be running at the time. There is a price for this feature, however. Latency tests tend to create a large amount of network traffic for short periods. They will also probably require a bit of human analysis to derive the most accurate conclusions.

These tests work on the premise that when a network interface card is not in promiscuous mode, it still has hardware filtering capabilities. Again, any packets not destined for a particular computer will be discarded by the network interface card. Even when network traffic increases significantly, those packets not destined for that particular computer, never get sent up the stack and avoid interrupting the CPU. Obviously, computers that are in promiscuous mode do not have hardware filtering capabilities. All of the packets from the added network traffic get sent up the stack, dramatically affecting computer performance due to the CPU having to process those extra packets. This is where the additional latency is detected.

The antisniffing application simply builds a baseline of all computers on the network by issuing ICMP echo requests (PING packets). These requests have microsecond timers that allow you to determine the average response times. Once this data has been obtained, the application saturates the local network with illegitimate traffic. While this flooding is taking place, the application again issues timing packets to determine the average response times and compares them with the baseline figures taken earlier. Computers that are in their default mode only show a slight increase in latency, while computers in promiscuous mode often show a latency increase of up to four times the others.

SPOOFING AND DENIAL OF SERVICE THREATS

We will be discussing spoofing and denial of service attacks at this point because they are often used in combination with each other. As we explain these attacks, we will try to quickly review the mechanics behind some of the protocols that are the usual tools of these attacks. See Chapter 3 for more comprehensive protocol descriptions and header format diagrams.

All of the things that we are about to discuss involve what is loosely described as a "man in the middle" attack. Anytime you monitor or place yourself in between any communication, you are committing a man-in-the-middle attack. Take a look at Figure 5-3.

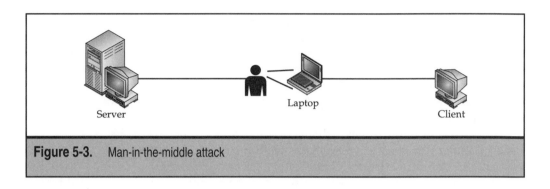

Figure 5-3.   Man-in-the-middle attack

SPOOFING

Spoofing is simply something masquerading as something else. There are several ways to spoof an identity, and a machine can do this on many different levels. The most common methods are

▼ ARP spoofing

■ IP spoofing

■ DNS spoofing

■ Web spoofing

▲ Telnet hijacking

As we pointed out in Chapter 2, these methods simply occur at different layers in the protocol stack. To stay secure, you need to protect yourself at every layer in the protocol stack. *Spoofing* means finding a way, at any layer, to convince another machine that it is the intended destination.

ARP Spoofing

The Address Resolution Protocol (ARP) is used by the Internet Protocol (IP) Network layer protocol to map logical IP network addresses to the hardware or physical addresses used by a Data Link layer protocol.

An Ethernet network uses frames that contain two hardware addresses that are generally referred to as MAC (Media Access Control) addresses. These Data Link addresses within the frames identify the source and destination of each frame sent over the network. (The destination address could also contain all 1's to identify the frame as a broadcast to be sent to all computers connected to the network.) Each computer network interface card (NIC) is allocated a globally unique 6-byte or 48-bit address. This means that the hardware address is dependent on what the manufacturer then "burns" into the PROM chips on the card. The MAC address therefore, in theory, is a globally unique and unchangeable address that is stored on the network card itself. These MAC addresses make it possible for data to be sent back and forth, independent of whatever application protocols are used on top of it. Ethernet builds frames of data, consisting of 1500-byte blocks. Each frame has an Ethernet header, containing the source MAC address and destination MAC address.

Wrapped inside this frame is the Network layer's IP address, which is logical and decided upon by the network administrator. IP is a protocol used by applications, independent of whatever network technology operates underneath it.

IP communicates by constructing packets that are similar to frames but that have a different structure. These packets cannot be delivered without the Network layer. In our case, they are delivered by Ethernet, which splits the packets into frames, adds an Ethernet header for delivery, and sends them onto the wire. However, at the time of construction, this Ethernet frame has no idea what the MAC address of the destination machine is, which it needs to create an Ethernet header. The only information it has available is the destination IP from the IP packet's header.

There must be a way for the Ethernet protocol to find the MAC address of the destination machine, given a destination IP. This is where ARP, the Address Resolution Protocol, comes in. ARP operates by sending out ARP request packets. An ARP request essentially asks, "Is your IP address x.x.x.x? If so, please send your MAC address back to me." These packets are broadcast to all computers on the LAN, even on a switched network. Each computer examines the ARP request, checks if it is currently assigned the specified IP, and sends an ARP reply containing its MAC address. To minimize the number of ARP packets being broadcast, the operating systems keep a cache of ARP replies. When a computer receives an ARP reply, it updates its ARP cache with the new IP address to MAC address association.

ARP is a stateless protocol. Most operating systems will update their cache if a reply is received, regardless of whether they have sent out an actual request. ARP spoofing involves constructing forged ARP request and reply packets. By sending forged ARP replies, attackers could convince a target computer to send frames destined for computer A instead to computer B. Once these frames have been redirected, computer A will have no idea that this redirection took place. The process of updating a target computer's ARP cache with a forged entry is referred to as *poisoning*. The attackers insert their computer in the communications path of two target computers and sniff the wire. The attackers' computer then forwards the frames between the two target computers so that the communications do not get interrupted. The attack is performed as follows (where X is the attacking computer, and A and B are targets):

▼ X poisons the ARP cache of A and B.

■ A associates B's IP with X's MAC.

▲ B associates A's IP with X's MAC.

All of A and B's IP traffic will go to X first, instead of directly between A and B. This method is extremely potent because not only can computers be poisoned, but routers as well. All Internet traffic for a host could be intercepted using this method when the malicious hackers' computer positions itself between the target computer and the LAN's router.

IP Spoofing

IP spoofing is an attacker compromising the routing packets to redirect a file or transmission to a different destination. This is also known as *source address spoofing.* The routing packets of most file transfers are transmitted in the clear, making it easy for an attacker to modify source data or change the destination of information. The method is also used to disguise the attacker's identity, to keep system administrators from identifying those who breach their security.

Some protocols are more susceptible to spoofing than others. The Internet Control Message Protocol (ICMP) is vulnerable because it passes information and error messages between communicating nodes on a network. The Internet Group Management Protocol (IGMP) is vulnerable because it reports error conditions at the user datagram level and contains routing and network information. The User Datagram Protocol (UDP) can also have its identification query function compromised.

The solution to preventing IP spoofing is securing transmission packets and establishing screening policies. Encryption can help prevent attackers from reading the packet's contents. Authentication can ensure that a legitimate source, and not a spoofed middleman, sent the contents of the packet. In either case, any attempt to tamper with the packets would leave some indication to warn system administrators.

The easiest way to protect yourself from such an attack is to have your firewall or router reject any external packets that appear to come from an internal IP address. IP spoofing in its most common form involves spoofing the IP address as well as spoofing the TCP connection by predicting sequence numbers. *Nonpredictable sequence numbering,* or making the SYN/ACK dialogue more difficult to predict, also reduces the chances of a session being intercepted and hijacked. The other method of IP spoofing involves altering the route of the packets using the source routing feature in the protocol. Most of these attacks involve interfering with the TCP three-way handshaking process, as shown in Figure 5-4.

The three-way handshake is used to ensure that a path between the hosts exists in both directions. Let's assume that a client on Host A wishes to open a connection with a server on Host B.

A sends a *SYN_segment* (Synchronize Segment) with an Initial Sequence Number (ISN) to *B* with the SYN (synchronize) flag set. The ISN identifies the first byte of data that will be sent for this particular connection.

B, upon receiving the SYN_segment, will send a *SYN+ACK_segment* (Synchronize Acknowledgement Segment) with both the SYN and ACK flags set. This indicates that *B* has received the first SYN, and it is ACKing it while simultaneously sending a SYN of its own that contains its own ISN.

When *A* receives the ACK, it knows that a two-way path exists. *A* then sends an *ACK_Segment* (Acknowledgement Segment) and begins sending data.

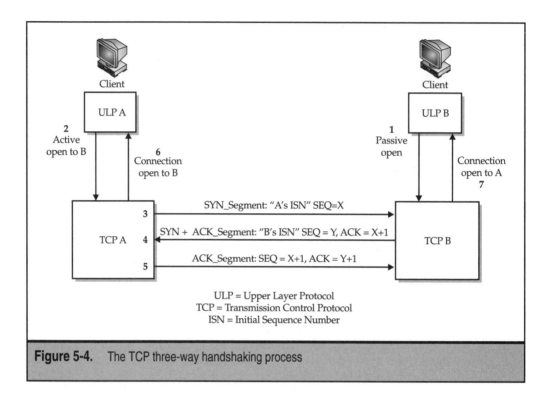

Figure 5-4. The TCP three-way handshaking process

When *B* receives the ACK for its SYN, it then knows that a two-way path exists, and *B* may then begin sending data to *A*.

Once both hosts have received ACKs to their SYNs, the connection is established.

A segment is a single TCP "packet." This first packet with the SYN flag set is sometimes jokingly referred to as "the original SYN."

IP Source Route Spoofing

Source routing is an IP option used today mainly by system administrators to check network connectivity. When an IP packet leaves a system, its path usually is determined arbitrarily by the routers along the way, depending upon their given configurations. Source routing provides a means to override the control of the routers.

There are two kinds of source routing: strict and loose. *Strict* source routing lets a manager specify the path through all the routers to the destination. Return responses use the exact same path in reverse.

Loose source routing is when a manager doesn't care how the packet gets where it is going as long as somewhere along the way it travels through a network segment address specified in the packet by the manager using the loose source routing option.

Loose source routing aids attackers when the attackers spoof one of the permitted source addresses using the loose-source-routing option to force the response to take a path that travels over the same network segment that the attackers' computer is con-nected to. This gives them the opportunity to examine, damage, or alter the contents of those packets before they arrive at their intended destination.

Imagine host badguy.example.com (real IP address 192.168.1.2) trying to communi-cate with one of your hosts while masquerading as goodguy.example.net (real IP address 172.16.3.4). It will have no difficulty generating IP datagrams containing the source address 172.16.3.4; these will reach your system because IP routing will take care of that. But how can the datagrams forming the other half of the conversation get back to badguy.example.com? Surely, they'll go to goodguy.example.net instead. This is where IP source routing comes in. If you look at an IP header, you will see that there is an options field. Two of those options allow the sender to specify the route the datagram will follow rather than letting intermediate routers decide, with the expectation that datagrams sent in response will follow the same route back. Thus, it's possible to steal someone else's IP address and still get packets back and forth. The route can be mandated and can involve the use of other routers or hosts that normally would not be used to for-ward packets to the destination. You could do this by masquerading as the trusted client of a given server in the following manner:

1. First, the attacker could change his machine's IP address to be the same as a trusted client machine.

2. Then this attacker would create a source route to the server by specifying the exact path that the IP packets should take to the server and from the server back to the attacker's machine, using the trusted client as the last hop in the route to the server.

3. A client request is sent by the attacker to the server using this source route.

4. The client request is then accepted by the server just as though it came directly from the trusted client.

5. The server, using the source route, forwards the packet to the attacker's host.

Because source routing is so easily abused, most routers, firewalls, operating systems, and some applications are usually configured to drop packets that have source-rout-ing-options traffic. There is rarely a legitimate need for source routing. You can get a basic idea of this by looking at the private IP subnet in Figure 5-5.

In the absence of source routing, IP spoofing is harder, but it can sometimes still be achieved using TCP sequence number prediction. This is a neat trick, since it involves badguy.example.com carrying on a conversation with you though he can't hear your replies, which is why it is often referred to as *blind spoofing*.

Nonblind Spoofing

Nonblind spoofing interferes with a connection that sends packets along your local subnet. Generally, one of the two hosts involved is on your subnet, or a situation exists where all data traffic is passing by your network device. With nonblind spoofing, the attacker has access to the TCP sequence number by simply accessing the reply packets on the local wire.

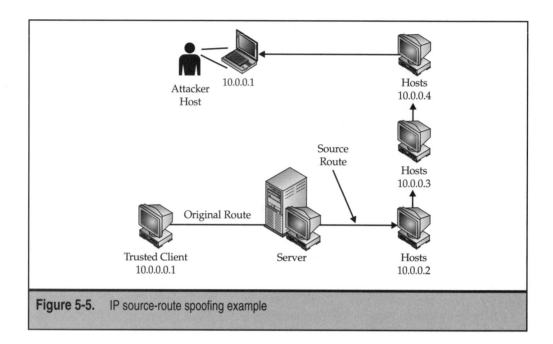

Figure 5-5. IP source-route spoofing example

Blind Spoofing the IP packets sent with the forged IP source address will reach the target fine. (Remember, IP is a connectionless-oriented protocol, meaning each packet is sent without knowing whether it will arrive.) However, the packets sent back by the target machine will be dropped. The attackers never get to see them. The intervening routers know where the packets are supposed to go—they are supposed to go to the target machine. As far as the Network layer is concerned, this is where they originally came from, and this is where responses should go. Of course, once the packets are routed there, and the information is sent up the protocol stack and reaches TCP, it is discarded. (The target machine cannot respond.) Now the attackers have to be clever and find a way to "know" what was sent and to "know" the response the server will be looking for. The attackers cannot see what the target host sends, but they can try to "predict" what it will send. That, combined with the knowledge of what it "will" send, is how the attackers will try to work around this blindness.

Remember, a lot of this is explained in more detail in Chapter 3, but we will go over some of this again. When you make a connection, data is sent in packets. Packets take care of low-level traffic and make sure the data arrives (sometimes with special error handling). The backbone of the Internet uses the IP protocol version 4. It is totally independent of all hardware protocols, making it perfect for such an endeavor.

TCP and UDP are higher-level protocols wrapped in IP packets. All of those packets contain a header and a payload (data). The packet's IP header contains the source address and destination address of the communicating hosts along with a value that identifies the type of protocol wrapped up in it, for example, 6 for TCP, 17 for UDP, and so on.

If you unwrap the IP packet, you will find a packet containing the protocol indicated in the value. If that value happened to be the number 6, you would have yourself a brand new baby TCP packet (congratulations, it's a TCP!!!).

TCP packets contain, among other things, port numbers of the source and destination host, sequence (SEQ) and acknowledge (ACK) numbers, and a number of different flags. The SEQ number is counted byte per byte and indicates the number of the next byte to be sent or that is sent inside this packet. The ACK number is the SEQ number that is expected from the other host. SEQ numbers are chosen at connection initiation.

UDP packets contain, among other things, port numbers of the source and destination host. UDP has no such thing as SEQ or ACK and is generally considered a weak protocol in terms of security.

As you know, TCP uses a SEQ/ACK system (Sequence nr./Acknowledge) to do what it is supposed to do. A connection is initiated like this in its simplest form, and this is known as the *three-way-handshake* (see the "IP Spoofing" section earlier in this chapter):

Packet 1: Client to Server

▼   Flags: SYN ("I want to initiate a connection")

▲   SEQ: clientnr

Packet 2: Server to Client

▼   Flags: SYN, ACK (ACK: The request is being acknowledged)

■   SEQ: servernr

▲   ACK: clientnr+1

Packet 3: Client to Server

▼   Flags: ACK

■   SEQ: clientnr+1

▲   ACK: servernr+1

If we go back to what we talked about earlier, in our spoofed case we will have the following:

Packet 1: T to X (actually A to X, but we spoof it)

▼   Flags: SYN

▲   SEQ: clientnr

Packet 2: X to T

▼   Flags: SYN, ACK

■   SEQ: servernr

▲   ACK: clientnr+1

Now what? Because we (host A) can't see packet 2, we cannot use "servernr" to calculate the required ACK (servernr+1) for packet 3. We have to predict it one way or another, so packet 3 can be as follows:

Packet 3: T to X (actually A to X, but we spoof it)

▼   Flags: ACK

■   SEQ: clientnr+1

▲   ACK: guessed_servernr+1

If guessed_servernumber is incorrect, we have failed to initiate a connection. A second problem is that we don't know if we have failed or not.
When host X sends its answer

Packet 2: X to T

▼   Flags: SYN, ACK

■   SEQ: servernumber

▲   ACK: clientnr+1

then host T will receive the packet! Host T doesn't know of any connection initiated to X, so it tells X to stop the connection initiation because it is bogus. It does that by sending a RST (reset) to host X. If the RST reaches host X, the connection we are trying to set up will not get established. Now we have to prevent T from doing such things. You could just wait until T is offline, or you could try a denial of service attack to take host T down. The most common method is to perform a SYN flood from A to T. You SYN flood the port on T that you are faking. When it is flooded, it will not be able to handle further incoming packets. This is to make it so it will not be able to see "packet 2" and send its show-stopping RST. You can read about SYN floods later in this chapter.

TCP Sequence Number Prediction

Now the attackers need to get an idea of where in the 32-bit sequence-number space the target machine's TCP is. The attackers connect to a TCP port on the target (SMTP is a good choice) just prior to launching the attack and complete the three-way handshake. The attackers then need to get an idea of what the round-trip time (RTT) from the target machine to the spoofing machine is like. (The process can be repeated several times, and an average of those RTTs is calculated.) The RTT is necessary to being able to accurately predict the next Initial Sequence Number (ISN). The attackers have the baseline (the last ISN sent) and know how the sequence numbers are incremented (128,000 per second and 64,000 per connect). They now have a good idea of how long it will take an IP datagram to travel across the Internet to reach the target (approximately half the RTT, because most times the routes are symmetrical). After the attackers have this information, they immediately proceed to the next phase of the attack. (If another TCP connection were to arrive on any port of the target before the attackers were able to continue the attack, the ISN predicted by the attackers would be off by 64,000 of what was predicted.) You might better visualize how this attack is done if you look at Figure 5-6.

When the spoofed segment makes its way to the target machine, several things can happen, depending on the accuracy of the attacker's prediction:

▼ If the sequence number is exactly where the receiving TCP expects it to be, the incoming data will be placed on the next available position in the receive buffer.

■ If the sequence number is less than the expected value, the data byte is considered a retransmission and is discarded.

▲ If the sequence number is greater than the expected value but still within the bounds of the receive window, the data byte is considered to be a future byte and is held by TCP pending the arrival of the other missing bytes. If a segment arrives with a sequence number greater than the expected value and not within the bounds of the receive window, the segment will be dropped, and TCP will send a segment back with the "expected" sequence number.

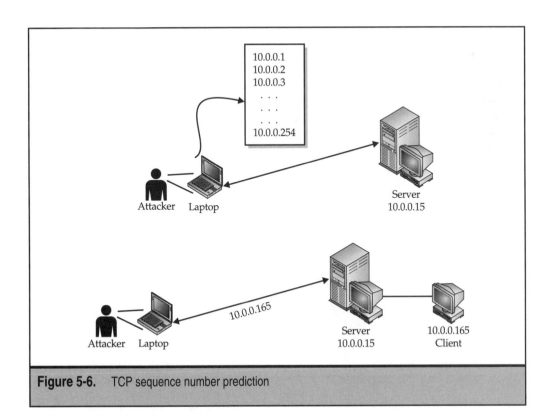

Figure 5-6. TCP sequence number prediction

DNS Spoofing

DNS stands for "Domain Name Service." Domain names are the familiar, easy to remember names for computers on the Internet (such as "internic.net"). They correspond to a series of numbers, called *Internet Protocol numbers,* that are used to route packets to their destination. Those numbers are usually represented in a "dotted quad" notation you will probably recognize as something like 111.122.133.144. Domain names are generally used as a convenient way of locating information and reaching others on the Internet. This service allows easily remembered alphanumeric names for humans to be mapped to the numerical addresses necessary for computers. For example, the name *ds.internic.net* has the three IP addresses, 198.49.45.10, 192.20.239.132, and 204.179.186.65. Most likely, each of these IP addresses will eventually be mapped onto different computers, although multiple IP addresses could also map onto a single computer. The reason for many-to-many name-to-address mappings is that they give system administrators a way to balance the network load on their machines. The name *ds.internic.net* is probably accessed hundreds of thousands of times every day by people all over the world. Take a look at Figure 5-7.

The best way to handle this kind of volume would be to have the mapping of this one domain name spread across multiple machines with multiple IP addresses. DNS is a distributed database that maps domain names to IP addresses. DNS offers a practical way around the difficulty of maintaining large host files in an organization. On a larger scale, computers with Internet access use DNS to resolve Uniform Resource Locators (URLs). This is so you don't need to know the IP address of a web server, only its name (URL).

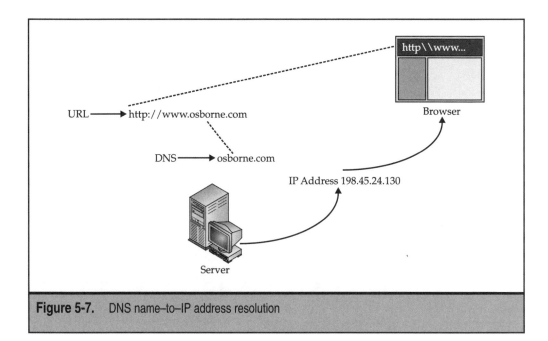

Figure 5-7. DNS name–to–IP address resolution

DNS names take the form <domain name>.<domain type>, like *internic.net*. Some common domain types include *.edu* for educational institutions, *.mil* for the military, *.org* for nonprofit organizations, *.gov* for government organizations, and *.com* for commercial institutions.

The Internet makes use of a network of DNS servers to form a distributed database of mappings between domain names and IP addresses. At the top of this network of servers are the root DNS servers, and under these root DNS servers are the top-level servers.

When a computer, known as a *DNS client,* wants to resolve a URL, it queries its *DNS server.* The DNS client uses a *DNS resolver* to locate its DNS server. If the DNS server is not authoritative for the destination domain, or if the DNS server doesn't have the information in its cache, it will not be able to answer the client's query immediately. When this happens, the DNS server will either act as a DNS forwarder or will issue a recursive query. When a DNS client receives a nonauthoritative response from a DNS server, it means that the DNS server has found the answer in its cache, instead of contacting the appropriate authoritative DNS server for that domain.

A DNS forwarder will forward the query to a second DNS server higher up in the tree of servers. This ends the contact between the DNS client and the first DNS server. Alternatively, if recursive querying is allowed, and it usually is, the DNS server will ask a root name server for the IP address of a host that's authoritative for the destination domain and will then contact the authoritative server and report back to the DNS client. So a *recursive* query is one where a server issues one or more queries to answer another query.

DNS spoofing involves forcing a DNS client to make a query to an impostor server and then tricking the client by sending a wrong answer from the impostor server. *DNS spoofing* is when a DNS server accepts and uses incorrect information from a computer that has no authority to give that information. DNS spoofing involves poisoning the cache by forging that data and placing it in the cache of the name servers. Spoofing attacks can cause serious security problems for DNS servers vulnerable to such attacks, for example, causing users to be directed to wrong Internet sites and web pages or causing e-mail to be routed to unauthorized mail servers.

Three ways to carry out a DNS spoofing attack are

▼   Spoofing the DNS responses

■   DNS cache poisoning

▲   Breaking into the platform

Spoofing the DNS Responses

Attackers can use the recursive mechanism to their own advantage, by predicting the request that a DNS server will send out and replying with false information before the real reply arrives. Each DNS packet has an associated 16-bit ID number that DNS servers use to determine what the original query was. In the case of BIND, the most common DNS server software, this number increases by 1 for each query, making the request easier to predict. This has been fixed in the later versions of BIND, where DNS packets are assigned random numbers.

Once false hostname and mapping information has been sent back to the requesting client, the attackers can then redirect future name resolution mappings, while exposing intended network data to the threat of capture, copying, corruption, and alteration. DNS involves a high trust relationship between client and server, and it is this trust that makes DNS vulnerable to spoofing. A cryptographic authentication mechanism can solve this problem.

To test whether a DNS server is vulnerable to this DNS spoofing attack, you can send queries to the target name server, assuming that its traffic will flow somewhere over your network link. Then you can analyze the queries to see whether it is possible to predict the next DNS query ID number of a DNS query packet. If the DNS query's IDs are predictable, you can then assume that it is possible to poison the server's cache with invalid data.

DNS Cache Poisoning

Another DNS vulnerability lies within its DNS caching procedure. To save time, if they receive a similar query again, some DNS servers will cache all local zone files as well as the results of all recursive queries they have performed since their last startup. (Zone files contain the information for all zones for which the DNS server is an authority.) The length of time that recursive query results are held in the DNS cache is called the *time to live* (TTL), and that time is configurable.

DNS cache poisoning involves sending a DNS server incorrect mapping information with a high TTL. The next time the DNS server is queried, it will reply with the incorrect information. DNS cache poisoning occurs when malicious or bad data received from a remote name server is saved or "cached" by another name server. This bad data is then made available to programs that request the cached data through the client interface. It is possible to limit the vulnerability to this DNS cache poisoning attack by reducing the time that information is stored in the cache (the TTL); however, this will affect the server's performance, as shown in Figure 5-8.

Breaking into the Platform

Yet another DNS spoofing attack involves breaking into the target network's DNS server. Once attackers have control of the underlying DNS platform, they have control of the network environment. This can be accomplished by intercepting a *zone transfer*.

A zone transfer is the transfer of the DNS database to a secondary server. It allows name servers that are authoritative for the same domain to stay synchronized with each other. The information in a zone transfer is very useful to attackers. This would include information such as IP addresses of important hosts. DNS servers should be configured to allow zone transfers only between primary and secondary DNS servers, because zone transfer attempts are often the first indication that a network is being probed. DNS communication uses both the Transmission Control Protocol (TCP) and the User Datagram Protocol (UDP). DNS queries are handled on UDP port 53, while DNS zone transfers are handled on TCP port 53. Most firewalls are configured to block TCP port 53 to prevent zone transfers.

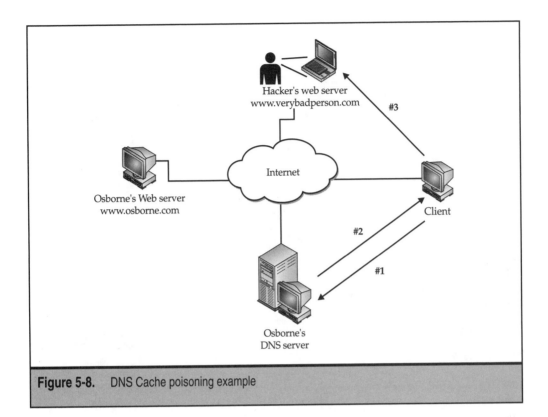

Figure 5-8. DNS Cache poisoning example

One way of preventing the damage that unauthorized zone transfers can cause is by using a split DNS system. This involves setting up both an internal DNS server for internal computers (like an internal mail exchange server) and an internal name server. Then an external DNS server is set up containing just the information needed by external computers like SMTP gateways and external name servers. It is more secure to have a separate server for incoming SMTP mail. This way, should the external mail exchange be compromised, the attackers will not automatically have access to the internal mail system.

Web Spoofing

Web spoofing is a situation in which attackers create what appears to be normal access to the Web when in fact it is just a bogus, and convincing, duplication of the Internet.

This fake Internet looks and acts like the real one, but the attackers are controlling all of the traffic between the victim's browser and the attackers' computer. The attackers' computer is now essentially acting as a proxy server for the victim. Since the attackers have access to any and all data going to or from the victim to the supposed Internet servers, their possibilities for mischief are unlimited. Take a look at Figure 5-9.

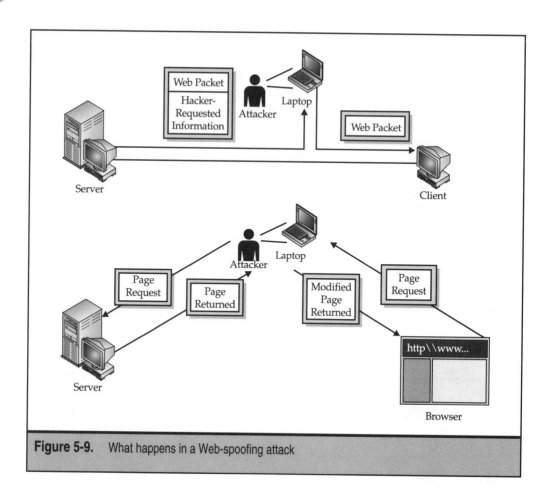

Figure 5-9. What happens in a Web-spoofing attack

The attackers can easily watch the traffic as it goes by, recording which pages the victim visits and the contents of those pages. When the victim fills out a form, the entered data is transmitted to the attacker's server first. Now the attackers can record all of the requested data as well as all of the responses sent back by the originally intended server. Since all e-commerce is transacted using forms, the attackers can steal any account numbers and passwords the victim enters.

This can be done even if the victim has a "secure" connection (using Secure Sockets Layer [SSL]) to the server. Although the victim's browser may show a secure connection on screen (such as an image of a lock or a key), it is really just a nice, secure connection to the attacker's server.

The attackers can also easily modify any of the data going in either direction between the victim and the intended Internet server. So now, instead of just copying any form data that goes by, the attackers can modify any form data being submitted by the victim. For example, if the victim is ordering a product online, the attackers can change the product number, the quantity, or the ship-to address. Instead of getting the cute new pair of baby booties being ordered for a baby shower, the victim gets the brand-new top of the line bowling ball he has always wanted and for a price that is hard to beat. (Well, the *price* might not be tough to beat, but the *cost* would be… given the fact it didn't cost the attacker anything.)

The attackers can also modify the data returned by a web server, for example, by inserting misleading or offensive material to trick the victim or to cause antagonism between the victim and the server. (I know what you're thinking, that's impossible because everyone on the Internet is above that sort of behavior. But it is possible.)

URL Rewriting

The attackers' first trick is to rewrite all of the URLs on a web page so that they point to the attackers' server rather than to some real server. Assuming the attackers' server is on the machine:

```
http://www.verybadperson.com
```

the attacker rewrites a URL by prepending

```
http://www.verybadperson.com
```

to the front of the URL. For example:

```
http://home.osborne.com
```

becomes

```
http://www.verybadperson.com/http://home.osborne.com
```

The victim's browser requests the page from

```
www.verybadperson.com
```

since the URL starts with

```
http://www.verybadperson.com
```

The remainder of the URL tells the attackers' server where on the Internet to get the real document.

Telnet Hijacking

A telnet session hijack occurs when attackers intercept the data transmission channel after a client has authenticated itself to the server. The attackers then resume the session, at which point they become the client. For this attack to work, attackers must be in the path of the data transmission by gaining access to one of the routers along the way or on the same network segment of either the client or the server. What makes this kind of attack attractive is that the hijackers don't need to know the user name and password to gain access to the host.

The attackers only need to determine the TCP/IP sequence numbers and acknowledgements. Placing the attacking host's network interface into promiscuous mode should allow attackers to determine the sequence numbers and acknowledgements, since telnet packets are sent in cleartext. By keeping track of telnet sessions being started, the attackers can choose which one of those sessions to hijack.

Next, the attackers choose a spoofed source and destination MAC address. The original client and server's network interfaces retain their correct MAC addresses. Since the attacking host has its network interface in promiscuous mode, it is able to process all of the packets on the LAN segment anyway.

Now the attackers have the option of deciding whether to drop the original connection at the client's host by sending a reset packet to it. The attackers can also choose to keep the client's session alive, allowing the attackers to see what the user is typing on his or her keyboard even while the session has been hijacked. When the attackers are ready to take over the telnet session, they just press any key. At this point, the attackers have taken over the telnet session and can start issuing commands on the server. The attacking software inserts the correct sequence numbers and the acknowledgements for the packets being sent in the hijacked session, allowing the attackers to continue the connection without detection. The original client can continue to type at his or her keyboard, but gets no response from the server. When the attackers are done issuing their commands, they can end the hijacked session at any time.

If the attackers choose not to drop the original client's connection, they can reset the connection or synchronize the communication channel. This gives back the original telnet session to the client undetected, as long as they were keeping track of the input and output characters transmitted, the sequence numbers of the packets, and which packets have been acknowledged during the hijack. Take a look at Figure 5-10.

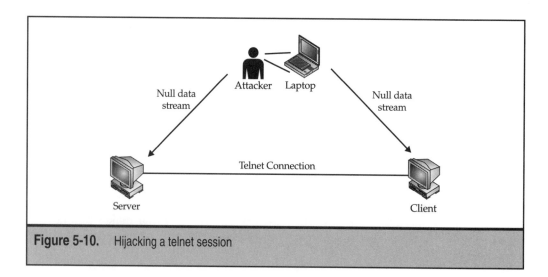

Figure 5-10.   Hijacking a telnet session

DENIAL OF SERVICE

Denial of service (DoS) is an attack designed to render a computer or network useless and incapable of providing normal services. The most common DoS attacks target network bandwidth or connectivity. *Bandwidth* attacks flood the network with so much traffic that all of the network resources are consumed, making it impossible for normal users to get through. *Connectivity* attacks flood a computer with so many connection requests that all available operating system resources are consumed, and the computer can no longer process normal user requests.

Another DoS attack variant is called a distributed denial of service (DDoS) attack. This kind of attack uses multiple computers to launch a synchronized DoS attack against one or more targets. Using client/server technology, the attacker is able to multiply (also known as amplify) the effectiveness of the denial of service significantly by using the resources of multiple unwitting accomplice machines as distributed attack platforms. You can get a basic idea of what I mean by looking at Figure 5-11.

Here are some of the more common methods used for denial of service attacks:

▼ **SYN flood**   SYN flooding is a network denial of service attack. It takes advantage of the way TCP connections are created. The SYN flood attack simply sends TCP connection requests faster than a machine can process them:

1. The attacker creates a random source address for each packet.

2. The SYN flag set in each packet is a request to open a new connection to the server from the spoofed IP address.

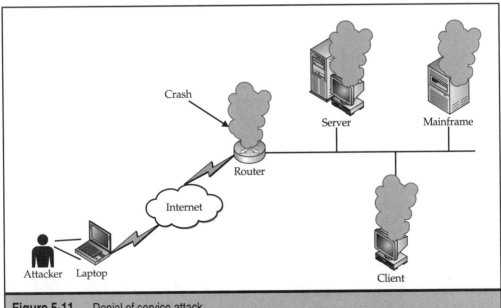

Figure 5-11.   Denial of service attack

3. The victim responds to the spoofed IP address and waits for confirmation that never arrives.

4. Waiting for these replies causes (usually after about 3 minutes) the victim's connection table to fill up.

5. Once the table is full, all new connections are ignored.

6. A SYN flood can be used for other types of attacks, such as disabling one side of a TCP connection in preparation for a session hijacking attack, or preventing authentication or logging between servers.

■ **Land attack**   The land attack uses IP spoofing in combination with an initial TCP connection. It sends a packet in which the SYN flag in the header is set during the opening of a TCP connection. Then the IP addresses in the underlying IP packet are altered. In a land attack, both the source and destination IP addresses are modified to be identical in the address of the destination host. Once the destination host receives the packet, it answers the SYN request. When answering, the destination host first creates a packet with the ACK flag set, then changes the destination address to the source address and vice versa. If both addresses are identical, guess what? The machine starts sending the packet back to itself. Now the machine is in an ACK war against itself. This continues, creating a denial of service situation.

■ **ICMP flood**   This denial of service attack sends such a large amount of ICMP echo request packets to the target machine that it can't respond quickly enough to handle the amount of traffic on the network. First, the attackers will spoof their source IP; otherwise, their machine would be a victim of its own attack since it would use resources not only to send the packet but also to receive packet responses. Now it continues to send packets, while the target has to use resources to receive *and* reply to the packets. Multiply this by a distributed attack across multiple machines, and the target will easily be brought down. Intrusion detection system (IDS) services on routers and firewalls can be programmed to limit ICMP echo requests or to drop them entirely from entering the subnet.

■ **Ping of death**   The ping of death attack sends ICMP echo request packets (pings) that are larger than the operating systems data structures storing this information can hold. Sending a single, large ping packet to many systems causes them to hang or even crash.

■ **Smurf attack**   The Smurf attack takes advantage of a feature in the IP stack known as *directed broadcasts.* A Smurf attack floods your router with ICMP echo request packets (pings). Since the destination IP address of each packet is the broadcast address of your network, your router broadcasts the ICMP echo request packet to all hosts on the network. The more machines on the network, the more ICMP echo request and response traffic generated. Should the attacker choose to spoof the source IP address of the ICMP echo request packet, the resulting ICMP traffic will not only clog up your network (the "intermediary" network), but will also congest the network of the spoofed source IP address known as the "victim" network by sending ICMP echo reply packets. To prevent a network from becoming an intermediary, turn off broadcast addressing on any network routers that allow it, or configure the firewall to filter the ICMP echo requests. To avoid becoming the victim of the attack itself, set the firewall or routers to filter ICMP echo replies, or limit echo traffic to a small percentage of overall network traffic.

- ■ **UDP flood**   The User Datagram Protocol (UDP) is a connectionless protocol. It does not need to establish a connection to transfer data. Since no connection setup is required before data is transferred, it is difficult to bring a host down by flooding the host with just UDP packets. A UDP flood denial-of-service attack is created when the agent host sends a packet to a random port on the target machine. When the target host receives a UDP packet, it determines what is listening on the destination port. If nothing is listening, it returns an ICMP packet to the forged source IP address, indicating that the destination port is unreachable. If enough UDP packets are sent to dead ports on the target host, not only will the target host go down, but computers on the same segment will also be disabled because of the amount of traffic.

- ■ **Teardrop attack**   Most networks have a maximum packet size that they can handle. This packet size is called the *maximum transfer unit* (MTU). If the underlying network cannot transport a given packet because it is too large, the packet has to be broken into smaller pieces. This is called *fragmentation*. Once the packet fragments have reached their final destination, the packet is reassembled from the fragments. Both fragmentation and reassembly are done in TCP/IP by the IP layer. When the IP layer receives an IP packet, it typically does the following:

  1. It verifies that the packet is at least as long as the IP header and makes sure that the header is contiguous.

  2. It runs a checksum on the header of the packet to check for any errors. If there are, the packet will be dropped.

  3. It verifies that the packet is at least as long as the header indicates. If it's not, the packet will be dropped.

  4. It checks to see if the packet is for this machine. If it is, it will continue processing the packet. If it's not, it will either forward it or drop it.

  5. It processes the IP options, if any.

  6. If the packet has been fragmented, it will keep the fragment in a queue until all other fragments have been received.

  7. It passes the packet to the upper levels.

- Use something like SSH instead, because it uses public-key encryption. (Encryption is covered in Section IV, "Secure Communications.")

- Do not allow telnet traffic beyond the firewall.

- **Simple Network Management Protocol (SNMP)** This protocol is used to collect management information Base (MIBs) from different network devices.

  - This is another unsafe protocol that also allows cleartext passwords.

  - Everything is done using community strings. Should that be compromised, then all SNMP-enabled devices with that string could also be compromised.

  - This should be filtered at the firewall.

- **Domain Name System (DNS)** Also called *name service*, this application maps IP addresses to the names assigned to network devices.

  - DNS can be attacked and should be kept in the DMZ behind your firewall.

  - Hackers will attempt to get a DNS server's zone files.

  - There are a couple of ways to do a zone transfer: you use NSLOOKUP, and/or a slave or secondary name server queries a primary name server to get its zone files.

  - Name poisoning is possible when a hacker is able to insert wrong or false DNS information into a zone transfer, usually between a master and a slave.

  - If the DNS server is compromised, then the name-to-IP mapping can be manipulated to be bogus, thereby poisoning your DNS facility.

- **Internet Control Message Protocol (ICMP)** ICMP performs a number of tasks within an IP Internetwork.

  - IP routing specifies that IP datagrams travel through an Internetwork one-router hop at a time.

  - The entire route is not known at the outset of the journey. Instead, at each stop, the next router hop is determined by matching the destination address within the datagram with an entry in the current node's routing table.

  - Each node's involvement in the routing process consists only of forwarding packets based on internal information.

  - IP does not provide for error reporting back to the source when routing anomalies occur. This task is left to another Internet protocol: the Internet Control Message Protocol (ICMP).

  - The principal reason for which ICMP was created is for reporting routing failures back to the source. ICMP also provides a method for testing node reachability across an internet (the ICMP Echo and Reply messages), a method for increasing routing efficiency (the ICMP Redirect message), a method for informing sources that a datagram has exceeded its allocated time to exist within an Internet (the ICMP Time Exceeded message), and other helpful messages.

■ All in all, ICMP is an integral part of any IP implementation, particularly those that run in routers.

■ ICMP messages generally contain information about routing difficulties with IP datagrams or simple exchanges such as timestamp or echo transactions.

■ **Routing Information Protocol (RIP)**   Routing is central to the way TCP/IP works. Network devices use RIP to exchange routing information.

▲ **Network File System (NFS)**   This is a system developed by Sun Microsystems that enables computers to operate drives on remote hosts as if they were local drives.

By exploiting these weaknesses, attackers can gain access to a computer with the permissions of the account running the application, which usually has permissions that can do a lot of damage.

One of the newest forms of Application layer attack exploits the openness of several new technologies used commonly on the Web. Examples include new HTML specifications, web browser functions, and other popular protocols like Internet Relay Chat (IRC). These attacks, which use Java applets and ActiveX controls, involve passing harmful programs over the Internet and into your network by loading them through the victim's browser.

SUMMARY

By now, you should have a good handle on the most common potential threats to your network. Be on the alert for packet sniffers, spoofing, telnet hijacking, denial of serves, and password attacks.

CHAPTER 6

Intrusion Detection Systems

Intrusion detection is a critical element in your electronic security infrastructure. When installed in conjunction with a firewall, an intrusion detection system (IDS) provides an additional layer of security that significantly decreases the risk of an attack going undetected. Firewalls have emerged as the primary tool to prevent unauthorized intrusion into systems, and you should view them as the first line of defense on any system connected to the Internet. However, firewalls are limited in their ability to detect many types of unauthorized behavior. Network IDSes provide an additional layer of security that significantly strengthens an organization's perimeter defense. Network IDSes are usually installed to identify weaknesses in a company's security policy, alerting the security administrator to enhance or revise the network's security defenses. The primary objective is to assure the firewall's policy is correctly configured so that unauthorized or suspicious connections are identified and wherever possible prevented. Remote control and distributed denial of service Trojans are very dangerous to the integrity of a network. This is especially true for telecommuters on VPNs. In this chapter, we will discuss the characteristics and concepts behind intrusion detection. We will also talk about the tools and methods available to help accomplish this.

WHY IS INTRUSION DETECTION NECESSARY?

You may have asked yourself, "Why would I need an intrusion detection system when I have a perfectly good firewall?" Many people think that firewalls are all-powerful, that they can recognize attacks and block them. Unfortunately, this is not the case. As pointed out throughout this book, the only thing worse than no security is a false sense of security.

Firewalls are discussed in more detail in Chapter 7, but for now, let's think of a firewall as a device that shuts off all services completely and then turns on only those services that are necessary for the operation of your network. In other words, the first thing the firewall does upon installation is stop all communication in both directions. That isn't exactly the kind of "productivity tool" that most businesses have in mind for their network. So, immediately after installation, the systems administrator will carefully decide the specific kinds of traffic the firewall considers acceptable and in what directions it will allow the information to travel. For example, a typical corporate firewall allowing access to the Internet would stop all UDP and ICMP traffic and all *incoming* TCP connections. It would, however, allow all *outgoing* TCP connections. This is to stop any malicious incoming connections from the Internet while still allowing internal users to connect in the outgoing direction.

You could also think of a firewall as a kind of electronic fence around your network with a carefully scrutinized, well-protected gate. As a rule, fences do not know when somebody is trying to break in.

Intrusion detection systems, on the other hand, know when someone is coming through the fence, how they are getting through, and who they are. An IDS even knows when someone is just "hanging around" and peering in through the slats. An IDS is dynamic—it constantly watches and learns to recognize attacks against the network that firewalls, which are very literal creatures, are unable to see.

Another major security factor that people tend to overlook is that firewalls are at the boundaries of your network. Approximately 80 percent of all malicious or suspect activity comes from inside the network. You need to do more than concentrate on "evil outsiders," because the most likely threat is from within. By monitoring all network activity and analyzing normal traffic patterns, an IDS can discover any anomalies pointing to possibly damaging behavior by outsiders and insiders alike.

Firewalls act as a barrier between your network (where the insiders are) and the Internet (where the outsiders are) by filtering traffic according to a given set of policies. It is possible to bypass the firewall, usually from within. This is not uncommon. Filtering out everything except that which is needed can block that which is often wanted, such as music, pornography (I know what you're thinking, "Pornography on the Internet?!? No way!") and other popular content. To circumvent this blocking, users may connect their computer over an unauthorized modem connection while they're still attached to the internal network. Any attacks over that connection would bypass the firewall completely and provide full access to malicious hackers.

Firewalls are subject to attack themselves. Attacks, methods, and easily downloadable software designed to circumvent firewalls have been widely publicized and are easy to come by. A common attack strategy is to utilize *tunneling* to bypass firewall protections. Tunneling is the practice of encapsulating a message inside of another message. That means that a dangerous payload that is normally sent in one protocol (that might be blocked by the firewall filters) could simply be placed inside a second message using a protocol that is not blocked.

WHAT ARE INTRUSION DETECTION SYSTEMS?

It won't do you much good to know *how* people are likely to try to break into your network if you don't have a way of knowing *when* an attack is taking place. It is almost as important to know if and when they are trying to do so as how. An intrusion detection system is the security equivalent of real-world cameras, alarms, and guards.

You might have locks on every door in your corporate headquarters, and you may even make sure that they are always locked. Remember the old saying "locks were made to keep honest people honest"? Locks can be picked or destroyed. But let's say, for instance, that you found one that couldn't be picked or destroyed. Before you get too happy, know that the doors and windows they are attached to can be broken open with the locks still intact and locked. Untouched locks won't keep anyone from going in though, will they? That old saying still holds true with respect to firewalls, VPNs, and the like. If, for example, you told a firewall to lock the SMTP (mail) "door," it wouldn't stop someone from coming in through the HTTP (web) "window."

If your company were to install alarms, cameras, and security guards, you could be informed when security was being breached, see where it was being breached, and send someone to that location to fix or stop what was being done to breach it. You could even take that security plan a step further—you could keep tapes *logging* the security breach events and keep a record of the individuals that were doing it. Armed with this information, you could respond more quickly or before it ever happened again, should that same individual appear on your security camera monitors in the future.

Intrusion detection systems do for network security what alarms, cameras, security guards, and tape logs do for physical location security. The difference is that notification devices would replace alarms, sensor equipment would replace cameras, countermeasure features would replace security guards, and network activity monitoring logs that examine packets and data at all or a select few of the layers in the OSI model would replace videotape library logs.

Intrusion detection systems are designed to alert you when your system is being probed or attacked. They generally collect information from different sources at strategic vantage points in the network, analyze the information, and then send alarms, drop data or entire segments, implement countermeasures, log events and activities, and initiate any number of other responses.

Many different intrusion detection systems are available, and almost as many different designs and implementations. Because there are so many different intrusion detection systems, it helps to have a model within which to consider all of them. The Common Intrusion Detection Framework (CIDF) loosely defines a set of components that make up an intrusion detection system. These components include:

▼ **A-boxes** Network activity *A*nalysis devices that can be specific hardware, software, or both.

■ **C-boxes** *C*ountermeasure mechanisms or response procedure equipment.

■ **D-boxes** Storage mechanisms (hard *D*isks), which are essentially the logging equipment.

▲ **E-boxes** These are considered *E*vent generators, also known as sensors.

A CIDF component can be hardware, a software package, or part of a larger system such as a firewall. You can see where these components fit in an intrusion detection system in Figure 6-1.

A-Boxes/Analysis

A-boxes analyze the data gathered from the event generators or sensors. They analyze things like statistical anomalies, virus signatures, common attack characteristics, and other criteria to decide when and if an attack appears imminent, is taking place, or has taken place. Analysis is a continual and dynamic necessity, because new attacks and vulnerabilities present themselves all the time. A large portion of intrusion detection research goes into creating new ways to analyze these anomalous events to extract relevant information. Various approaches for analyzing these anomalies have been considered by IDS designers, even to the degree of studying the similarities based upon historical military models and biological immune system deficiencies and vulnerability models.

C-Boxes/Countermeasures

C-Boxes represent the devices that provide countermeasures. Many intrusion detection systems are designed only as alarms. However, most intrusion detection systems provide for some form of countermeasure capability, such as shutting down open TCP connections,

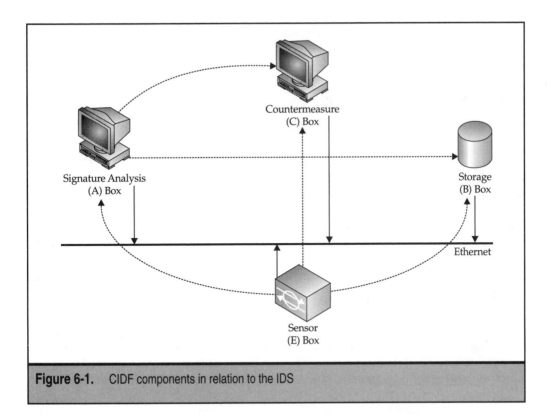

Figure 6-1.   CIDF components in relation to the IDS

modifying router filter lists, and network segmentation. By employing countermeasure capability, an IDS can be used to prevent further attacks from occurring after initial attacks are detected.

D-Boxes/Logging

E-boxes and A-boxes tend to produce a lot of data. This data must be stored and made available later to system administrators so that it can be compared against baseline data gathered during normal operations. Again, this is a constant and dynamic process, and requires devices that can store and log this information. This is what D-boxes are all about. They represent the devices that provide the means to store and log security activity data and make it available later.

E-Boxes/Sensors

The purpose of an E-box is to provide information about events to the rest of the system. An *event* can be complex set of circumstances like failed logon attempts to a given file by a certain person at a given time, or it can be a low-level network protocol occurrence. It doesn't necessarily have to indicate malicious intrusion. E-boxes are the sensory organs of an intrusion detection system. Without E-box inputs, an IDS has no empirical information from which to draw conclusions about security events.

IDS DETECTION METHODS

There are three methods that are commonly used in IDS implementations: *misuse detection*, *anomaly detection*, and *signature recognition*. They can be used exclusively or all three can be combined.

Misuse Detection

Misuse detection involves encoding information about specific behaviors known to be indicators of intrusion and then filtering this data for these indicators.

Misuse detection involves the following:

▼ A good understanding of what "misuse" is

■ A reliable record and baseline of user activity

▲ A reliable way to analyze these records

Misuse detection is best used for detecting known use patterns. One of its weaknesses is that you can only detect what you know about. You can, however, leverage the knowledge you have gained from previous misuse to spot new exploits of old intrusion methods.

Misuse detection essentially checks for "activity that's bad" by comparing all activity with abstract descriptions of undesired activity. This approach attempts to assign rules to describe known "undesired" usage. The rules might be based on past attacks or activity thought to exploit known weaknesses, as opposed to describing historical "normal" usage. Rules may be written to recognize a single event that represents a threat to system security or a sequence of events that might indicate possible prolonged attacks. How well these rules work depends on how active and diligent the systems administrators are.

The rules need to focus on defining specific descriptions and instances of types of attack behaviors to flag. They describe an attribute of an attack or class of attacks, and may require the recognition of sequences of events. A misuse information database provides a good way to address newly identified attacks before overcoming the vulnerability on the target system. Usually, misuse rules tend to be specific to the target machine.

Anomaly Detection

Anomaly detection compares observed activity against expected normal usage profiles that can be developed for users, groups of users, applications, or system resource usage. Anything that falls outside the parameters of this normal behavior is considered an anomaly.

The most common way people approach network intrusion detection is to look for statistical anomalies. You may also hear this referred to as "statistical analysis." This method involves finding variations from normal patterns of behavior. The idea behind this approach is to measure a baseline of such things as disk access, CPU utilization, user logon activity, and file access. You then calculate the averages in frequency and variability. Once you have these figures, the system can send alarms or directed e-mails whenever a deviation from this baseline is detected. The advantage to this approach is that it can sense the anomalies without having to know the underlying cause behind them.

Let's say, for example, that you monitor the traffic from individual workstations. The system notes that at 2 A.M., a lot of these workstations start logging into the servers and carrying out tasks. You might consider this interesting to note and might take action on it, such as allowing access to this system only during certain windows of time.

Signature Recognition

Most IDS products are based upon examining network traffic and looking for well-known patterns of attack. This means that for every known hacker technique, a *signature* for this technique is then programmed into the system.

This signature technique can be as simple as running a pattern-matching comparison scheme. It might be something as simple as character-string (order of text) matching. One classic example of this is to examine every packet on the wire and search for the character string "/cgi-bin/phf?," which might indicate somebody attempting to access this common vulnerable CGI script on a web server. Some IDS systems are built from large databases that contain hundreds or even thousands of such strings. They just sniff the network and trigger alarms and/or log events for every packet they see containing one of these strings.

ASSESSMENT ARCHITECTURES

IDS products are generally based on one of two types of architectures. The architecture you would choose depends entirely on the kind of network traffic that would be running on your network. However, using a hybrid approach (a combination of both) is by far the most effective.

Host-Based Assessment

Host-based intrusion detection involves loading software on the system to be monitored. The loaded software uses log files and/or the system's auditing agents as the data source. *Network-based* intrusion detection systems, on the other hand, monitor the traffic on the entire network segment as a data source. Both network-based and host-based intrusion detection sensors have pros and cons, and in the end, you'll probably want to use some combination of both. The person responsible for monitoring the intrusion detections needs to be an alert, competent systems administrator (someone like yourself) who is familiar with the host machine, network connections, users and their habits, and all software installed on the machine. This doesn't mean that you need to be an expert on the software itself, but you should have a feel for how the machine is supposed to be running and what programs are legitimate. Most security breaches can be detected and contained by sharp systems administrators who notice something "different" about their machines or who have noticed a user logged on at a time that is unusual for that user.

Host-based intrusion detection involves not only looking at the communications traffic in and out of a single computer, but also checking the integrity of your system files and watching for suspicious process activity. To get complete coverage at your site with host-based intrusion detection, you need to load the intrusion detection software on every computer. There are two primary classes of host-based intrusion detection software:

▼ Host wrappers or personal firewalls

▲ Agent-based software

Either approach is much more effective at detecting trusted, insider attacks than a network-based approach. Both are relatively effective for detecting attacks from the outside. You can see an example of a host-based intrusion detection system in Figure 6-2.

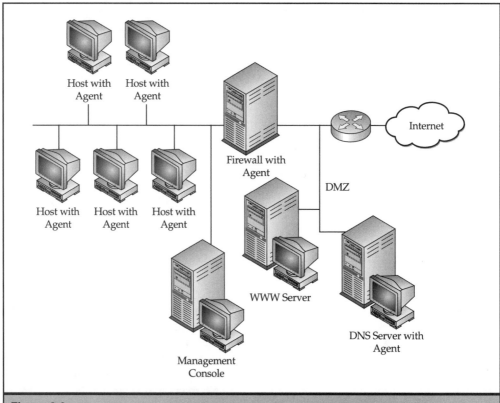

Figure 6-2. Example of a host-based intrusion detection system architecture

Host wrappers or personal firewalls can be configured to look at all network packets, connection attempts, or login attempts to the monitored machine. They can also monitor dial-in attempts or other non–network-related communication ports.

Host-based intrusion detection is only as good as its logging. If a machine has only a handful of users, perhaps only the connections from the outside and the integrity of the system files need to be monitored, whereas a machine with a lot of users or network traffic may need more stringent monitoring. Programs can be written to analyze log files and alert the systems administrator via e-mail or pager when something is thought to be unusual. System logging output should be sent to a remote site or modified so that the log files are put into nonstandard places to prevent hackers from covering their tracks. *Host-based intrusion detection systems* (HIDS) methods deploy a monitor on each system. HIDS tend to be a more scalable solution than a *network-based intrusion detection system* (NIDS), but it is also harder to manage. Intrusion is easier to detect at the system level, and the accuracy rate is better than with NIDS.

HIDS products, as their name implies, reside on each individual host or server. Applications-based IDS products are a subset of HIDS. HIDS are reactive rather than proactive, scrutinizing event logs, critical system files, and any other auditable resources, looking for unauthorized changes or suspicious patterns of behavior, such as numerous incorrect password attempts. One of the particular problems addressed by HIDS is Trojan horses, where a hacker will attempt to deliver and run a program inside the firewall to undermine file integrity, either to cause damage or perhaps to weaken security in preparation for a more opportune time of intrusion. Since the HIDS is a reactive system, the HIDS will only detect the Trojan horse after it has become active.

HIDS also addresses internal data theft, usually characterized by such anomalous behavior as an attempt to gain access to sensitive data from an unusual location. Why, for instance, would someone in sales be trying to access the human resource files of the company's directors?

HIDS monitor specific files, logs, and registry settings on a single computer and can send alarm notifications upon any access, modification, deletion, and/or copying of the monitored object. The role of a HIDS is to flag any tampering for a specific computer. The HIDS can also be set up to automatically restore altered files when changed to ensure integrity.

A derivation of HIDS is *centralized host-based intrusion detection systems* (CHIDS), which serves the same purpose, but does the analysis centrally by having monitored files, logs, and registry settings sent to the centralized host management system for analysis. The main difference between these systems is that a CHIDS is more secure. It monitors all necessary information from the host and sends it to a centralized location. If the host is compromised, the alerting and forensic analysis can still take place uninterrupted. The catch is that centralized analysis requires substantially more network bandwidth to send that data to the centralized host.

A HIDS will monitor policy and make reactive decisions locally, and will send alerts to a centralized host only when warranted. This uses up much less network bandwidth. As implied, the drawback of a HIDS is that should the host be compromised, there will be no alarm notification nor any forensic data left to determine what happened or what was lost.

Network-Based Assessment

Network-based intrusion detection systems (NIDS) analyze network traffic for attacks that exploit the connections between computers and the data that can be accessed via a network connection. NIDS can detect the broadest range of attacks on corporate information assets, which may include denial of service (DoS) or distributed denial of service (DDoS) attacks. DoS and DDoS attacks are aimed at stopping the enterprise or its customers from accessing corporate IT assets. The role of the network IDS is to flag and sometimes stop an attack before it gets to information assets or causes damage. NIDS are more effective than HIDS and their derivatives for monitoring both inbound and outbound network traffic in general.

NIDS search for signs of intrusion through TCP analysis. They monitor traffic in real time, examining data packets to spot denial of service attacks or dangerous payloads before they reach their destination. They also match activity patterns against known attack signatures, updating their signature database as new attacks are discovered. Although NIDS are capable of raising alerts and terminating the connection, they have a number of shortcomings. Most of them work on what is called the "promiscuous model," meaning that they examine every single data packet. Use of system resources is consequently heavy, to the extent that NIDS usually require a dedicated host. Even then, they can still be overwhelmed by heavy traffic. NIDS are also unable to see across switches or routers, so a sensor has to be installed on every network segment. They are unable to analyze encrypted traffic, common on virtual private networks (VPNs), but still provide the most cost-effective intrusion detection on unencrypted networks up to 100 Mbps.

Unlike host-based systems, network-based systems can use proactive, invasive monitoring techniques to detect possible vulnerabilities. You can draw from a database of attack scenarios and run them against target systems within the network. Analysis can be run to determine whether weaknesses or holes exist that can be exploited. Regular port scanning techniques can be run on a regular basis to check not only for weaknesses but also whether proper auditing and logging are up to par. Network-based systems are usually dedicated platforms involving two components:

▼ A sensor for analyzing network traffic

▲ A management device

The sensors in a network-based IDS capture traffic on the network being monitored and analyze the traffic using configured parameters. The sensors analyze packet headers to determine source and destination addresses, the type of protocol being used, and the packet payload. Once the sensor sees a violation, it can take a number of actions like sending alarms, logging the event, resetting the data connection, or dropping future traffic from that host or network. You can see an example of a network-based intrusion detection system in Figure 6-3.

Network-based intrusion detection involves looking at the packets on the network as they pass by the sensors. The sensor can see only the packets that happen to be carried on

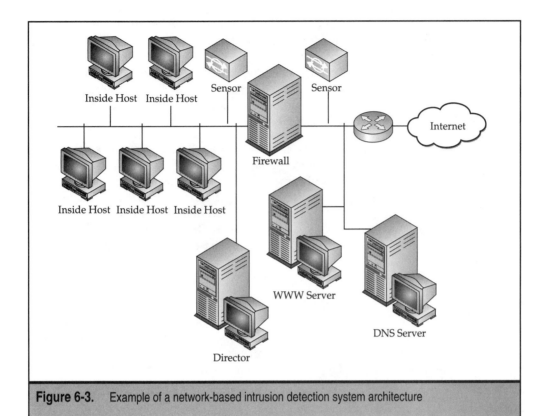

Figure 6-3.   Example of a network-based intrusion detection system architecture

the network segment it's attached to. Packets are considered to be of interest if they match a signature. There are three primary types of signatures. These are

▼   **String signatures**   String signatures look for a text string that indicates a possible attack. An example string signature for a UNIX-based system might be `"cat "+ +" > /.rhosts"`, which, if successful, might make a UNIX system extremely vulnerable to network attack. To refine the string signature to reduce the number of false positives, it may be necessary to use a compound string signature. A compound string signature for a common web server attack might be `"cgi-bin"` AND `"aglimpse"` AND `"IFS"`.

■   **Port signatures**   Port signatures simply watch for connection attempts to well-known, frequently attacked ports. Examples of these ports might include Telnet (TCP port 23), FTP (TCP port 21/20), SunRPC (TCP/UDP port 111), and IMAP (TCP port 143). If these ports are not used on your network, then incoming packets to these ports would be considered suspicious.

▲ **Header condition signatures**   Header signatures watch for dangerous or illogical combinations in packet headers. The most famous example is WinNuke, where a packet is destined for a NetBIOS port and the Urgent pointer or the Out Of Band pointer is set. This can result in the "blue screen of death" for Windows systems. Another well-known header signature is a TCP packet with both the SYN and FIN flags set, signifying that the requesting computer wishes to start and stop a connection at the same time.

Sometimes the detection method is simply anomaly based, where everything that looks normal is ignored. The advantage is that the system is able to recognize previously unseen attacks because anything not normal is considered to be an intrusion, but it also needs to be trained when changes are made to the network, such as the installation of a new application. The disadvantages of anomaly-based detection are that it will not be able to spot hostile or intrusive activities if the pattern of behavior appears to be normal. Also, because they are not referring to any kind of database of attack signatures, they cannot name attacks by type.

Hybrid Intrusion Detection Systems

Hybrid intrusion detection systems combine HIDS technology with the ability to monitor the network traffic coming in or out of a specific host with NIDS technology that monitors all network traffic. This is the ideal method of intrusion detection, but it is more expensive and requires more attention and more systems administrator expertise.

Placement of IDS Devices

Effective use of a network IDS requires two things:

▼ Proper placement of the IDS devices

▲ Policies associated with the type of traffic on the Internet connection

Placement of the network IDS to support a firewall depends on the intended objectives for intrusion detection. Ideally, two sensors should be used and placed on both the outside and inside of the firewall. Take a look at Figure 6-4 to get an idea of some IDS placement possibilities.

One sensor monitors the segment between the router and firewall. The second sensor monitors the segment between the firewall and internal network. Deployed in this configuration, the two sensors perform complementary functions acting as a firewall leak detector. The policy configuration of each sensor is a bit different. The outside sensor is used to monitor for attack attempts and network probes of an organization's Internet connection, while the inside sensor is used to detect certain types of events that have made it through the firewall or are coming from the internal network. The systems administrator should make sure the implementation doesn't just focus on external activity. Insiders can have distributed denial of service, Trojan horse programs, worms, and attack programs installed on internal computers. If this is the case, there is a potential for legal liability if an organization is duped into being the beachhead for the source of the attack.

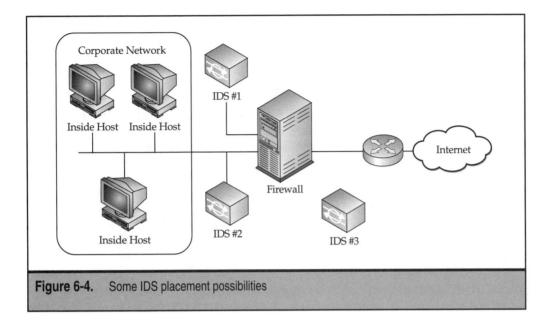

Figure 6-4. Some IDS placement possibilities

Some other common points of placement are

▼ **Network hosts** Even though network intrusion detection systems have traditionally been used as probes, they can also be placed on hosts (in nonpromiscuous mode). Let's say that you have a switched network where a disgruntled employee is on the same switch as the president. If these client machines were all running Windows 98, they would be completely defenseless since there are no logging capabilities. Without those capabilities, no data would be available to be fed into a traditional host-based intrusion detection system. This disgruntled employee could run a network-based password cracker for months without fear of being caught.

■ **Network perimeter** IDS can be very effective when installed on the network perimeter, such as on both sides of the firewall, near the dial-up server, and on links to partner networks. These links tend to be lower bandwidth than LAN connections and allow the IDS to better keep up with the traffic.

■ **WAN backbone** Another good spot is the corporate WAN backbone. A frequent problem is attacks from "outlying" areas into the main corporate network. Again, since these WAN links tend to be low bandwidth, it's easier for the IDS systems to keep up.

▲ **Server farms** Servers are often directly connected to switches, and that makes monitoring more difficult since each server is ostensibly on its own network segment. Another problem is that IDS systems cannot keep up with the speed and high volume of traffic generated on LANs. For extremely important servers, you may be able to install dedicate IDS systems that monitor just the individual server's link. Also, application servers tend to have lower traffic than file servers, so they are better suited for IDS systems.

Setting Intrusion Detection Policies

It is important to tune the sensor's policies for the type of systems inside the firewall. For example, if the devices such as user workstations and servers are based on Microsoft Windows, there is little need to test for UNIX exploits. In addition, if there are no externally available web or FTP servers on the inside of the firewall, there is no requirement to check for those categories of exploits.

The outside network sensor should be configured to detect a variety of probing activities, firewall exploits, suspicious activity, and protection of external or DMZ devices at the Internet connection. You must take into account that most exploitation activity will be on the Internet side, so this is where the most care must be taken in tuning this network sensor to protect your network. Be careful when implementing all of a vendor's signature sets—you don't want to get bogged down detecting signatures that have no impact on internal systems. When you activate detection of port probes, the objective is to learn that when probing activity increases significantly, it could indicate an imminent attack. You also need to keep in mind that activating port probe signatures can generate heavy event traffic.

Implement custom signatures to monitor external connection rules. Nearly all firewalls have some rules that allow for external access to services on the internal network. Normally, these are narrowly defined to include specific addresses and ports. Custom-developed IDS policies should be developed to monitor and log all activity on those external access rules.

Activation of suspicious behavior signatures should be tuned to protect the firewall, routers, and externally connected devices such as web, mail, DNS, and FTP servers that are often attached to DMZ segments.

Inside the firewall, the objective is to detect suspicious activity. Detection of Trojans, back doors, unauthorized access exploits, and outward-bound attack signatures is important. Properly configured firewalls will filter many exploits, but are typically weak against Trojan and backdoor activity.

This is why you will want to activate all Trojan and backdoor detection signatures on your IDS. A good network IDS will be able to detect hundreds of Trojans and their variants by analyzing packet contents and not just their default TCP ports. Remember that many infected computers could belong to home users attached via modems, DSL, and cable modems. A VPN connection does not protect against a Trojan horse in a system outside your perimeter.

HOW TO KNOW WHEN YOU HAVE BEEN ATTACKED

The whole point of having an IDS is to know when you are being attacked. You will need to carefully and consistently do the following activities to successfully do this.

Baselining Your Network

The first and most basic step you need to take to protect your network is to establish a baseline database of your file system so that you can regularly compare this baseline with the current and future state of the file system. If any unexpected files or directories show up, there is a possibility that the system has been breached. You should probably think of performing these comparisons on an hourly basis for your web server systems files and directories, and on a one- or two-time-a-day basis for systems files and directories on the rest of your servers. If you do this, you will be able to detect and correct abnormal conditions on your system as they happen.

Always remember that baselining is necessary under *any* circumstance. Although very powerful intrusion detection products may exist that can dynamically adjust themselves to catch some attackers before they begin to alter your file systems, they are still nowhere near as foolproof as baselining and file integrity verification. Even the best dynamic intrusion detection software can only detect attack signatures that it has been preprogrammed to recognize.

The first step is to make sure that you know exactly what your file system is supposed to look like. If new or altered files start showing up in your file system, a *root kit* may have been installed by a malicious hacker.

A root kit is a set of tools that attackers will use to stay on your system by hiding their activities and any changes they might make to your system. They will even try to replace key security modules that you might be using to detect their activity with hacked copies of those modules that will leave that activity unnoticed. One common component of a root kit is a program or set of programs that will erase all evidence of the attacker's activities from any log files on a regular basis.

A good baseline database for file integrity verification will discover a root kit immediately. You can get rid of the root kit by going back to your latest good backup, or by reinstalling the compromised security modules. Make your first baseline of your file systems when your system is first configured and definitely before it has been connected to the Internet.

No amount of verification will help you if your system was already compromised before you made your first baseline of the file systems. You have to be *sure* that you have started with a clean system that you are *sure* has never been compromised. Just because nothing out of the ordinary has happened yet, doesn't necessarily mean you have a clean machine.

Make sure you keep this in mind: losing your baseline database is like losing the key to your safe. If you are a little shaky about the whole "good thing/bad thing" theory, then just know that that would be a bad thing. It would be prudent to keep consistent backups of the baseline database. It also would be prudent to make those backups on multiple types of media and to keep those backups in different physical locations as well as keeping one of those copies in a fire safe offsite. (Losing the key to *that* safe would *really* be like losing the key to your safe.)

You could encrypt the baseline database, and that would keep attackers from viewing or altering the contents of that baseline database, but it wouldn't keep them from deleting the database all together. It is also important that your file integrity analysis software itself not be replaced with a Trojan horse. One way you could do this would be to install one copy of the file integrity analysis software on each of your servers and have those servers check each other or even check the software against a clean copy from an original read-only CD.

File Integrity Analysis

File integrity analysis is another kind of security that complements intrusion detection systems. Integrity analysis involves the use of message digests like MD5 or some other kind of cryptographic checksum methods to check critical files, system files, and other objects. (We will be going over message digests and cryptography in some detail in later chapters, so don't worry about these terms just now.) Once it has done this, it compares them to their original reference values and then flags any differences or changes.

The reason file integrity is so important is that smart attackers often alter system files. This happens at three stages of the attack:

▼ First: The attackers alter system files, which is the goal of the attack, like the insertion of a Trojan horse.

■ Second: The attackers attempt to leave back doors in the system so that they can reenter the system later.

▲ Third: The attackers attempt to cover their tracks so that system owners are completely unaware of the attack.

Not only is file integrity analysis used to determine whether attackers have altered system files or executables, but it can also help in determining whether vendor software patches and any other desired changes have been applied. This is also very important when doing forensic examinations of a system after an attack to help diagnose the attack type to add to the attack signature database.

Systems Log Analysis

Any attackers who know what they're doing do not want to leave any evidence of their activity in the systems log. The first order of business after breaking into your network is to try to alter and/or delete their activity in the log entries.

At the very least, you should check your logs daily so that the amount of data does not become overwhelming. Logging all activity on a network produces an impressive amount of mundane data. If this sounds like a lot of work, it is. However, it is nothing compared to going over a week's worth of data. Scripts can be written or even purchased that can help you in this task.

There are generally considered to be six types of systems logs that should be monitored. Those types are

▼ User activity:

■ Repeated failed login attempts

■ Logins from unexpected locations

■ Logins at unusual times of day

- Unusual attempts to change user identity
- Unusual processes run by users
- Unauthorized attempts to access restricted files
- Process activity:
  - Processes that are run at unexpected times
  - Processes that have terminated prematurely
  - Unusual processes such as those that were not caused by normal authorized activities
- System activity:
  - Unexpected shutdowns
  - Unexpected reboots
- Network connections:
  - Connections to or from unusual locations
  - Repeated failed connection attempts and their origination and destination addresses and ports
  - Connections made at unusual times
  - Unexpected network traffic, such as that which violates any rules of your firewall configuration or unexpected traffic volume
- Network traffic monitoring:
  - Sweeps of your network address space for various services that would indicate attempts to identify hosts on your network and the services they run
  - Repeated half-open connections that could signify IP spoofing attempts or denial of service activity
  - Successive attempts to connect to unusual services on your network's hosts
  - Transactions originating outside your network with destinations also outside your network, signifying traffic that should not be traversing your network
  - Attempted sequential connections to specific services, signifying someone trying to run network probing tools against your networked systems
- ▲ Application-specific log files:
  - Unusual uses of outgoing modems
  - Excessive or unusual file transfers
  - Unusual entries from mail utilities
  - Excessive mailings
  - Unusual database transactions

Other good ideas would be to check your web server's access logs to examine all web server access of any type as well as its error logs to see all failed access attempts and other errors. Also consider sending your system's log entries to a read-only file, a secure remote machine, or a printer port to make it harder for attackers to alter or delete these records.

HONEY POTS

A *honey pot* can help protect your network from unauthorized access. The honey pot contains no data or applications critical to the company, but has enough interesting data to lure hackers. A honey pot is a computer on your network whose sole purpose is to look and act like a legitimate computer. It actually is configured to interact with potential hackers in such a way as to capture details of their attacks. Honey pots are also known as a sacrificial lamb, decoy, or booby trap. The more realistic the interaction, the longer the attackers will stay occupied on honey pot systems and away from your production systems. The longer the hackers use the honey pot, the more will be disclosed about their techniques. This information can be used to identify what they are after, their skill level, and the tools they use. All this information is then used to better prepare your network and host defenses.

Honey pot systems are decoy servers or systems set up to gather information regarding an attacker or intruder into your system. It is important to remember that honey pots do not replace other traditional security methods, but rather provide an added level of protection.

Honey pots can be set up inside your network, outside the network, in the DMZ, or in as many locations as you like. They are essentially variants of standard intrusion detection systems but with more of a focus on information gathering and deception.

An example of a honey pot system installed in a traditional Internet security design can be seen in Figure 6-5.

The purpose of a honey pot is simply to have one machine that is easy to break into. Its aim is to lure would-be attackers into a system long enough to see what they might be after, who they are, and how they are getting in. Most attackers, especially these days, get very excited when they finally see a system "stupid enough" to let them in. This is partly because they tend to be kids (not necessarily in terms of age, but characterized by a consistent lack of maturity, like the author of this book) and also since security is much more common than it was at the inception of the Internet. The general consensus is that once attackers breaks into a system, they will come back for more. During these return visits, additional information can be gathered, monitored, and saved.

Generally, there are two popular reasons for setting up a honey pot:

▼   You can learn how intruders probe and attempt to gain access to your systems. Since a record of the intruder's activities is kept, you can gain insight into attack methodologies to better protect your real production systems.

▲   You can gather forensic information required to aid in the apprehension or prosecution of intruders. This is the sort of information often needed to provide law enforcement officials with the details needed to prosecute.

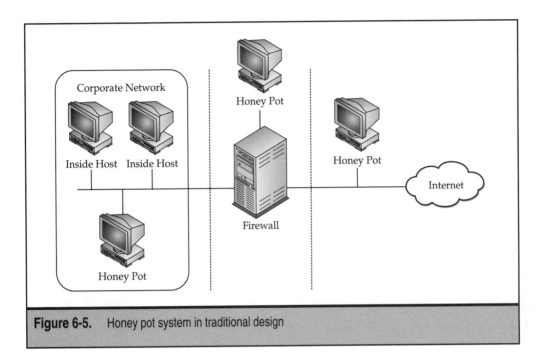

Figure 6-5.   Honey pot system in traditional design

When setting up a honey pot, certain goals have to be considered. Those goals are

▼ The honey pot system should appear as generic as possible. If you are deploying a Microsoft NT–based system, it should appear to potential attackers that the system has not been modified; otherwise, they could disconnect before much information is collected.

■ Be careful about the kinds of traffic you allow attackers to send back out to the Internet. You don't want to become a beachhead (launching point) for attacks against other systems on the Internet. This is a good reason for installing your honey pot on the DMZ or inside the firewall. Remember, you could be legally liable in some areas should your system be used this way and at the very least, would be extremely unpopular with the victims.

▲ Make your honey pot an interesting site by placing dummy information on it, or make it appear as though the intruder has found an intranet server. Make your honey pot appear legitimate so that your attackers will spend enough time investigating that you can gather as much information as possible.

Keep in mind that hackers tend to take things personally. Should they discover that you are setting them up, they might rally against the organization that has set these traps and make them a public target for other hackers. Examples of this sort of activity can be easily found on any of the popular hacker sites or publications.

Another big benefit to having a honey pot system as opposed to just an IDS by itself is that many intrusion detection systems have a problem distinguishing hostile traffic from harmless traffic. Honey pots are systems that should *never* be accessed. This means that *all* traffic to a honey pot system is extremely suspect.

One of the most common things hackers do is scan the Internet doing "banner checks." The honey pot can be set up to provide a banner that looks like a system that can easily be hacked, then to trigger if somebody actually does the hack. For example, the POP3 service reports the version of the software. Several versions of well-known packages have buffer-overflow holes. A hacker connects to port 110, grabs the version info from the banner, then looks up the version in a table that points to which exploit script can be used to break into the system.

A honey pot can be used to augment the deployment of an IDR system. Some of the problems with commercial IDR include inability to detect low-level attacks, techniques, or tools that are new or not previously known, or use of techniques that may appear as legitimate user activity. To a certain extent, the honey pot is also subject to missing new attacks. However, the honey pot is uniquely capable of letting you know that hackers are in your network doing things they have no business doing. The honey pot alone may spot them because as far as other security measures (including IDS) are concerned, they are legitimate users.

CHAPTER 7

Firewalls

Now that you've learned about protocols, packets, layers, ports, and sockets, you can get an idea of what firewalls are and how they work to secure your network. Firewalls are not a cure all. Firewalls examine the data coming into and leaving the network at any number of the OSI layers. Humans don't exist in any of those layers. Remember, the biggest threat to network security isn't errant data—it is people. Firewalls are analogous to locks and keys in the real world. The saying, "locks keep honest people honest" applies equally well when it comes to firewalls. You can have the most sophisticated firewalls, DMZs, and perimeter security in the world, but one simple modem attached to an employee's PC inside your network can bring the whole system to its knees without your knowledge—or even the employee's knowledge. Always keep this in mind. That being said, let's talk about firewalls. In this chapter, I discuss what a firewall is and how the four main types of firewalls work. I also provide an introduction to firewall architecture.

Firewalls, or "gateways," are tools that provide for security and consist of four basic types. Those types are

▼ Packet-filtering gateways or firewalls

■ Circuit-level gateways or firewalls

■ Application-level gateways or firewalls

▲ Stateful-inspection gateways or firewalls

They can be used as stand-alone items or in combination with some or all of the other types. The design you choose depends on the resources you have available to you, the level of security you require, the degree of convenience you wish to maintain, and any number of other factors unique to your situation.

All firewalls rely on information generated by protocols that function at different layers of the OSI model. Knowledge of these layers is one of the keys to understanding the different types of firewalls. As a rule, the higher the OSI layer at which your firewall examines these packets, the greater the protection provided by that firewall. You can see how the different types of gateways and protocols compare to the OSI reference model in Figure 7-1.

Throughout this chapter, I use the words "gateway" and "firewall" interchangeably. These two words are used interchangeably in the real world and in the marketplace as well.

WHAT IS A FIREWALL?

A firewall, or gateway, is essentially a software or hardware device that examines and filters external information coming from untrusted sources outside of your network into your internal private network or computer system.

A good analogy to a firewall would be similar to that of the checkpoint you must pass through to get to your gate at the airport. Everyone has to walk through metal detectors, and, at some airports, show some form of identification. Meanwhile, your carry-on bags go through an X-ray device that reveals the contents of each bag. As an added precaution,

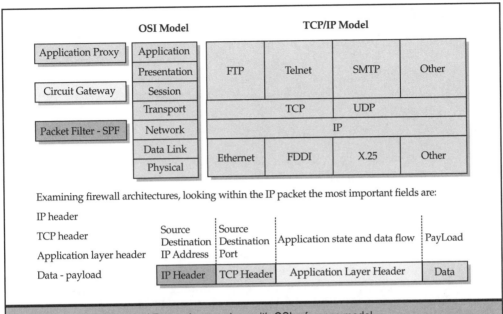

Figure 7-1.   Gateway and Protocol comparison with OSI reference model

should the individuals reading the X-ray not be able to recognize the contents, they will move you to the side and inspect the bag's contents.

Firewalls do basically the same thing. Instead of baggage, we have packets. The firewall "opens up" each packet to see who the packet is from, where the packet is going, what its contents are, the action of the packet, and other criteria that are considered acceptable.

Neither the airport security checkpoint nor the firewall is foolproof. The airport security personnel could be poorly trained, the X-ray machines could be malfunctioning, other doors to the gate area may not be properly monitored, and destructive devices could be nonmetallic, harmless-looking objects carried by individuals with fake IDs. Likewise, packets of data can be made to look as though their origin (though false) is authentic, their contents may be other than they seem, and the service claimed may not be as labeled. Malicious hackers use these tactics every day.

The airport analogy is just a simplification of the overall purpose of a firewall. The main purpose of a firewall system is to control access to or from a protected network. A firewall can be a personal computer, a router, a dedicated piece of hardware, or some other form of communication. It can protect you from outside intrusion from over the Internet while permitting employees to access Internet services such as WWW and e-mail. A firewall can also be used as an access control device that only allows access to the Internet for certain people within the organization and denies access to others. Many firewalls even have features that can control, authenticate, and secure users who may want to access the company's internal data from the Internet or even data from another company.

TYPES OF FIREWALLS

Firewalls generally consist of four different types. Those types are

▼ Packet filtering

■ Circuit level

■ Application level

▲ Stateful MultiLayer Inspection (SMLI)

Each type can be used alone or with one or more other types. The design you choose depends on the resources you have available to you, the level of security you require, the degree of convenience you wish to maintain, and other factors unique to your situation.

Packet-Filtering Gateways

Packet-filtering gateways are the most rudimentary form of firewalls. They work by examining packets as they come into and leave your network. Two related terms are *ingress*, meaning packets entering the network, and *egress*, meaning packets leaving the network. A packet-filtering gateway is a standard feature of any good router.

A packet-filtering firewall is usually a router or computer running software that can screen incoming and outgoing packets. A packet-filtering firewall accepts or denies packets based on information contained in the packets' TCP and IP headers. For example, most packet-filtering firewalls can accept or deny a packet based on the packet's contents, which consist of the following:

▼ Source address

■ Destination address

■ Protocol type

■ Source port

▲ Destination port

All routers, even those that are not being used as packet filters, routinely check the content of these packets to determine where to send the packets they receive. However, when being used as a packet-filtering firewall, a router goes one step further. Before forwarding a packet, the firewall compares the contents of the packet against a table containing rules that dictate whether the firewall should deny or permit packets to pass. This table is known as an Access Control List (ACL).

A packet-filtering firewall scans the ACL until it finds one that agrees with the information in a packet's full association. If the firewall encounters a packet that does not meet one of the rules, the firewall applies the default rule. Those rules should be explicitly defined in the firewall's table. For strict security, the default rule should be to drop any packet that does not meet the criteria of the other rules.

You can define packet-filtering rules that indicate which packets should be accepted and which packets should be denied. For example, you could configure rules that instructed the firewall to drop packets from specific untrusted hosts, which you would identify in the table by their IP addresses. You could also create a rule that permitted only incoming e-mail messages traveling to your mail server and another rule that blocked incoming e-mail messages from an untrusted host that had flooded your network with several megabytes of data in the past.

In addition, you can configure a packet-filtering firewall to screen packets based on TCP and UDP port numbers. Configuring a firewall in this way enables you to implement a rule that tells the firewall to permit particular types of connections, like Telnet and FTP connections, only if they are traveling to appropriate trusted Telnet and FTP servers, respectively. This rule usually depends on a generally accepted TCP/IP network convention regarding *well-known ports*. Servers and their clients generally run the most common TCP/IP applications over these well-known ports. However, those servers and clients are not *required* to use the well-known ports, and malicious hackers will be the first to take quick advantage of that convention. I discuss this in more detail in the "How Packet Filtering Works" section just ahead.

Most packet-filter implementations are done at *border routers*—routers that tend to sit at the border between your internal network and the outside world.

Stateless and Stateful Packet Filters

Most packet filters are *stateless*, which means that each packet is evaluated on its own merit and on a packet-by-packet basis. Stateless (also known as *static*) packet filters don't maintain "call setup" information between the client and server. You will recall in earlier chapters that we talked about connection-oriented protocols, such as TCP, and the applications that use them, such as FTP.

In FTP, the client sends a PORT command to the server machine. The command includes parameters to tell the server which socket (the combination of IP address to connect to and port to open at that address) to connect to, which is some arbitrary higher-numbered port on the client machine. This is what creates a *state* of communications. If we wanted to keep track of the state of this connection, the outgoing PORT command of an FTP session would have to be noted and remembered so that an incoming FTP data connection can be verified against it. Stateless packet filters don't do this. A client requests information on one TCP port, causing the server to return data at some arbitrary TCP port above the well-known range. In this scenario, which is typical of FTP, all of those higher-numbered ports above 1023 must be left open. Any clever malicious hacker could simply create Trojan horses that wait for packets entering at ports above 1023 and ride in on top of them.

Stateful packet filters "listen" to all communications and store these conversations in memory. Armed with this knowledge, the stateful packet filter would only pass those responses that are returning on expected ports from expected IP addresses. Ports will be opened only if the stateful device determines that these packets are part of conversation from a machine originating the request. If the packet matches a conversation stored in memory, it allows it to pass through. When the conversation is completed, the port closes so that if a packet is returned without the expected response, the packet will be dropped.

How Packet Filtering Works

A packet-filtering router inspects every packet and determines where to deliver it, based on the source, destination, and port addresses. A *packet filter* is a software feature set on the router that lets you control IP addresses, protocol types, TCP/UDP ports, and source routing information. You then use the packet-filtering software to configure the ACL and the criteria by which the router will decide to pass or drop the packets and define which devices users can access inside or outside the firewall based on the specific services identified by their given TCP or UDP port. This filtering may be performed as the packets enter the router, exit the router, or a combination of the two.

Most routers come from the factory set to let everything through by default. However, once you set the rules—the Access Control Lists (ACLs)—to tell the router that you wish it to inspect these packets, the packet-filtering router becomes very literal, indeed. For example, let's say you create an ACL that tells that packet-filtering router to let all packets with the source address 192.10.15.20 through. The packet-filtering router says, "Roger that, Houston, let all packets from that address through" and then promptly drops everything on the planet that doesn't come from that address.

This is known as an "implicit deny" default characteristic, and it is a common feature on some of the devices on the market today. Be sure you know what characteristic is the default feature on the products you work with. This characteristic is always in place whether you specify it or not at the end of every ACL. It is also *very literal* with respect to the order in which you place the statement on your ACL. The ACL is used to instruct the packet filter what to do. If your ACL is set to instruct it to "let *all* packets through, and *then* deny any packets from the source address 192.10.15.20," it does the first thing you said before moving on to the next request (the deny). So it is very important that you tell it, "OK, sorry, deny packets from source address 192.10.15.20, and *then* let all packets through." Figure 7-2 shows a representation of where packet-filtering gateways relate to the OSI reference model.

Again, for a packet-filtering gateway, decisions are usually based upon examining the packet's IP header and TCP header:

▼ Source address
■ Destination address
■ Protocol type
■ Source port number
▲ Destination port number

Apart from filtering by source and destination addresses, you can also filter the protocols or port numbers. It is possible to filter an entire suite of protocols such as SNMP,

SMTP, UDP, ICMP, TCP, and so on, but this tends to be a bit too general. Let's say you have your packet filter set to examine data based on the following criteria:

▼ The physical network interface that the packet arrives on

■ Source address from where the data is (supposedly) coming from (source IP address)

■ Destination address that the data is going to (destination IP address)

■ The type of protocol (TCP, UDP, ICMP)

■ The Transport layer source port

■ The Transport layer destination port

▲ Whether this packet is a start of connection request

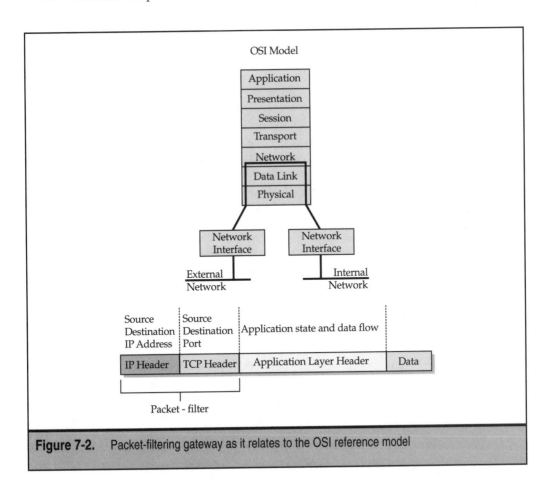

Figure 7-2. Packet-filtering gateway as it relates to the OSI reference model

You could, for example, decide to set up a policy blocking the SMTP mail port 25. While port 25 is indeed the normal mail port, there is no way we can control the use of that port on an outside host. Malicious hackers could easily access any internal machine and port by originating their call from port 25 on the outside machine.

A better policy would be to permit *outgoing* calls to port 25. That is, we want to permit our hosts to make calls to someone else's port 25 so that we can be sure that it is, in fact, mail delivery that is taking place.

A TCP conversation consists of packets flowing in two directions. Even if all of the data is flowing one way, acknowledgement packets and control packets must flow the other way. We can accomplish what we want by paying attention to the direction of the packet, and by looking at some of the control fields. In particular, an initial open request packet in TCP does not have the ACK bit set in the header; all the other TCP packets that follow do. So, packets with the ACK bit set are part of an ongoing conversation. Packets without the ACK bit set represent connection establishment messages, which we will permit only from hosts inside our network. The idea is to ensure that a host machine outside of your network cannot *initiate* a connection, but can *continue* one.

ICMP is a different story in that the ICMP protocol layer does not utilize port numbers for its needs, so it is difficult for packet filters to apply any kind of real security policy to this form of network traffic. When an error message is received that contains errors or that it doesn't recognize, it will just discard the message without any notification. To apply an effective security policy for ICMP, the packet filter must have some way of maintaining a log to ensure that an ICMP reply message was recently requested from an internal host and that it contains information that was the same as that sent in the recent datagrams to defend from denial of service attacks.

Similar problems exist with UDP. UDP utilizes port numbers, but since it is connectionless, does not maintain any connection "state." To apply some type of security, the packet filter would also need to maintain a log to determine whether subsequent packets were in response to previous packets from within.

These capabilities are usually only available on products with stateful packet-filtering capabilities. Since many products can only filter packets based on simple IP addresses, they may not understand how to process information in the higher-level protocols, such as FTP. Some of the more sophisticated packet filters are able to detect IP, TCP, UDP, and ICMP. Using a packet filter that includes the TCP/UDP port-filtering capability, you can permit certain types of connections to be made to specific computers while prohibiting other types of connections to those computers and similar connections to other computers.

Headers also contain other information deeper inside the packet for options like source routing. Source routing and fragmentation are a favorite technique of malicious hackers. The problem is that source routing makes address spoofing (pretending to be from an address that is not yours) very easy. Once the malicious hackers can spoof a source IP address, they can do whatever that address has been authorized to do. Source routing overrides any normal routing tables. Usually, you can easily send random TCP packets with a bogus source address, and you won't see the reply. While that's not a perfect defense against an attack, it makes it much harder. With source routing, however, the malicious hackers manually specify the hops that the packet should take on its way back to them. When a TCP packet arrives via source routing, the destination host is supposed to turn the route around for response packets so that all of the reply packets arrive at the malicious hackers' machine. This is not something you want, trust me.

The upside to stateless packet filters is that they work at close to wire speeds. This just means that it is not very processor intensive. Generally, the best solution is a combination of packet-filtering routers used in conjunction with circuit and/or application devices, which we will be talking about very soon.

Keep in mind that packet filtering is not limited to routers and firewalls. Most operating systems come with some form of packet-filtering capabilities that can be implemented at the host level and that can provide a lot more protection. This is what is known as *end-system filtering* because it is the last stop for the packet. Remember, a lot of your biggest threats can come from within, so it is a good idea to implement this feature end-system filtering as a last line of defense. Think of it as a backup firewall should your perimeter security be compromised.

Packet Filtering Advantages and Disadvantages

The primary advantage of using a packet-filtering firewall is that it provides some measure of protection at relatively low cost and causes little or no delay in network performance. If you already have a router with packet-filtering capabilities, setting up a packet-filtering firewall will cost no more than the time it takes to create the packet-filtering ACLs.

Although the cost of a packet-filtering firewall is attractive, this firewall alone is often not secure enough to keep out hackers with more than a passing interest in your network. Configuring packet-filtering rules can be difficult, and even if you manage to create effective rules, a packet-filtering firewall has limitations. For example, suppose you created a rule that instructed the firewall to drop incoming packets with unknown source addresses. While this rule would make it more difficult for malicious hackers to access some trusted servers, it wouldn't stop them from substituting the actual source address on a malicious packet with the source address of a trusted client.

A packet-filtering firewall usually checks information only at Layer 3, the Network layer, so sneaking packets through this type of firewall is relatively easy. A malicious hacker simply needs to create packet headers that satisfy the firewall's ACL rules to squeak those packets by.

Network and Port Address Translation (NAT and PAT)

NAT stands for "Network Address Translation." NAT was originally designed to address problems being created by the exponential growth of the Internet and the inevitable shortage of public IP addresses that would eventually follow. In March 1994, the organization responsible for the public addressing scheme of the Internet, the Internet Assigned Numbers Authority (IANA), had an idea about conserving the unique IP addressing space on the Internet (public space) by setting aside (reserving) a large set of addresses to be used in corporate private networks (private space). These addresses could be used arbitrarily and at the complete discretion of those corporations. To make this solution viable, the IANA made sure that this pool of reserved addresses would not be routable on the public Internet. The solution was called *private addressing* and was defined in Request For Comments (RFC) 1597. The Internet Engineering Task Force (IETF) records the decisions it comes to in documents called RFCs. The standards upon which the Internet is formed are based on these RFCs.

These private address ranges are

▼ 10.0.0.0 through 10.255.255.255

■ 172.16.0.0 through 172.31.255.255

▲ 192.168.0.0 through 192.168.255.255

In May 1994, the idea of Network Address Translation (NAT) was defined in RFC 1631. NAT is a process whereby any internal private addresses (such as those defined in RFC 1597) could be mapped or "translated" by an intermediate device (usually a router) to an external assigned public IP address. The intention was that, regardless of the addressing scheme being used internally by a given organization, it would not conflict with any of the publicly assigned IP addresses used on the Internet.

In the computer industry there are many terms that are used so often that their true meanings become blurred. "VPN" is probably the best example of that, but "NAT" is certainly a close second. A lot of terms with their own definitions like IP Masquerading, Network Port Address Translation, NAT Overload, Port Address Translation (PAT), and Static and Dynamic NAT have inadvertently been fused into the universal heading of NAT. The popular conception of "NAT" is a many-to-one IP address translation whereby any number of private inside (local) IP addresses is mapped to a single public outside (global) IP address. That is a gross simplification of the definition, but be aware that it is the most common overall conception.

We will discuss the real definitions of those terms shortly. I should also make clear at this point that many people have a similar misconception that NAT (the popular definition) in and of itself can be used as a network security solution. While it does, by its very nature, make the computers inside your network harder to attack, it should not be used by itself as a "firewall" for your network. Remember that as we discussed in earlier chapters, the only thing worse than no security is a false sense of security.

We will start off with a loose but familiar analogy: telephones. NAT is similar to a corporation's private branch exchange (PBX). Suppose you are a large corporation, let's say McGraw-Hill, and you want to provide a way for people to call the thousands of people working there without having to remember or look up employees' individual numbers. With a PBX (which is essentially like your own mini phone company or a voice equivalent of an intranet) you publish a single number in the phone book for McGraw-Hill, say, something like 1-800-555-1000. You can call that number and ask for anyone from the president (who will undoubtedly be in a meeting…) to a mail-room clerk and reach him or her. Your call can go through in a number of ways (just like NAT does, as you will soon see). For example, you can call the -1000 number and have a receptionist say, "McGraw-Hill" and then transfer you to whomever you ask for. You could also know the direct number, simply dial 1-800-555-1001, and bypass the receptionist.

Dialing out can work the same way. If McGraw-Hill has 1000 employees, each with a telephone number, they will need to make outgoing calls. The company knows it is very unlikely that every one of those people will phone out at the same time, so they may have only purchased a hundred phone lines from the telephone company (telephone companies like to refer to these lines as "hunt groups"). When the president dials out from his extension at 1001, his telephone hunts for a free phone line and may be put on line 31 of those hundred phone lines if 30 people have already dialed out. He could just as easily be put on line 12 when he dials out later or put on any other line for that matter (up to 100), depending on how many lines are being used at any one moment. All 100 lines could also be occupied simultaneously, in which case he might get a busy signal and have to try later. For the sake of our comparison to NAT, let's assume that McGraw-Hill chose to purchase only one outside line. Then every one of those 1,000 internal connections could only "map" to that one outside line when dialing out. (That does not make for very good "hunting.") Now, anyone trying to dial out at the same time would have to wait until that connection was available.

That said, let's take a look at the different implementations of address translation that work in much the same way. Many methods can approach problems and situations similar to those that would be encountered in the telephone analogy.

Static NAT   Static NAT is a one-to-one address mapping between the inside private addressing scheme and the outside public addressing scheme. An example would be the use of an unregistered private local IP address like those mentioned in RFC 1597 to a registered public global IP address on a one-to-one basis. This would be analogous to the example where McGraw-Hill purchased only one number from the telephone company for its 1000 employees to dial out.

Dynamic NAT   In Dynamic NAT, only the address (not the port) is translated. In this case, usually the number of external public global IP addresses is less than the number of private local IP addresses being hidden behind the NAT router. Each time a request is made from a host on the private network, The NAT router chooses an external IP address that is currently unused and then performs the translation. McGraw-Hill's 1000 internal extensions would be analogous to the private internal IP addresses, and the 100–phone line "hunt group" would be analogous to a 100-address block of public global IP addresses from an ISP.

Network Port Address Translation (NPAT), Port Address Translation (PAT), IP Masquerading, and NAT Overload   This is the popular conception of NAT that I spoke of earlier. This is in reality NAT combined with PAT, or what is often referred to as *NPAT*. This implementation can provide some level of protection without having to do any special setup or firewall policy implementations and is certainly better than no security at all. The reason for this is that NPAT only allows connections that have originated from machines inside your network. NPAT allows an internal client to connect to an outside FTP server, but prevents an outside client from connecting to an internal FTP server, because it would have to originate the connection.

You could, however, make some internal servers available to the outside world by using inbound mapping techniques. Doing that simply involves mapping certain well-known TCP ports, like 21 in the case of FTP, to specific internal addresses specified by the administrator.

By using this method, you can have more control of access to your FTP or web servers from outside your network.

Remember when we talked earlier about denial of service (DoS) attacks like SYN floods, ping of death, and the like? Although these attacks do not, on the surface, appear to compromise the security of servers inside your network, they can cause those servers to crash. Once those servers have crashed, malicious hackers can use this as a diversion for subsequent attacks like web server "spoofing" as well as many other spoofing attacks, and those are serious security breaches, indeed! Network address translation can go a long way in providing protection from such attacks.

The TCP/IP protocols include a multiplexing facility so that any computer can maintain multiple simultaneous connections with a remote computer. This multiplexing facility is the key to single-address NPAT.

To multiplex several connections to a single destination, client computers label all packets with unique *port numbers*. Each IP packet starts with a header containing the source and destination addresses and port numbers. This combination of numbers completely defines a single TCP/IP connection. The addresses specify the two machines that are communicating at each end. The two port numbers are there so that the actual connection between those two machines can be uniquely identified. We talked about this in earlier chapters when describing the port and socket concepts.

Each separate connection originates from a unique source port number in the client, and all reply packets from the remote server for this connection contain the same number as their destination port so that the client can send them back to its correct connection. This is the way the computer will know how to put all the individual packets of all the responses back together again.

NPAT changes the source address on every outgoing packet to its published single public address. It then also has to change the Source Port into a unique number so that it can keep track of each client connection. NPAT does this by using a port-mapping table to remember the numbers it substituted for the ports of each client's outgoing packets. The port-mapping table translates the client's actual local IP address and Source Port as well as its substituted source port number to a destination address and port. NPAT simply reverses this process for returning packets and then sends them back to the intended clients.

▼ When any remote server responds to an NPAT client, incoming packets arriving at the NPAT gateway will all have the same destination address, but the destination port number will be the unique substituted source port number that was assigned by the NPAT gateway. The NPAT gateway then looks in its port-mapping table to determine the actual client's destination address and port for which the packet was intended, and replaces these numbers before passing the packet on to the local client.

■ This process is completely dynamic. When a packet is received from an internal client, NPAT looks for the matching source address and port in the port-mapping table. If the entry is not found, a new one is created, and a new mapping port is allocated to the client like so:

- An incoming packet is received on a non-NPAT port.
- Look for the source address and source port in the mapping table.
- If found, replace Source Port with the previously allocated mapping port.
- If not found, allocate a new mapping port.

▲ Replace source address with NPAT address; replace source port with mapping port.

Packets received on the NPAT port undergo a reverse translation process like so:

▼ An incoming packet is received on an NPAT port.
- Look up the destination port number in the port mapping table.
- If found, replace the destination address and port with entries from the mapping table.

▲ If not found, the packet is not for us and should be rejected.

Each client has an idle timeout associated with it. Whenever new traffic is received for a client, its timeout is reset. When the timeout expires, the client is removed from the table. This is to make sure that the port-mapping table is kept to a manageable size. Most NPAT implementations can also track TCP clients on a per-connection basis and remove them from the table as soon as the connection is closed. This would obviously not apply to UDP traffic because it is not connection oriented. Since the NPAT's port-mapping table keeps complete connection information, like the source and destination address and port numbers, it can examine and validate any or all of this information before passing incoming packets back to the client. These capabilities can provide for respectable firewall protection against Internet-based attacks on your internal private network.

Circuit-Level Gateways

Circuit-level gateways relay TCP connections. When an outside user's computer (source) attempts to connect to a TCP port on a given system inside the network (destination), the source first connects to the port on the circuit-level gateway. It is the circuit-level gateway that will then connect to the intended destination on the internal network. During the conversation, the circuit-level gateway simply copies the bytes from the outside system to the destination system and vice versa. The circuit-level gateway acts like a wire or pipe, copying bytes between the inside connection and the outside connection. However, because the connection appears to originate from the gateway itself, it conceals information about the protected network. A circuit-level gateway monitors TCP handshaking between packets from trusted clients or servers to foreign hosts and vice versa to determine whether a requested session is legitimate (also referred to as a TCP Intercept). Figure 7-3 shows a representation of where circuit-level gateways relate to the OSI reference model.

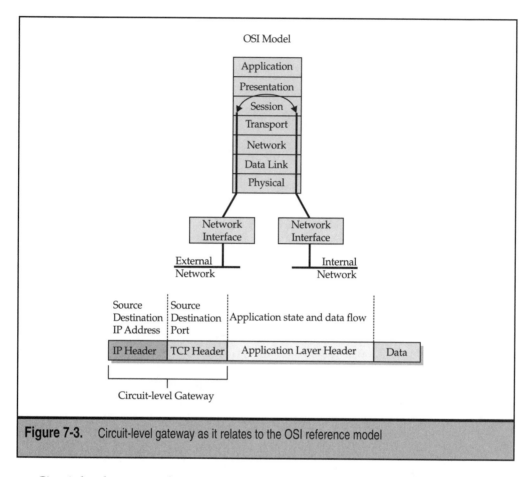

Figure 7-3.   Circuit-level gateway as it relates to the OSI reference model

Circuit-level gateway decisions are usually based upon examining the packet's IP header and TCP header:

▼   Source address

■   Destination address

■   Application or protocol

■   Source port number

■   Destination port number

▲   Handshaking and sequence numbers

How Circuit-Level Translation Works

To see whether a requested session is legitimate, a circuit-level gateway goes through a series of steps. See the following example. (Return to the Protocol Description section in Chapter 3 to help clarify some of these steps.)

1. A trusted client requests a service, and the gateway accepts this request, assuming that the client meets basic filtering criteria (such as whether DNS can locate the client's IP address and associated name).

2. Acting on behalf of the client, the gateway opens a connection to the requested foreign host and then closely monitors the TCP handshaking process that follows the connection. This handshaking process involves the exchange of TCP packets that have their SYN (synchronize) or ACK (acknowledge) bits set. These packet types are legitimate only at certain points during the session.

3. The first packet of a TCP session has the SYN bit set, indicating a request to open a session. This packet contains a random initial sequence number. For example, the trusted client might transmit a SYN packet with 1000 as the initial sequence number. Then the return packet from the foreign host will have the ACK bit set, acknowledging the receipt of the client's SYN packet. In this example, the foreign host would transmit an ACK packet numbered 1001, which is the next number in the sequence established by the trusted client.

4. The foreign host also transmits a packet with an initial sequence number, let's say, SYN 2000, for its side of the connection. The trusted client would then transmit an ACK 2001 packet, which is the next number in the sequence, acknowledging receipt of the foreign host's SYN packet and marking the end of the TCP handshaking.

A circuit-level gateway determines that a requested session is legitimate only if the SYN and ACK bits are set, and determines whether the sequence numbers involved in the TCP handshaking process between the trusted client and the foreign host are logical and legitimate. Once a circuit-level gateway determines that the trusted client and the foreign host are authorized to participate in a TCP session and verifies the legitimacy of this session, the gateway will allow establishment of the connection. From this point on, the circuit-level gateway simply copies and forwards packets back and forth. It doesn't need to filter them any longer.

The circuit-level gateway maintains a table of established connections, allowing data to pass when session information matches an entry in the table. When the session is completed, the gateway will then remove that particular entry in the table and then close the circuit for that session. So with a circuit-level gateway, all outgoing packets appear to have originated from that gateway, preventing direct contact between the trusted network and the foreign host. The circuit-level gateway's IP address is the only active IP address and the only IP address that the foreign host is aware of. This helps to protect trusted networks from spoofing attacks by foreign hosts.

Circuit Level Advantages and Disadvantages

A circuit-level gateway is very similar to a proxy server. Some would argue that it is, in fact, a proxy server. Although the term *proxy server* suggests a server that runs proxies (which is true of a circuit-level gateway), the term technically means something different. A proxy server is a firewall that uses a process called *address translation* to map all of your internal IP

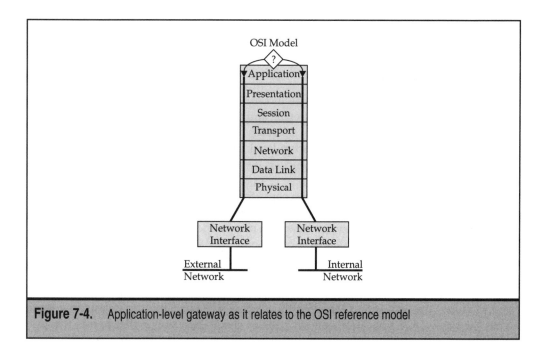

Figure 7-4. Application-level gateway as it relates to the OSI reference model

untrusted host. Remember, the proxies that an application-level gateway differs in two respects with regard to circuit-level gateways:

▼ The proxies are application specific.

▲ The proxies examine the entire packet and can filter packets at the Application layer of the OSI model.

Application Level Advantages and Disadvantages

A key benefit of application-level gateways is that most of them log activities and note significant events, intrusion detection, and other anomalies by sounding alarms, sending pages, e-mails, and so on. They maintain logs that record the source and destination addresses of packets associated with attempted or successful entries into your system, the time these attempts were made, and the protocol used.

An application-level gateway is one of the most secure firewalls available, but it does take away from the transparency and speed provided by other gateway types. Ideally, an application-level gateway would be as transparent as it is secure. Users on the trusted network would not notice that they were accessing Internet services through a firewall. The reality, however, is that users will often experience delays or need to log in several times before they are connected to the Internet or an intranet via an application-level gateway. Another common problem is that you will experience "timeouts" when trying to access virtually anything on the Internet. This goes back to deciding how granular you

wish your security to be when compared with the inconvenience and lack of productivity that can be a consequence of such granularity.

Stateful Multi-Layer Inspection (SMLI) Gateways

Stateful multilayer inspection firewalls combine all of the characteristics of the packet-filter, circuit-level, and application-level firewalls. They examine packets at the Network layer to determine whether session packets are legitimate and inspect the contents of those packets at the Application layer as well.

Unlike application-level gateways, SMLI firewalls allow direct transparent connections between a client and host. They rely on algorithms to recognize and process Application layer data instead of running application-specific proxies. They allow users to connect directly to the Internet by maintaining dynamic state tables on every connection in order to determine what information should be sent back to the end user.

A good analogy for this is a guard at a checkpoint noticing and remembering some defining characteristics of anyone leaving through the checkpoint and allowing them back in only if those defining characteristics remain identical. So, in the case of data, instead of examining the actual contents of each packet, it remembers the bit patterns of those packets and compares those bit patterns to packets that are already considered to be trusted. This is how the transparency is achieved.

Now let's say that an outside service is accessed. The SMLI server remembers characteristics about the original request like the port number, the source address, and the destination address. This process of remembering is known as *saving the state*. When the outside system responds to a request, the SMLI server compares the received packets with the saved state to decide whether to allow packets to pass.

SMLI firewalls offer a high level of security, good performance, and transparency to end users. Stateful inspection firewalls tend to be expensive and more complicated to implement. By virtue of this complexity, they could end up being less secure than some of the simpler solutions if not implemented properly. Initial analysis is done at the lower levels of the OSI model, but unlike the application-level gateways, stateful inspection uses business rules defined by the user rather than using specifically designed proprietary application information. This means that SMLI firewalls cannot apply different rules to different applications, so you will have to decide which is more important to your organization's security policy. The best solution is to combine these different gateway types at different points in your network, depending on the level and type of security you require. Figure 7-5 shows how an SMLI firewall compares with the OSI reference model.

How SMLI Works

SMLI firewalls screen the data being communicated at one or more layers anywhere from the Network layer up to the Application layer—or all of those layers for that matter—to check the data's integrity. You could think of it this way: SMLI firewalls have the data "checked" as opposed to "processed." Application-level gateways process the data.

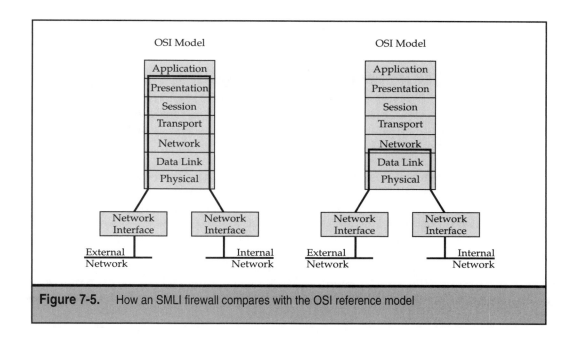

Figure 7-5. How an SMLI firewall compares with the OSI reference model

Incoming packets are checked in the context of previously received data by keeping track of session states. Session states are a combination of communications phase and endpoint application states. If there should be any unexpected changes in those session states, the firewall drops the rest of the session. Also, this type of firewall can check for new session requests in the context of active sessions allowed through the firewall. By doing this, SMLI firewalls can then dynamically configure their rule sets to changing network conditions and to the way some multisession applications behave.

The lack of application-level processing capabilities makes this type of firewall unaware of any application-specific features beyond basic session-state behavior. However, SMLI firewalls are less processing intensive when compared with application gateways and therefore are much faster.

Suppose your security policies require specific knowledge of a particular application to perform functions such as event logging, application-specific authentication, network address translation, or to disable certain application features that are considered a risk. Then you should consider using an application-level gateway in place of, or for best results, in combination with, an SMLI firewall.

FIREWALL ARCHITECTURES

The purpose of a firewall is to block unauthorized access between external untrusted networks (like the Internet) and your internal private networks. The results of firewall compromise can be considered *zones of risk*. In the case of a private network connected directly

to a public network, the entire private network constitutes a zone of risk and is subject to attack. With the implementation of even the simplest packet filtering firewall into the network separating the public and private networks, the zone of risk is reduced to being the packet-filtering firewall (usually a router) itself. This reduces your concern and takes your focus away from a direct attack on the internal network and instead focuses it on a single entity, the firewall. You can see an example of this simple direct connection to the Internet in Figure 7-6.

Screening Router Architecture

The simplest firewall architecture usually uses the packet-filtering capabilities of the router connected between the Internet and your internal private network. Normal routers simply determine whether and where a packet should be routed. When you use the router as a packet filter, too, you introduce an additional level of security by allowing the router to also determine if a packet should be routed or dropped. It does this by examining the packet's source, destination, and port numbers. It can then decide if the contents are consistent with the port request being made.

While screening routers can be used as a firewall, the amount of protection they offer is at best rudimentary. Once the router's defense has been compromised, your entire network is then open to attack. In addition, most routers do not have any, or at best, very little, logging and auditing capabilities. Figure 7-7 shows a typical example of how a screening router firewall architecture might be implemented.

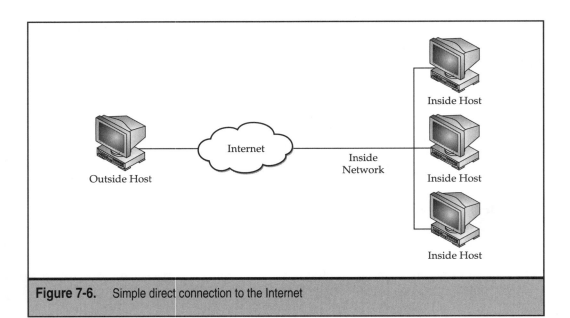

Figure 7-6.   Simple direct connection to the Internet

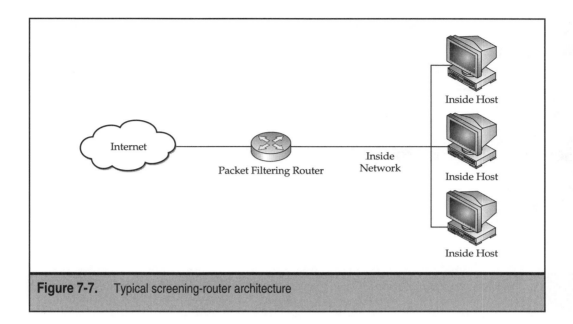

Figure 7-7. Typical screening-router architecture

Screened-Host Firewall Architecture

A screened-host firewall architecture is also sometimes referred to as a *bastion host* architecture. The way this works is that a firewall with a connection to both the public and private networks. While it is capable of routing packets from one network to another, this functionality is typically disabled. Instead, a proxy service or stateful inspection implementation manages the evaluation of packets and then establishes the proxy that transports them to their destination. Many of these architectures allow users to authenticate themselves at the firewall and to establish direct connections with the requested host inside the network. These architectures can create a weakened state of security on the internal private network simply because they offer an opportunity for individuals to gain unauthorized access to the firewall and compromise the internal private network by the very nature of it sharing the same wire or network segment. Figure 7-8 shows a typical example of how a screened-host firewall architecture might be implemented.

Dual-Homed Firewall Architecture

A dual-homed firewall architecture is a better alternative to packet-filtering router firewalls. It generally consists of a gateway system with two network interfaces to separate the internal private network subnet from the DMZ (Demilitarized Zone) subnet. Between the DMZ and the connection to the Internet itself would be a packet-filtering router to provide additional protection. This is where your intermediate screened subnet or DMZ is used for locating your publicly available services such as web servers and FTP servers.

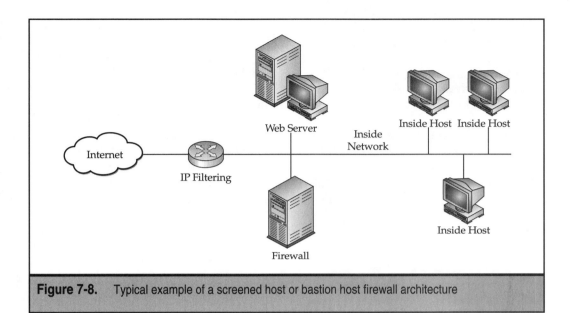

Figure 7-8.   Typical example of a screened host or bastion host firewall architecture

Unlike the packet-filtering firewall, the dual-homed gateway completely blocks traffic between the Internet and the internal private network. Access and services are controlled by the proxy features in the gateway sitting between the DMZ and the internal private network. It is a simple firewall and also quite secure.

This type of firewall architecture can be used to implement a second design policy such as denying all services unless they are specifically permitted so that no services can pass except those for which specific proxies exist. You would also want to make sure that the ability of the gateway to accept source-routed packets is disabled so that no other packets could be passed by the gateway to the internal private network subnet. It can be used to achieve a high degree of privacy since routes to the internal private network subnet need to be known only to the firewall and not to outside hosts on the Internet. Those hosts on the Internet have no way to send packets directly to hosts on your internal private network. The names and IP addresses of site systems would also be kept hidden from Internet systems because the firewall would not pass DNS information.

A simple setup for a dual-homed gateway would be to provide proxy services for Telnet and FTP, and centralized SMTP services in which the firewall would accept all site mail and then forward it to internal systems. Because it uses a host gateway, the firewall can be used to require users to present authentication tokens or other advanced authentication measures. The firewall can also keep track of all logon attempts. In addition, the firewall can log any attempts to probe your network, to warn you about possible intruder activity.

Dual-homed gateways could be a disadvantage to some organizations since all services are blocked by the gateway except those for which proxies exist, making access to other services impossible. Systems whose services require such access would need to be placed on the DMZ side of the gateway, and that would not be as secure. You need to make sure that the security policies of the gateway system are very secure so that if anyone should make it past the packet-filtering capabilities of the router directly connected to the Internet, they would be stopped by the gateway. Figure 7-9 shows an example of how a dual-homed firewall architecture might be implemented.

An even more secure and flexible way of approaching your network security would be a combination gateway architecture you might want to call a "dual dual-homed" firewall architecture. (I think I may have just "spilled the beans" with respect to my ability to start a naming convention of any kind…oh, be nice!) This architecture would allow you to use whatever combination of gateways necessary for your particular security policy needs, regardless of how unusual they might be. You could use combinations such as application-level/SMLI gateways, application-level/packet-level/SMLI, or any others you can think of to meet your needs. You can see what is meant by this in Figure 7-10.

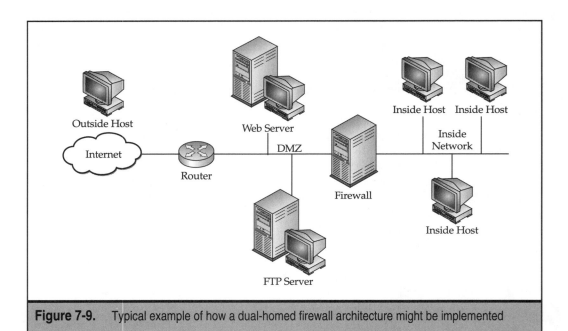

Figure 7-9.   Typical example of how a dual-homed firewall architecture might be implemented

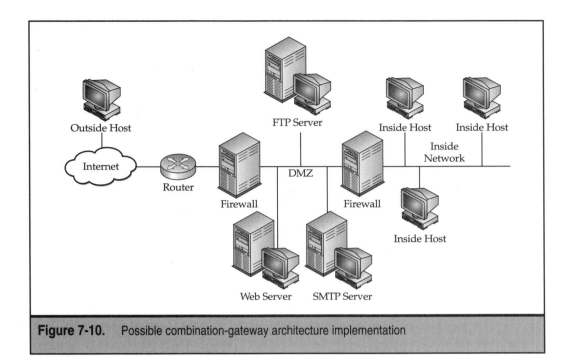

Figure 7-10.   Possible combination-gateway architecture implementation

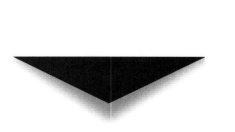

PART II

Virtual Private Networks

CHAPTER 8

VPN Basics

In earlier chapters, we talked about firewalls and saw that they were an important security feature for company networks. Firewalls, for the most part, are for preventing unauthorized users and data packets from leaving and entering the corporate network. However, once those packets pass through the firewall and into the Internet, they become fair game and completely visible to anybody who chooses to look at them. It is for this reason that this book is being written.

Another reason is because the term "VPN," or "Virtual Private Network," is probably the most recklessly used term in the networking industry. It is being portrayed as a panacea for a broad set of problems and solutions, when the objectives themselves have not been properly thought out.

This confusion has resulted in a situation where the press, telecommunications industry, networking industry, vendors, and consumers of networking technologies alike generally use the term VPN to describe almost any product that connects to the Internet. Every company gets a fair chance to claim that its existing product is actually a VPN. But no matter what definition you choose, the basic idea is to create a private network via tunneling and/or encryption over the public Internet.

This chapter will discuss the fundamentals of what a VPN is and some of the components and methods used to implement it.

WHAT IS A VPN?

VPN is simply an acronym for Virtual Private Network. You may also have seen it represented as VPDN, for Virtual Private Dial-up Network. So much for the acronyms. A VPN is a "virtual" network that is kept private by "tunneling" private data through the underlying infrastructure of the public Internet. A VPN uses the Internet or another public network service as all or part of its WAN backbone.

In a VPN, inexpensive local connections to an Internet service provider (ISP) replace PSTN or ISDN dial-up connections for remote users and replace expensive leased lines (such as ATM or Frame Relay connections) for LANs at remote sites.

A VPN also allows a private intranet to be securely tunneled through the Internet, making secure e-commerce and extranet connections with other business partners, suppliers, and customers possible. Tunneling allows two corporate sites to connect in a secure manner by using the infrastructure of the Internet transparently as what appears to be the VPN's own private network, in just a single hop. A VPN can also provide protection for departments within a corporation sharing the same LAN segment. This practice is not uncommon and is often used for interdepartmental data transfer.

VPN security is achieved by using end-to-end authentication and encryption, thus saving money by replacing long-haul links with local links, and reducing the cost of maintaining and constantly having to upgrade their customer premise equipment (CPE) telecommunications devices.

Many different types of tunneling architectures can be used, and we will discuss each of them in more detail in later chapters. After reading those chapters, you will have a much better understanding of what is actually taking place in each of those architectures. For now, let's take a look at Figure 8-1 to get a general idea (*very* general, mind you!) of what takes place.

Tunneling is a technology that enables one network to send its data over another network's connections. Tunneling works by encapsulating a network protocol within packets carried by the second network, which is accomplished by encapsulating the data inside a protocol that is understood at the entry and exit points of the other network. The entry and exit points are the endpoints of the tunnel. The system at the entry point encapsulates the data, while the system at the exit point removes this data from the encapsulation.

Tunneling creates circuit-like connections across the packet-oriented Internet. Instead of using a packet-oriented protocol such as IP, which might send packets across a variety of routes before they reach their common destination, a tunnel represents a dedicated virtual

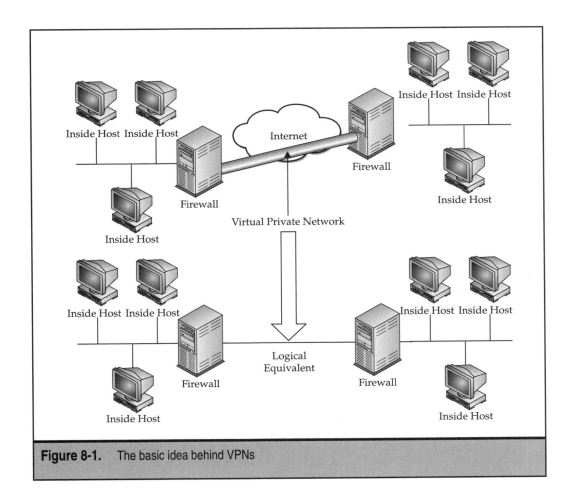

Figure 8-1.   The basic idea behind VPNs

circuit between two endpoints. However, since it works across a shared-network infrastructure, tunneling gives enterprises a cost-effective solution between packet and leased-line communications.

To create the virtual circuit, a special tunneling protocol must encapsulate each source network packet into a new packet that contains the connection management intelligence necessary to set up, tear down, and manage the tunnel.

The great thing about tunneling is that it doesn't matter *what* kind of protocol is carried. It could be Apple Computer's AppleTalk, Novell's IPX, Compaq's DECnet, IBM's SNA, NetBEUI, or anything else, routable or not!

You could even use non-Internet routable reserved IP addresses set aside in RFC 1918 (for example, 10.0.0.0, 172.16.0.0, or 192.168.0.0) internally, obviating the need for Network Address Translation (NAT), discussed in Chapter 7. Now if that's not a deal, I don't know what is. This is a big factor in what makes VPN so secure.

VPNs allow telecommuters, remote employees such as salespeople, or even branch offices, to connect in a secure way to a corporate server located at the edge of the corporate LAN. VPNs make these connections by using the switched-circuit architecture common in the infrastructure provided by public telephone companies and long-distance carriers. From the user's perspective, the VPN is a direct LAN connection between the user's computer and a corporate server. The nature of the intermediate connection is irrelevant to the user because it appears as if the data is being sent over a dedicated private link.

VPN technology also allows a corporation to connect to branch offices or to other companies over a public infrastructure while ensuring that all of its communications are secure. This is also what is known as an *extranet*. A VPN connection across the Internet can logically operate as a WAN link between the sites.

In both cases, the secure connection across the internetwork appears to the user as a private network communication even though the communication, in reality, occurs over a public infrastructure—hence the name *Virtual* Private Network.

VPN technology is designed to address modern issues raised by the efforts to move toward increased telecommuting, widely distributed global operations, and highly coordinated partner operations. It facilitates a way for workers, clients, customers, business partners, and even competitors to connect to central resources, communicate with each other, and efficiently manage inventories while meeting today's fast-paced production and economic requirements.

INTERNETS, INTRANETS, AND EXTRANETS

In any business related to information technology, you are going to come across a lot of definitions of what a VPN is. VPNs are the hot commodity these days, and everyone is offering some version of what their VPN is. They will tell you whatever it takes to get you to buy their product, so be wary of those who would sell you anything.

Before we get into any great detail regarding the history of VPNs, we should discuss the technologies that take advantage of the capabilities provided by VPNs. They are

▼   Internets

■   Intranets

▲   Extranets

Let's start by defining these technologies. We will break these down in order, and this will give you a better appreciation of what VPNs are when we talk about their history. Take a look at Figure 8-2 first, so you can get a visual idea of the upcoming descriptions.

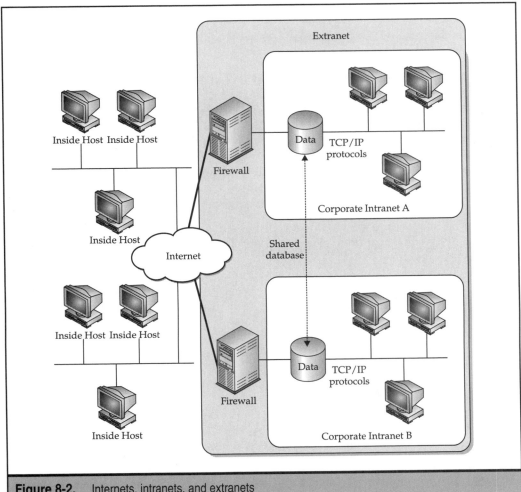

Figure 8-2.   Internets, intranets, and extranets

Internets

The Internet, or "World Wide Web" as it is commonly referred to, has rapidly become a global information phenomenon. This growth has exceeded virtually everyone's predictions and at blinding speed. This rapid growth, ease of use, and ubiquity is also making international cultural interaction and social exchange as easy as clicking a button and will change forever the way we interact and do business. The Internet is quite literally a worldwide network of networks. The Internet has given rise to a similar term, "internet," spelled with a lowercase *i*. The technical definitions are actually case sensitive.

▼ An *internet*, with a lowercase *i*, is any collection of networks that are networked or connected together over a common infrastructure.

▲ The *Internet*, with an uppercase *I*, is the World Wide Web, or global Internet that you have come to know and love.

The beauty of the Internet is that, thanks to TCP/IP, all of this ubiquitous international interaction can be accomplished regardless of the underlying technology being used with respect to operating system, network topology, protocols, or even software applications. The whole reason you are reading this book (whether you know it or not) is that the very properties of TCP/IP that make the seamless communication between disparate systems possible are also the reason we need VPNs. TCP/IP is a very "trusting" protocol suite, and this is what makes it such an attractive playground for hackers.

Intranets

Over the last few years, a great deal of hype has surrounded intranet technology. Thank goodness VPNs came around to take over that role; intranets were starting to get tired of all that hype anyway.

Intranets are essentially when a company treats its internal network infrastructure as though "it" was its very own World Wide Web. In other words, you could think of intranets like an "Internet" or "World Wide Web" involving and accessible exclusively by the employees of that particular organization.

Intranets are so popular for the same reasons the Web became so popular. Anyone and everyone knows how to use it, they like to use it, and by virtue of that, they require no training, which improves productivity, which is what businesses like because... Well, listen, this is not an Economics 101 text book, so you're not going to get any of that knowledge here, not if I can help it!

By emulating the Internet, intranets are also based on the same protocol suite, TCP/IP. However, by virtue of their autonomy, they are not subject to the same perils of the Internet, because only employees are given access to an intranet.

Extranets

Extranets are simply an extension of the intranet concept and a subset of the Internet. The Internet allows any and all networks and entities to connect with each other and is not governed by any one organization in particular. An intranet, by contrast, is a group of networks governed by one and only one organization. An extranet is a combination of the Internet and an intranet. It involves a collaboration of organizations or companies that share a common interest and are willing to connect each of their *intranets* into a miniature isolated version of the global *Internet* in order to share their resources (within arbitrarily specified limits set by their respective systems administrators). An extranet can be used for things like inventory, purchase orders, product catalogs, and whatever else might improve the organizations' logistical productivity.

Extranets can be set up using private-network infrastructures that used circuit-switched technologies like leased lines and, more recently, Frame Relay networks provided by traditional telephone carriers. In fact, until recently, this was the only way extranets were set up. This was a very expensive enterprise, to say the least, costing many thousands of dollars a month in tariffs regardless of whether they had actual data transmission. The tariff cost is nothing when you also consider the need for highly trained network engineers (like you!) to maintain the equipment.

With the recent advent of the global Internet, extranets can be set up using public, packet-switched technologies for a fraction of the price. This is where VPNs shine. Their usefulness will become more evident as we progress.

HISTORY OF VPNS

Prior to the Internet, companies that had geographically dispersed locations would create wide area networks (WANs) to tie their local area networks together by leasing private lines from public telephone carriers. These might include T carriers (T1/T3), ATM, Frame Relay, ISDN, and even X.25 or PSTN (Public Switched Telephone Network). Mobile users would use the PSTN, or possibly an ISDN connection, to dial in to access servers maintained by these companies at given locations around the world. These networks were considered private because they were, for all practical purposes, the only ones using these leased lines.

The technology for implementing VPNs has existed for some time. Its origin can be found in the *virtual circuit*. The basic structure of the virtual circuit is to create a logical path from the source to the destination. This path may incorporate many hops between routers for the formation of the circuit. The final, logical path or virtual circuit acts in the same way as a direct connection between the two ports. In this way, two applications could communicate over a shared network.

The viability of a VPN is exactly the same as any other private network. This means that information passed between parties on the network is protected from attackers. The difference between a VPN and another private network is that a VPN operates using a public network infrastructure like the Internet. The VPN occupies its own private address space on the public network, making it inaccessible to other users on that network. In addition, the data in transit is encrypted so that even if that data were to be intercepted, the attacker would not be able to read it.

The best way to bring some things to light is by first learning what precipitated their necessity. Let's start by taking a look at what took place prior to the invention of VPNs. We'll start by talking about the olden days, when companies used private networks.

Private Networks

The most familiar telecommunications technology is telephone circuit-switching networks. These networks have a fixed and predetermined capacity, because a telephone call requires a dedicated link, and no other user can take advantage of the network facilities involved in the link until the call is terminated. In contrast, Internet services are produced with packet-switching networks, which can theoretically handle an unlimited number of connections over the same facility, because they operate by splitting data into small packets that contain all the necessary information to reach their destination, where they are reassembled into the original data. This splitting allows the network to send packets of different origins in interleaved order through the same link, at the expense of a longer transmission time for the same amount of data when the number of connections grows.

Traditional WAN-based private networks used leased-line circuits from each site back to a corporate headquarters. These leased-line circuits were priced according to distance, making them expensive for geographically dispersed locations. These traditional WANs often used a hub-and-spoke model, requiring traffic going between branch offices to travel through the corporate headquarters. Despite these disadvantages, traditional WANs offer the highest level of security and network performance. Figure 8-3 shows a basic idea of how a WAN works.

Ironically, private networks share the same telecommunications infrastructure as the Internet. However, private network lines cost considerably more than standard telephone lines because they are often configured for higher speeds and greater bandwidth. Private networks usually require access systems involving added infrastructure equipment. These are usually located at the corporate headquarters. Access systems include communication servers (more commonly known as *access servers*), modem banks, and telephone termination equipment. Access to the private network from remote locations

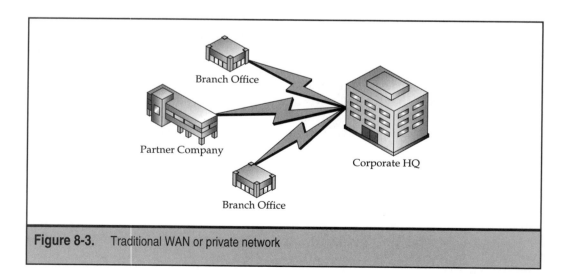

Figure 8-3. Traditional WAN or private network

can be accomplished using standard dial-up or digital telephone lines. These can include technologies such as Integrated Services Digital Network (ISDN) or Digital Subscriber Line (DSL) lines. Private networks can be maintained by the company itself or through an ISP like UUNET, Sprint, ATT, or MCI WorldCom. Take a look at Figure 8-4 to see the more modern implementation of a private network.

Higher performance, speed, and security are obvious advantages of a private network. Another advantage that has been a big obstacle for VPN viability is that bandwidth is guaranteed. These problems have been addressed with newer VPN technologies over IP-based networks using Multi-Protocol Label Switching (MPLS), which will be talked about in future chapters.

Private networks can also support a variety of protocols such as Frame Relay, Asynchronous Transmission Mode (ATM), and time-sensitive protocols like SNA.

The disadvantage is that these networks are extremely expensive. A large enterprise planning to connect many offices located globally via a fiber-optic telecommunications backbone can spend millions of dollars on these private networks. Another problem is that these networks quickly become difficult to manage, especially when adding any new locations, business partners, suppliers, or other online concerns. Traditional WANs require highly specialized equipment. This also means that highly specialized, technical in-house personnel are needed to manage these networks.

The technical differences between circuit switching and packet switching arise from how they treat congestion. In a telephone or circuit-switching network, when the last free link has been used to establish a call, all further attempts by any user to call another user will fail until one or more of the ongoing calls are terminated, freeing up a line. Corporations pay for the full-dedicated bandwidth of the network whether they use it or not.

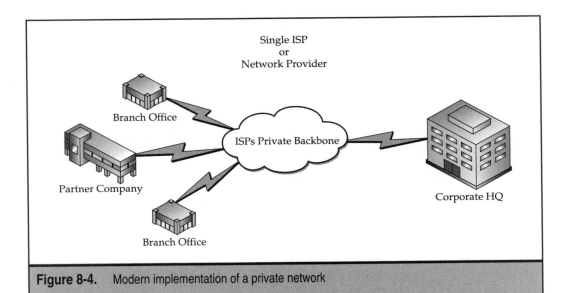

Figure 8-4.   Modern implementation of a private network

In a packet-switching network, however, no user is to be denied access no matter how many connections are taking place, but transmission times will increase with every additional connection.

Public Networks

With the advent of the Internet (you've probably heard of this recent phenomenon), the rules changed. There were many compelling reasons for these companies to have a presence on this new "Information Superhighway" by connecting themselves to Internet service providers (ISPs) that had direct connections to the Internet. You can see an example of a public network represented in Figure 8-5.

This posed an interesting new problem, namely, privacy. By virtue of having their private WANs connected to the Internet, the network—and the data on it—was vulnerable to anyone and everyone on this new Internet who had the initiative to tamper with it. Unfortunately there was, and continues to be, no shortage of those individuals. Along with VPNs, this was also the impetus behind the creation of the firewalls in use today. VPNs make use of many existing technologies such as the virtual circuit and the security firewall.

That VPNs make use of a public network makes them highly cost-effective. With certain VPNs the companies requiring security facilities can make use of any service provider, which will be oblivious to the Virtual Private Network it is hosting. Alternatively, the service provider can support the VPN so that the company need not maintain its own private network. In addition, if the company chooses to manage its own Virtual Private Network and does not lease the service from a service provider, the maintenance and administration of a VPN is considerably simpler than that of private leased lines.

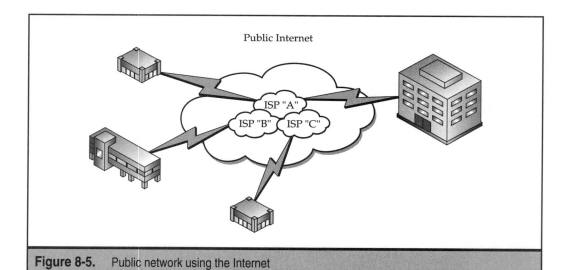

Figure 8-5.   Public network using the Internet

WHAT IS TUNNELING?

Tunneling can be defined as a way to transfer data between two similar networks over an intermediate network. Tunneling encloses one type of data packet into the packet of another protocol. Before the encapsulation takes place, the packets are encrypted so that the data is unreadable to anyone monitoring the network. These encapsulated packets travel through the Internet until they reach their intended destination. Once there, the packets are separated and returned to their original format. The protocol of the encapsulating packet is understood by the network and by both the points where the packet enters and exits the network.

Tunneling generally involves three different protocols:

▼   **Carrier protocol**   This is the protocol used by the network that the information is traveling over, usually TCP/IP.

■   **Encapsulating protocol**   This is the protocol that the original data is packaged in, like GRE, IPSec, L2F, PPTP, or L2TP.

▲   **Passenger protocol**   The original or native data that is being carried from the network where the originating host resides like IPX, AppleTalk, or IP.

Tunneling enables you to place a packet that uses a protocol not used by the Internet (such as IPX) inside an IP packet and to send it securely over the Internet. You can even place a packet that uses a private, nonroutable, IP address inside a packet that uses an assigned IP address to tunnel your private network through the Internet.

TUNNELING PROTOCOLS

VPNs are generally based on one of two types of tunneling protocols:

▼   **Layer 2 tunneling protocols**   Layer 2 protocols correspond to the Data Link layer and use frames as their unit of exchange. PPTP, L2TP, and L2F are Layer 2 tunneling protocols. These protocols encapsulate the data in a Point-to-Point Protocol (PPP) frame to be sent across an internetwork.

▲   **Layer 3 tunneling protocols**   Layer 3 protocols correspond to the Network layer and use packets. IP over IP and IP Security (IPSec) Tunnel Mode are examples of Layer 3 tunneling protocols. These protocols encapsulate IP packets in an additional IP header before sending them across an IP internetwork.

Layer 2 Tunneling Protocol Characteristics

For Layer 2 tunneling protocols like PPTP and L2TP, the tunnel endpoints must first agree to the communications session as well as to any necessary configuration variables, like encryption, assignment of addresses, and compression. The data is then sent across the tunnel using, in most cases, a datagram-based protocol like IP. The tunnel is then managed using a tunnel maintenance protocol.

Layer 2 protocols like PPTP and L2TP are based on the well-defined PPP protocol and as such inherit a suite of useful features. These features address the basic VPN requirements as follows:

▼ **User authentication**   Layer 2 tunneling protocols inherit the user authentication schemes of PPP, including the EAP methods discussed next.

■ **Token card support**   Using the Extensible Authentication Protocol (EAP), Layer 2 tunneling protocols can support several authentication methods like one-time passwords and smart cards.

■ **Dynamic address assignment**   Layer 2 tunneling supports dynamic assignment of client addresses based on the Network Control Protocol (NCP) negotiation mechanism of the PPP protocol.

■ **Data compression**   Layer 2 tunneling protocols support PPP-based compression schemes.

■ **Data encryption**   Layer 2 tunneling protocols support PPP-based data encryption mechanisms.

■ **Key management**   Layer 2 protocols rely on the initial key generated during user authentication and then periodically refresh it.

▲ **Multiprotocol support**   Layer 2 tunneling supports multiple payload protocols, which makes it easy for tunneling clients to access their corporate networks using IP, IPX, AppleTalk, and so on.

Layer 3 Tunneling Protocol Characteristics

In the case of Layer 3 tunneling protocols, all of the configuration issues are set up out of band, usually manually. For these protocols, there may be no tunnel maintenance phase. For Layer 2 protocols (PPTP and L2TP), however, a tunnel must be created, maintained, and then terminated.

Once the tunnel is established, it can then send the tunneled data. The tunnel client or server uses a tunnel data transfer protocol to transfer the data. Essentially, when the tunnel client sends a payload to the tunnel server, the tunnel client first appends a tunneling protocol header to the payload. Then the client sends the encapsulated payload across the internetwork, where it is routed to the tunnel server. Once the tunnel server receives the packets, it removes the tunneling protocol header and then forwards the payload inside to its intended destination. Data sent between the tunnel server and the tunnel client behaves the same way.

Layer 3 protocols address the basic VPN requirements as follows:

▼ **User authentication**   Many Layer 3 tunneling schemes assume that the endpoints are already known and authenticated prior to tunnel establishment. An exception to this is IPSec ISAKMP negotiation, which provides mutual authentication of the tunnel endpoints. (Note that most IPSec implementations support machine-based certificates only, rather than user certificates. As a result, any user with access to one of the endpoint machines can use the tunnel. This potential security weakness can be eliminated when IPSec is paired with a Layer 2 protocol such as L2TP.)

- **Token card support** Layer 3 tunneling protocols can use methods similar to those used by Layer 2 tunneling protocols. IPSec defines public key certificate authentication in its ISAKMP/Oakley negotiation.

- **Dynamic address assignment** Layer 3 tunneling schemes assume that an address has already been assigned before initiating the tunnel.

- **Data encryption** Layer 3 tunneling protocols can use methods similar to those for Layer 2. IPSec defines several optional data encryption methods that are negotiated during the ISAKMP/Oakley exchange.

- **Key management** IPSec negotiates a common key during the ISAKMP exchange and also periodically refreshes it.

- ▲ **Multiprotocol support** Layer 3 tunneling protocols, such as IPSec tunnel mode, usually support target networks using the IP protocol.

A TUNNEL FROM THE PAST

To help grasp the concept of tunneling, it might prove helpful to discuss some recent history in the development of the tunneling protocols being used today. In the mid-1990s it was becoming apparent that the concept of VPNs had the potential of becoming a very lucrative market. At that time a couple of obscure little companies that you have probably never heard of (Cisco and Microsoft) were working on two different tunneling protocols.

Microsoft (for the most part) was developing the Point-to-Point Tunneling Protocol (PPTP), an OSI Layer 2 protocol that encapsulates PPP packets into IP packets.

Cisco, on the other hand, was busy working on the Layer Two Forwarding (L2F) protocol that would, unlike PPTP, use typical Data Link protocols like ATM, Frame Relay, and so on. I was actually hoping *you* might be able to extract which OSI layer the L2F protocol occupies. (Feel free to e-mail me should you get stuck here.) Then, considering the size of these two behemoths, they realized that perhaps the best thing to do would be to combine these two protocols into what is now known as the Layer Two Tunneling Protocol (L2TP), and again e-mails are welcome. These protocols are generally considered to be best for host-to-host tunnels as opposed to LAN to LAN.

To address this problem and to avoid the problems that come with proprietary solutions, a working group of the Internet Engineering Task Force (IETF) set about finding a way to make the IP protocol itself more secure. This Layer 3 solution was known as Secure IP or what is now referred to as IPSec. IPSec defines two headers in IP packets to handle authentication and encryption. The IP *Authentication Header* (AH) is for authentication, while the *Encapsulating Security Payload* (ESP) is for encryption purposes. These, in conjunction with other technologies that we will touch on, such as the *Generic Routing Encapsulation* (GRE) protocol, are the most common methods used in tunneling today.

Another alternative that has been making progress is SOCKS. This is a Session layer (OSI Layer 5) protocol that supports all IP transported applications, including TCP and UDP. Because it is a Session layer protocol, it operates independently of any application. It has been in development since about 1990 and is now an IETF standard (RFC 1928). SOCKS does not encapsulate traffic and therefore does not compete with standard tunneling protocols. However, SOCKS brings other benefits to VPN environments using these protocols like user-level authentication and encryption, continuous access control for any existing connections, unidirectional security, and application-specific security. Version 5 includes encryption negotiation. Any firewall can be configured to pass SOCKS traffic transparently.

As protocols like IPSec, L2TP, and SOCKS mature, their header option standards will become more concrete. Once those options, like company specific Atribute Value Pairs AVPs and the like, are more stable tunnels will be able to interconnect much more easily. All of these technologies can now be used in conjunction with each other to increase compatibility between systems, to help create transparency for the user, and to make VPNs extremely secure.

TUNNEL TYPES

Before we talk about the actual tunneling protocols, it is important to understand the kinds of tunnels that are created with these protocols. There are two types of tunnels:

▼ **Voluntary tunnel** This tunnel is initiated by a remote PC. It is called a *voluntary tunnel* because the end device is in control of the tunnel creation. In this situation, the remote PC is the tunnel endpoint. Voluntary tunnels are also known as *client-initiated tunnels*.

▲ **Compulsory tunnel** This tunnel involves an intermediate device. This device is usually a Network Access Server (NAS), located at some ISP Point of Presence (POP), that is VPN aware. Any client can dial in with a non-VPN protocol, usually PPP, and connect to the ISP NAS. At this point, the NAS creates a compulsory tunnel with some host VPN server. With a compulsory tunnel, the remote PC is not the tunnel endpoint. The ISP NAS that sits between the remote PC and the host VPN server is the tunnel endpoint and acts as the tunnel client. Compulsory tunnels are also referred to as *NAS-initiated tunnels*.

As a rule, voluntary tunnels tend to be the most popular type of tunnel. However, sometimes a combination of the two is best (L2TP is a good choice) because not all computers that you may wish to give access to your VPN will always have the required VPN client software necessary for a pure voluntary tunnel. It is important that you understand that the two terms for each tunnel type are completely synonymous (that is, voluntary/client-initiated and compulsory/NAS-initiated). We reiterate this because different companies will use these terms interchangeably (as will we!) depending on their marketing approach, and when you hear them used, you should be aware of this.

Voluntary Tunnels

Voluntary tunnels are those created by a remote PC initiating a secure connection with a host VPN. If you plan to use voluntary tunnels, an appropriate *and* compatible tunneling protocol must be installed on the remote PC or client computer. Voluntary tunnels must also have some kind of an IP connection. It can be a dial-up connection or can even be directly attached to the LAN locally. Here are two possible situations:

▼ Let's say you are dialing in to the network. First, you need to establish a dial-up connection to the internetwork before the computer can set up a tunnel. This is the most common situation. A good example of this is the dial-up Internet user, who must dial an ISP and obtain an Internet connection before a tunnel over the Internet can be created.

▲ For a LAN-attached PC, the client already has a connection to the internetwork that can provide routing of encapsulated payloads to the chosen LAN tunnel server. This would be the case for a client on a corporate LAN that initiates a tunnel to reach a private or hidden subnet on that same LAN for, say, a human resource department.

All that is required is some form of IP connection. Some clients use dial-up connections to the Internet to establish the IP connection. This is a preliminary step in preparation for creating a tunnel and is not part of the tunnel protocol itself. Figure 8-6 shows a voluntary tunnel.

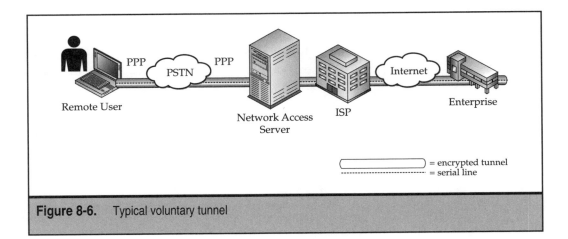

Figure 8-6.   Typical voluntary tunnel

Compulsory Tunnels

A number of ISPs' network access servers include the ability to create a tunnel on behalf of a dial-up client. The computer or network device providing the tunnel for the client computer is variously known as a Front End Processor (FEP) in PPTP, an L2TP Access Concentrator (LAC) in L2TP, or an IP Security Gateway in IPSec. For the rest of this section, we will use the term *LAC* since it is most common; however, what it's called usually depends on the tunneling protocol being used on the NAS. Generally what happens is that a corporation may have an agreement with an ISP to deploy a nationwide set of LACs. The LACs establish tunnels across the Internet to a host VPN server connected to the corporation's private LAN. This arrangement allows calls from geographically dispersed locations to go through a single Internet connection located at the corporate network. This type of tunnel is called a *compulsory* tunnel because the client is *compelled* to use the tunnel created by the LAC. After the connection is made by the remote PC, all of the network traffic to and from the remote PC is automatically sent through the tunnel that sits between the LAC and the corporate host VPN server. Any LAC can be configured to tunnel all dial-up clients to a specific tunnel server or even to tunnel individual clients based on the username or destination.

Unlike the separate tunnels created for each voluntary client, multiple dial-up clients can share a tunnel between the LAC and the host VPN server. If a second client dials into the NAS to reach a destination for which a tunnel already exists, there is no need to create a new instance of the tunnel between the LAC and the host VPN server, because all of the data from the new client is sent over the existing tunnel. Whenever multiple clients are connected over a single tunnel, the tunnel is not terminated or "torn down" until the last user of the tunnel disconnects.

We have covered just what tunnels are and the different variations that exist. In the next chapter, we will talk about the different VPN architectures that exist and learn where and how these tunnel types are applied.

SUMMARY

In this chapter, we introduced you to public and private networks and how they evolved into VPNs. We talked about the history of VPNs. We also talked about Internets, intranets, and extranets. You were introduced to the concept of tunneling and the different types that exist in Layer two and Layer three. In the next chapter, we will discuss the different VPN architectures that exist and their optimum implementations.

CHAPTER 9

VPN Architectures

Now we have talked about Internets, intranets, extranets, and the concept of tunneling. We have also discussed public and private networks. Given all of these networking technologies and designs, you can see why VPN definitions have been so vague. In this chapter, we discuss the components and characteristics of the networks we've talked about and the architectures used to implement them. This chapter should give you a clearer picture of the contemporary definition of today's VPNs.

VPN COMPONENTS AND GENERAL REQUIREMENTS

The two main concerns when implementing a VPN are security and performance. The original version of TCP/IP (version 4 or IPv4) was not designed with either of these concerns in mind. Nobody anticipated the sheer number of users and the types of applications that would be used on the Internet. Nor, in the beginning, was the need for strong security or guaranteed performance evident. The newer version of TCP/IP being developed is known as IPv6 or IPng (IP next generation) and has security measures built in, but that is another book all together.

To successfully use the Internet for VPNs and to serve as a viable substitute for T-carriers, Frame Relay, or other WAN links, a way is needed to ensure comparable security and network performance—and not just comparable price.

The standards for security on IP networks have evolved to a point where the Internet can be used to create VPNs. Work on providing guaranteed performance is at an earlier stage of development, but Internet service providers (ISPs) and long-distance carriers, with the help of advances like Multi-Protocol Label Switching (MPLS), are quickly working to rectify that. This can be accomplished to some degree now by using a common carrier or global ISP for your VPN infrastructure.

Meeting VPN Requirements

Four general requirements provide the necessary security for VPNs across any public network. Those four requirements are

▼ **Authentication** This makes certain the data is authentic and that it has originated from where it should. Authentication is the process of verifying that the sender is who he or she says. Strong authentication is essential to verify the identities of site locations as well as individual users.

■ **Access control** This ensures that only authorized users are given access to the network. Access control determines the amount of freedom a user has. It also controls the access of employees and other outside users to applications and different areas of the network. Be sure that all users have full access to the applications and information they need, but nothing more.

■ **Confidentiality** This keeps anyone from examining or copying data during transmission across the Internet. Encryption algorithms are used to scramble the data so that only those who have the encryption key can read the information. The security of the encrypted data grows as the key becomes longer. Once the encryption key length is selected and implemented, the next step is to ensure that the keys are protected through a key management system. *Key management* is the process of distributing the keys, refreshing them at specific intervals, and revoking them when necessary. Public Key Infrastructure (PKI) is the technology used to do this.

▲ **Data integrity**   This makes certain that no tampering or altering of the data occurs during transmission across the Internet. This is done using *hash algorithms.* One-way mathematical functions are applied to the data to maintain the integrity of the message.

These requirements can be met using the following methods:

▼ Various password-based systems like Terminal Access Controller Access Control System (TACACS and TACACS+) and Remote Authentication Dial-In User Service (RADIUS)

■ Protocols like Challenge Handshake Authentication Protocol (CHAP) and the less sophisticated Password Authentication Protocol (PAP)

■ Hardware- and software-based tokens

■ Digital certificates

■ Encryption algorithms like DES, triple DES, IDEA, RC2, RC4, RC5, and Blowfish

▲ Hash algorithms like MD2, MD4, MD5, RIPEMD-160, and SHA

We will be discussing those methods in more detail in later chapters, but for now you can get an idea of the tools used to meet the general VPN requirements.

Making VPN Connections

In the past, WANs were made by leasing dedicated permanent connections from communications service providers and running them between each site in their organization. The data traveling across these connections was devoted to the traffic from these organizations exclusively.

Unlike the private networks created by the leased lines used in traditional corporate networks, VPNs do not maintain permanent links between the endpoints that make up the corporate network. The private network being created is now *virtual.* This means that the network creation is dynamic, with connections set up according to organizational needs. It also means that the network is formed logically, regardless of the physical structure of the underlying network.

When a connection between two sites is required, it is created. As soon as the connection is no longer needed, it is torn down. This makes the network resources and bandwidth available for other uses. The connections used to create a VPN do not have the same physical characteristics as the hard-wired connections used on the leased lines.

To extend the traditional WAN concept to these organizations, we need a way to tunnel the information through the Internet. Unlike the private leased-line connections, all traffic from many organizations travels over the same public Internet connection. Various protocols are available to create these tunnels.

As we discussed in earlier chapters, tunneling allows these organizations to encapsulate their data in IP packets. This method hides the underlying routing and switching infrastructure of the Internet from both the sender and receiver. The encryption used by these protocols also protects the data from examination and copying by outsiders.

Tunnel Endpoints

Tunnels can consist of two types of endpoints: either an individual computer or a LAN with a security gateway, which might be a router or firewall. Only two combinations of these endpoints, however, are usually considered in designing VPNs.

The first case, that of *client-to-LAN tunnels,* is the type usually set up for mobile users who want to connect to the corporate LAN. The clients, that is, the mobile users, initiate the creation of the tunnel on their end in order to exchange traffic with the corporate network. To do so, they run special client software on their computer to communicate with the gateway protecting the destination LAN.

In the second case, *LAN-to-LAN tunneling,* like that of intranets and extranets, a security gateway at each endpoint serves as the interface between the tunnel and the private LAN. In such cases, users on either LAN can use the tunnel transparently to communicate with each other.

This brings us to the next section, which describes the architectures used to implement the types of tunneling we have discussed.

VPN ARCHITECTURES

In this section, we discuss the kinds of VPN architectures that are used most commonly today: access VPNs, intranet VPNs, and extranet VPNs. Access VPNs differ in many ways from intranet and extranet VPNs. The differences lie in the tunneling methods used by these architectures. We'll start with access VPNs.

Access VPN

Access VPNs are sometimes referred to as "remote access VPNs" or "Virtual Private Dial-up Networks" (VPDN). These are generally created for users who wish to access the corporate network from a remote location such as their home, a satellite office, or anywhere in the world. Remote users accessing the corporate network prior to VPN technology had to use the Public Switched Telephone Network (PSTN). They would dial into Access Servers using long-distance charges. This approach required individual modem connections and phone numbers that had to be maintained locally on the corporate campus by internal IT staff. Remote access VPNs obviate the need for all of these things. With the advent of the Internet, ISPs maintain local access numbers virtually everywhere. This means that remote users as well as the corporate campus need only make local phone calls to access the Internet over which these VPN tunnels are created. What's more, they are no longer locked into these slow analog phone lines. Virtually any technology can now be used, including high-speed xDSL, ISDN, and cable technologies, in addition to analog connections. Even wireless technologies can be used. There are basically two kinds of access VPNs: the client-initiated access VPN and the Network Access Server (NAS)-initiated access VPN.

Client-Initiated VPN

Client-initiated access VPNs are generally intended for users who need to access a home gateway through more than one provider network. The main characteristic of a client-initiated access VPN is that the user decides when and where to establish a VPN across a public network and handles the software directly. Because the decision is left up to the user, this type of VPN is also called a "voluntary" VPN. The NAS does not establish the tunnel, so this type of VPN can easily span many ISPs to reach a home gateway without additional configuration. Encryption is established between the user and the home gateway for end-to-end privacy.

A client-initiated access VPN requires IPSec support on both the user PC and corporate home gateway. Standards compliance is necessary because tunnels may span multiple backbone types, and IPSec as a Layer 3 standard is media independent.

The client-initiated VPN can use only one VPN connection at a time. Connecting to different VPN gateways requires that a user log out, reconfigure client software, and log in again.

The typical sequence of establishing a VPN tunnel and starting a session is the following:

1. The remote user initiates a PPP connection to the ISP's Point of Presence (POP) where the NAS is.

2. The ISP's NAS accepts a call and assigns an IP address to the user.

3. Client software initiates an IPSec tunnel to the home gateway. The client PC sends a certificate that contains a key and security identification number to the IPSec-compliant firewall over the Internet.

4. The corporate (home) gateway authenticates the remote user and then either accepts or declines the tunnel-creation request.

5. The corporate gateway confirms the acceptance of the call and creates the VPN tunnel.

6. The corporate gateway then exchanges PPP negotiations with the remote user. At this point, encryption using the Diffie-Hellman Digital Encryption Standard (DES) algorithm establishes a "shared secret" unique key between client and firewall. Then, an encrypted IP connection is established between the endpoints that encrypts the entire datagram and is transparent to any of the network systems in between.

7. All data is tunneled between remote user and the corporate gateway from end to end.

8. The remote user terminates the tunnel upon logout and closes the connection.

9. All encryption keys are deleted.

To sum it all up, client-initiated VPNs create a tunnel all the way from the corporate Access Server to the remote user itself. This requires the remote user to have some kind of VPN client software installed that is compatible with whatever VPN technology is being used by the corporate Access Server. This is a two-step process. The remote user must first access the local ISP NAS using PPP to connect to the Internet. Once this connection has been established, the remote user (client), using PPTP or L2TP, creates a tunnel all the way through to the corporate Access Server. When using this method, all communication between the corporate network and the remote user is completely encrypted from end to end. You can see how this works in Figure 9-1.

Advantages   Some of the advantages of a client-initiated VPN are

▼ The VPN tunnel is created end-to-end all the way from the remote user to the corporate gateway.

■ The remote user can establish a VPN tunnel from anywhere regardless of the ISP or network provider.

▲ The VPN can span multiple provider networks transparently.

Disadvantages   Some of the disadvantages of a client-initiated VPN are

▼ Corporate security policies are more difficult to enforce.

■ This architecture does not scale well.

■ The remote user must obtain IP addresses from the ISP, so private address schemes like those in RFC 1597 are ineligible.

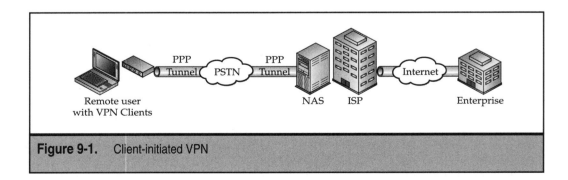

Figure 9-1.   Client-initiated VPN

■   The remote user needs client software, which can be an administrative concern when supporting many users.

▲   Supporting hardware/software products must be IPSec compliant.

Network Access Server (NAS)–Initiated VPN

The main characteristic of a NAS-initiated access VPN is that tunnels are initiated from the ISP's Network Access Server (NAS) to the home gateway. Tunnels are established consistent with customer security policy requirements, and the connection is transparent to users. The fact that the VPN tunnel is automatically created upon connection to the ISP is why it is called a "compulsory" VPN. Encryption is established between the ISP's NAS and the corporate gateway over an unencrypted PPP session between the remote user and the ISP's NAS.

NAS-initiated access VPNs require no specialized client software, allowing greater flexibility for companies to choose the access software that best fits their requirements. NAS solutions use robust tunneling protocols such as L2F or L2TP. IPSec provides encryption only, in contrast to the client-initiated model, where IPSec enables both tunneling and encryption.

The typical connection sequence in a generic NAS-initiated access VPN is as follows:

1. The remote user initiates a PPP connection to the ISP's NAS.

2. The ISP's NAS accepts the call and identifies the customer affiliation of the user, based on domain name, and accesses the approved customer profile to identify the corporate (home) gateway.

3. The ISP's NAS dynamically assigns the user an IP address.

4. The ISP's NAS initiates the tunnel to the corporate gateway.

5. The corporate gateway uses standard PPP authentication protocols such as PAP or CHAP to authenticate the user via a TACACS+ or RADIUS access control server at the corporate premises, authenticates the remote user, and then either accepts or declines the tunnel creation.

6. The corporate gateway confirms acceptance of the call and the tunnel is created.

7. The corporate gateway exchanges PPP negotiations with the ISP's NAS to establish IPSec or application-level encryption between the ISP's NAS and the corporate gateway.

8. An internal or private IP address can be assigned by the corporate gateway at this stage, using NAT. The ISP network does not see this address, because it is encapsulated.

9. The data is then tunneled end-to-end from the ISP's NAS and the corporate gateway during the communications session.

10. The user may choose to establish connections with multiple gateways simultaneously, like another corporate gateway, the Internet, and another secured web site.

11. The ISP's NAS terminates the tunnels upon logout and then closes the connection.

12. All encryption keys are deleted.

To sum up, with the NAS-initiated VPN, the remote user simply dials into the local ISP NAS over a standard PPP connection, at which point the ISP NAS creates the encrypted tunnel with the corporate Access Server. All communication over the Internet infrastructure to the corporate Access Server is encrypted. However, the connection from the remote client PC to the local ISP NAS is not. The difference is that no special client software is required on any of the remote clients. The connection over the PSTN is far less vulnerable than the actual Internet, but technically is still a liability. You can see how this works in Figure 9-2.

Advantages   Some of the advantages of a NAS-initiated VPN are

▼ No client software is required, which makes administration easier.

■ The service provider establishes the tunnel to preauthorized points for consistent policy enforcement.

■ It is easier for remote users since everything is transparent from their perspective.

■ It is much more scalable.

■ ISPs can provide value-added services.

■ The corporation can authenticate users locally.

■ The corporation can use private IP addresses using NAT.

▲ The remote user can initiate multiple VPN connections.

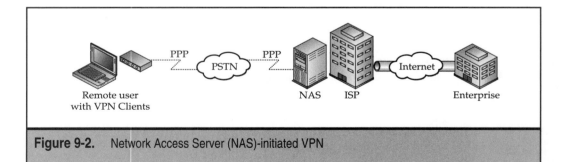

Figure 9-2.   Network Access Server (NAS)-initiated VPN

Disadvantages   Some of the disadvantages of a NAS-initiated VPN are

▼   It requires a more sophisticated network design.

■   The tunnel is not end-to-end from the remote user. Data is only encrypted between the ISP's NAS and the corporate gateway.

▲   It requires ISP services and is not as easy to use from "anywhere."

Intranet VPN

Intranet VPNs, also known as "router-to-router" or "site-to-site" VPNs, essentially provide the same type of function as access VPNs. These VPNs use the Internet to replace traditional wide area networks (WANs) used in connecting local area networks (LANs) at remote sites. These types of VPNs can be built using the Internet, or even over a part of the ISP's private IP infrastructure. For example, companies with offices around the world traditionally would provision expensive leased-line technologies such as Frame Relay, T1 or T3, or even ATM provided by public exchange carriers. Intranet VPNs avoid this requirement by setting up secure tunnels at the border routers of each of these locations. These IPSec or GRE tunnels actually "pass through" the public Internet transparently via the local ISPs. There are many categories: small office/home office (SOHO) sites, branch office to main office, enterprise, or a combination of all of these. New technologies such as DSL, cable, and wireless can now provide high-speed access at very affordable prices.

All of this is generally accomplished using Layer 3 tunneling techniques like GRE and/or IPSec. We discussed the concept of tunneling in earlier chapters, but to recap quickly: The tunnel is virtual and isn't tied to any specific "passenger" or "transport" protocols. Tunneling simply provides a way to encapsulate packets inside of a transport protocol.

Tunneling consists of three primary elements, which can be seen in Figure 9-3:

1. The *passenger* protocol. This is the protocol you are encapsulating, like AppleTalk, IP, or Novell's Internetwork Packet Exchange (IPX).

2. The *carrier* protocol, such as the Generic Routing Encapsulation (GRE) protocol or IPSec protocol.

3. The *transport* protocol, such as IP, which is the protocol used to carry the encapsulated protocol.

An important fact to keep in mind when implementing tunnels for any VPN environment is that IPSec encryption only works on IP unicast frames. GRE is capable of handling the transportation of multiple protocols including IP unicast *and* IP multicast traffic between two sites that only have IP unicast connectivity. Tunneling allows for the encryption and the transportation of multiprotocol traffic across the VPN since the tunneled packets appear to the IP network as an IP unicast frame between the tunnel endpoints. If all connectivity must go through the corporate gateway router, tunnels also enable the use of private network addressing schemes (like those in RFC 1597) across an ISP's backbone without having to use Network Address Translation (NAT).

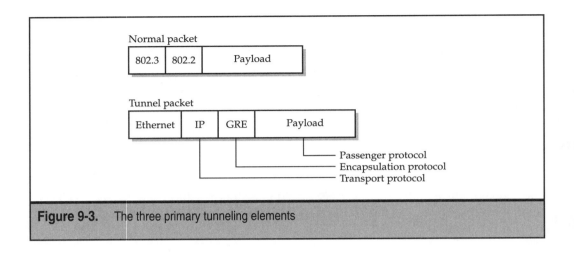

Figure 9-3. The three primary tunneling elements

Network redundancy also requires careful consideration in any decision to use GRE tunnels, IPSec tunnels, or tunnels using IPSec over GRE. GRE can be used with IPSec to pass routing updates between sites on an IPSec VPN. GRE can do this because it encapsulates the cleartext packet; then, IPSec in *transport mode* (discussed in the next paragraph) encrypts the packet. This makes IPSec over GRE able to pass routing updates, which are usually multicast, over an encrypted link. Remember, IPSec by itself cannot do this, because it does not support multicast.

IPSec can be configured in *tunnel mode* or *transport mode*. IPSec tunnel mode is an alternative to using a GRE tunnel. In IPSec tunnel mode, the entire original IP datagram is encrypted, and it becomes the payload in a new IP packet. This mode allows a network device, such as a router, to act as an IPSec proxy. That is, the router performs encryption on behalf of the hosts. The source router encrypts packets and forwards them along the IPSec tunnel. The destination router decrypts the original IP datagram and forwards it to the destination system. Tunnel mode protects against traffic analysis; with tunnel mode, an attacker can only determine the tunnel endpoints and not the true source and destination of the packets passing through the tunnel, even if they are the same as the tunnel endpoints.

In IPSec transport mode, only the IP payload is encrypted, and the original IP headers are left intact. This mode has the advantage of adding only a few bytes to each packet. It also allows devices on the public network to see the final source and destination of the packet. With this capability, you can enable special processing in the intermediate network based on the information in the IP header. However, the Layer 4 header will be encrypted, limiting the examination of the packet. When passing the IP header in the clear, transport mode allows an attacker to do some traffic analysis. An example of IPSec tunnel and transport modes can be seen in Figure 9-4. We will be discussing IPSec in much more detail in Section V.

Some ISPs are starting to provide a complete routing service, where the corporate intranet tunnels are maintained by the ISPs completely. This arrangement obviates the

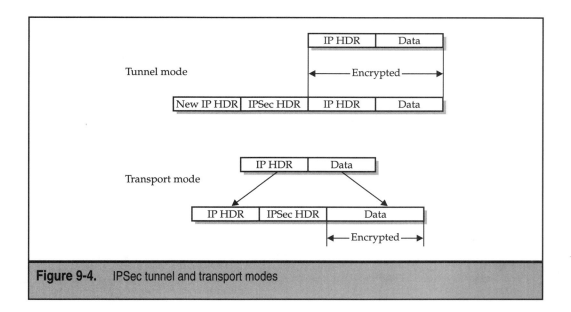

Figure 9-4.   IPSec tunnel and transport modes

need for complex router configuration at the corporate sites so that these companies need not maintain extensive IT staffing at the smaller branch offices.

The things that need to be taken into consideration revolve around Quality of Service (QoS). QoS has to do with prioritizing certain types of traffic such as SNA or even voice, which require little or no delay, over standard e-mail or file-transfer communications. The nature of the Internet is geared toward packet switching technologies, which are connectionless and can't always guarantee this QoS, as opposed to circuit switching technologies, which are connection oriented. However, new ways around these problems exist, like using priority queuing techniques, and tag switching or Multi-Protocol Label Switching (MPLS). We will be talking about MPLS and QoS in Section VI. Traditional WAN carriers can offer a VPN service similar to a Frame Relay service with QoS based on committed information rate. Figure 9-5 shows a common implementation of an Intranet VPN.

Extranet VPN

Extranet VPNs use many of the same technologies as Intranet VPNs. Extranets differ from intranets in that they allow access by remote users or business partners outside of their company's corporate internal network. This kind of VPN links customers, suppliers, partners, or communities of interest (COINs) to the company's corporate intranet using a shared infrastructure over dedicated connections. Extranets can also be thought of as intranets that provide limited access to these customers, suppliers, and partners while simultaneously allowing authorized access for telecommuters and remote offices. This is a powerful tool in business-to-business relationships. Figure 9-4 shows a common implementation

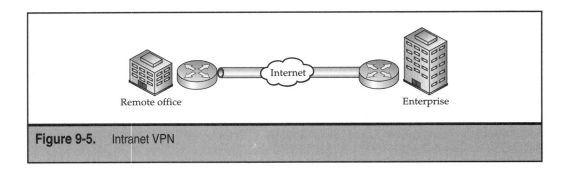

Figure 9-5.   Intranet VPN

of an extranet VPN design. By using digital certificates, clients establish a secure tunnel over the Internet to the enterprise. A certification authority (CA) issues a digital certificate to each client for device authentication. VPN Clients may either use static IP addressing with manual configuration, or they can use dynamic IP addressing with Internet Key Exchange (IKE) protocols or IKE configurations. The CA checks the identity of remote users and then authorizes these remote users to provide access to the corporate network.

TYPES OF VPNS

There are several types of VPNs, and they can be implemented in a number of ways. In this section, we will talk about some of these VPN types.

Software-Based VPNs

Software-based VPNs are ideal in situations where both endpoints of the VPN are not controlled by the same organization. This is also the case when different firewalls and routers are used within the same organization. At the moment, stand-alone VPNs offer the most flexibility in how network traffic is managed. Many software-based products allow traffic to be tunneled based on address or protocol. Tunneling specific traffic types is advantageous in situations where remote sites may see a mix of traffic. Software-based VPNs may be the best choice in situations where performance requirements are not critical, like remote users connecting over dial-up links.

A software-based VPN will be less expensive to implement since you have most of the hardware already in place and just need to add the software. Software-based solutions do not scale well, and performance will degrade progressively as more users are added. If you anticipate adding a great number of users over time and/or have many locations and tunnels that need support, you may be better off with a hardware-based VPN.

Table of Contents

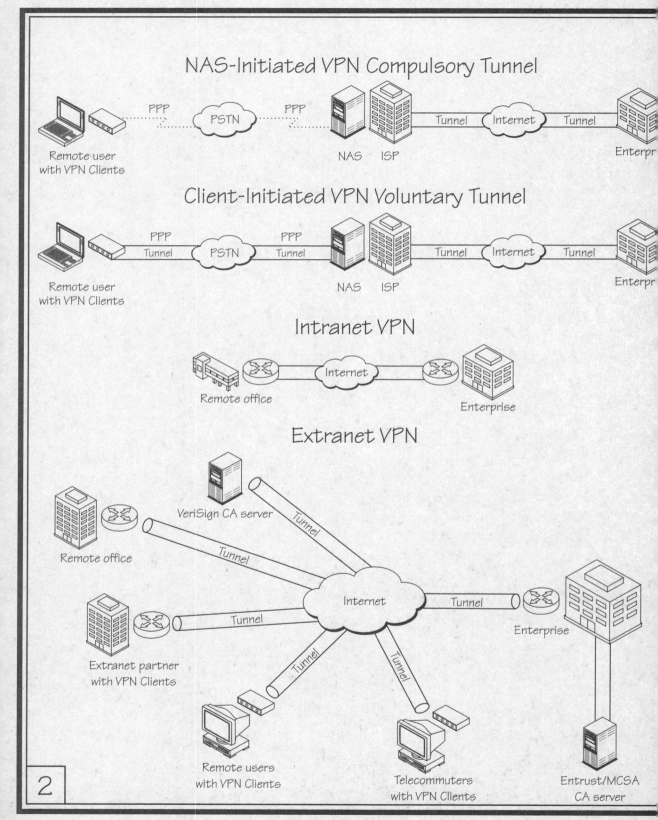

NAS-Initiated VPN Compulsory Tunnel

Remote user
with VPN Clients

PPP

PSTN

PPP

NAS

ISP

Tunnel

Internet

Tunnel

Enterpr

Client-Initiated VPN Voluntary Tunnel

Remote user
with VPN Clients

PPP
Tunnel

PSTN

PPP
Tunnel

NAS

ISP

Tunnel

Internet

Tunnel

Enterpr

Intranet VPN

Remote office

Internet

Enterprise

Extranet VPN

VeriSign CA server

Remote office

Tunnel

Tunnel

Extranet partner
with VPN Clients

Tunnel

Internet

Tunnel

Enterprise

Tunnel

Tunnel

Remote users
with VPN Clients

Telecommuters
with VPN Clients

Entrust/MCSA
CA server

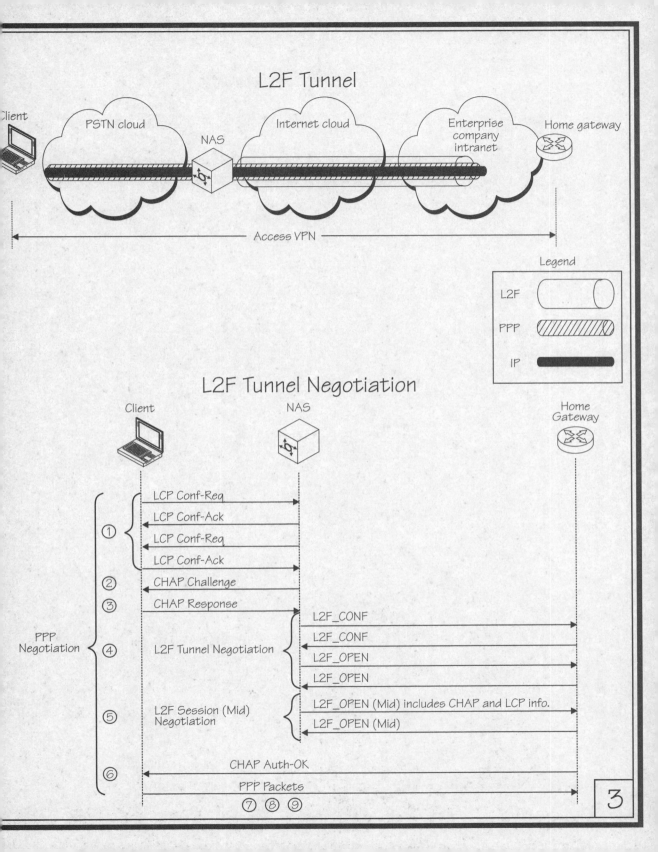

PPTP Data Packet

Data-link Header	IP Header	GRE Header	PPP Header	Encrypted PPP Payload (IP Datagram, IPX Datagram, NetBEUI Frame)	Data-link Trailer

PPTP Control Packet

Data-link Header	IP	TCP	PPTP Control Message	Data-link Trailer

PPTP Tunnel Negotiation

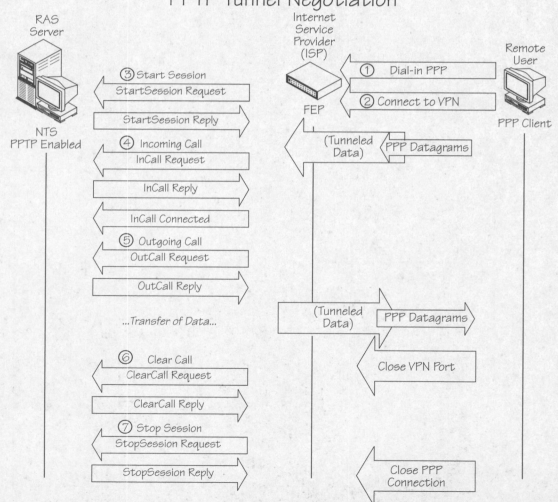

L2TP Tunnel Protected with IPSec

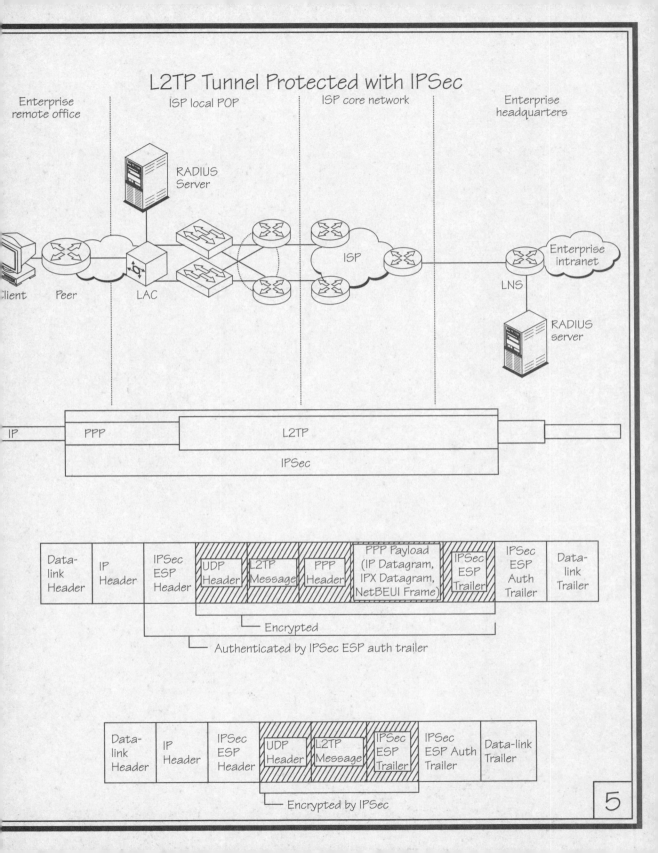

Enterprise
remote office

ISP local POP

ISP core network

Enterprise
headquarters

RADIUS
Server

ISP

Enterprise
intranet

Client   Peer   LAC   LNS   RADIUS
server

IP	PPP	L2TP	
		IPSec	

Data-link Header	IP Header	IPSec ESP Header	UDP Header	L2TP Message	PPP Header	PPP Payload (IP Datagram, IPX Datagram, NetBEUI Frame)	IPSec ESP Trailer	IPSec ESP Auth Trailer	Data-link Trailer

└─ Encrypted

└─ Authenticated by IPSec ESP auth trailer

Data-link Header	IP Header	IPSec ESP Header	UDP Header	L2TP Message	IPSec ESP Trailer	IPSec ESP Auth Trailer	Data-link Trailer

└─ Encrypted by IPSec

5

Complete IKE & IPSec Phase 1 & Phase 2 Exchange

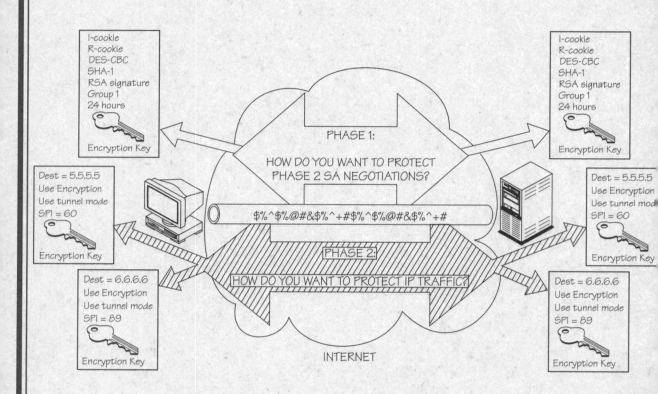

I-cookie
R-cookie
DES-CBC
SHA-1
RSA signature
Group 1
24 hours

Encryption Key

I-cookie
R-cookie
DES-CBC
SHA-1
RSA signature
Group 1
24 hours

Encryption Key

PHASE 1:

HOW DO YOU WANT TO PROTECT
PHASE 2 SA NEGOTIATIONS?

$%^$%@#&$%^+#$%^$%@#&$%^+#

PHASE 2:

HOW DO YOU WANT TO PROTECT IP TRAFFIC?

INTERNET

Dest = 5.5.5.5
Use Encryption
Use tunnel mode
SPI = 60

Encryption Key

Dest = 5.5.5.5
Use Encryption
Use tunnel mode
SPI = 60

Encryption Key

Dest = 6.6.6.6
Use Encryption
Use tunnel mode
SPI = 89

Encryption Key

Dest = 6.6.6.6
Use Encryption
Use tunnel mode
SPI = 89

Encryption Key

IPS AH & ESP Tunnel/Transport Modes

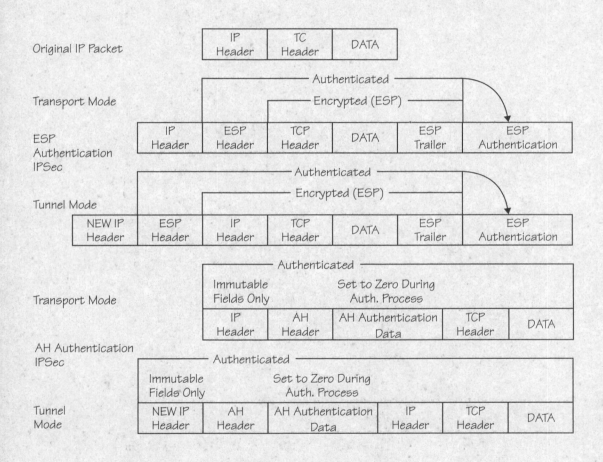

Encapsulation Security Payload (ESP) Header

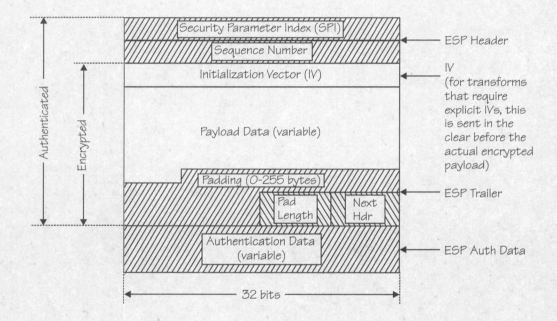

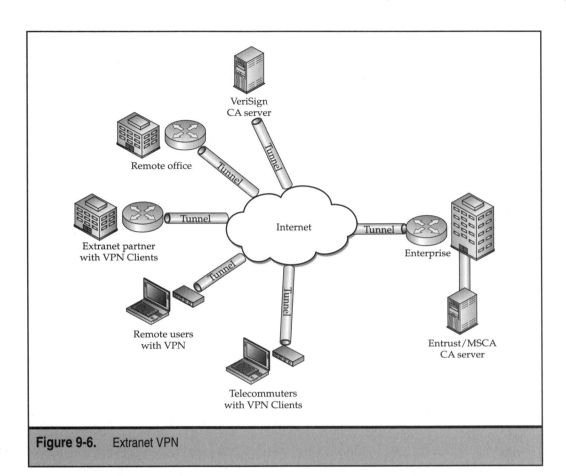

Figure 9-6.   Extranet VPN

Hardware-Based VPNs

Most hardware-based VPN systems are VPN concentrators or VPN-capable routers. They are secure, easy to use, and require very little effort to install; they are the closest thing to "plug-and-play" equipment available. They provide the highest network throughput since they don't need to waste CPU overhead like systems running operating systems or other applications.

Hardware-based systems may not be as flexible as software-based systems. Software-based products allow traffic to be tunneled based on address or protocol, while hardware-based products tend to tunnel all traffic regardless of protocol. A hardware-based VPN is probably more expensive on the front-end, but has the ability to support large numbers of tunnels or high levels of encryption without sacrificing system performance.

Firewall-Based VPNs

Firewall-based VPNs take advantage of the firewall's security mechanisms, including restricting access to the internal network. They are feature laden and have intuitive graphical user interfaces (GUIs) as well. Those features can include:

▼ Address translation

■ Authentication

■ Alarm and notification systems

▲ Comprehensive logging

Most commercial firewalls also "harden" the host operating system by removing dangerous or unnecessary services that provide additional security for the VPN server. However, operating-system protection provides much more security since very few VPN systems address operation-system vulnerabilities. Performance may be a concern, especially if the firewall is already under heavy load, but some firewall vendors offer very good hardware-based encryption processors to minimize the impact of VPN management on the system. The only thing that you would need to consider in this situation is a single point of failure.

IP-Based VPNs

IP-based VPNs will undoubtedly be the future of VPNs and will ultimately be the solution for the convergence of all communications in total. This will not be limited to just data, but voice and video as well. IP-based VPNs will allow any organization to fully exploit the ubiquity of the Internet and the economic benefits it provides. Unlike traditional dedicated WAN technologies whose scalability is cumbersome and complex, IP-based VPNs will scale very easily. Another major feature of these VPNs will be the ease of deployment that is unmatched by previous networking solutions and technologies.

These VPNs will be very flexible. They could include VPN equipment that is customer owned and managed completely on premises and privately operated.

They could be supplied and managed by the ISP with equipment located either at their public Point of Presence (POP) or at their customer's premises.

They could also be a combination of both. The bottom line is that IP-based VPN solutions will be very scalable to meet the needs of large and growing corporate communication requirements, but will also provide ease of deployment, management, and operation at the best possible cost.

Regardless of the implementation, an IP-based VPN solution must include a range of selectable Qualities of Service (QoS), intelligent and dynamic services, traffic engineering, and comprehensive security capabilities.

Advantages of Using the Public Internet Infrastructure

If dedicated leased lines and multiple-interface connection costs were of no concern, these dedicated interfaces and high-capacity permanent virtual circuits (PVCs) might very well be the choice for many organizations. The reality is that nobody can afford such a luxury, especially when traffic volumes are low or long distances are involved. Sharing network resources by using the public Internet is definitely more cost-effective than using dedicated leased lines.

The Internet can be accessed from virtually anywhere in the world with operations and administration supplied by the local ISP. The complexities of things like network interconnection across ISP boundaries and/or international borders can be hidden from the user by using an Internet-based infrastructure.

Another consideration in any network solution is scalability. *Scalability* involves the constant increasing of capacity for large, dynamic populations with varying bandwidth requirements. For instance, it would be impossible for a bank to have dedicated high-capacity links to every other bank it deals with. Even large companies that do maintain their own private intranet could never provide dedicated links to all of their customers and business partners. When e-commerce is involved, a mix of public and private networks is the only viable solution.

Provisioning Options

Obviously, the benefits of using VPN technologies must be balanced against the costs and complexities of implementing the technology. For some companies, owning and fully controlling the network resource will be paramount, while for others, the opportunity to outsource various pieces in order to address some of these complexities will be a priority.

A VPN may be provisioned in one of two ways:

▼ **ISP network-based VPN (edge-to-edge configuration)**   This where the VPN is built on "edge devices" that are located at the ISP's Point of Presence (POP). In this case, multiple customers share the same edge equipment. Secure connections are established on behalf of each customer from network edge to network edge. The ability to service many customers from a single POP can provide economies of scale to an ISP-based VPN in terms of maintenance, implementation, and management. The main advantages to this approach are lower installation costs and less complexity. The trade-off is that service levels and security would not be extended all the way to the customer premises.

▲ **Customer premises-equipment (CPE) based VPN (CPE-to-CPE configuration)**   This places the VPN access point at the customer premises. By having the full control of all the equipment involve you protect the access link by providing the customer end-to-end security and service-level control. A CPE-based VPN can be built by installing a secure VPN gateway with an existing customer router or by building upon a fully integrated VPN router. This VPN implementation can be readily deployed on a global basis, without requiring significant upgrades to the ISP's network infrastructure. Control over this VPN environment can be shared between the customer and the ISP. A potential problem for CPE-based VPNs is their distributed nature. This would make large-scale VPN management more complex.

Management Options

Management of the VPN and its operation could be handled in one of these ways:

▼ **ISP-managed VPN** This VPN would be owned and operated by the ISP. The service itself could be ISP based or CPE based. The important points are that the provider owns the equipment and takes responsibility for equipment management, service provisioning, service quality, and other aspects as dictated by the service definition.

■ **Customer-managed VPN** This VPN would be owned and operated by the customer, which means it is transparent to the ISP. This type of VPN is typically deployed as a private, end-to-end overlay on the Internet. All parts of this network are covered by the privacy and quality controls built into the VPN by the customer's network administrators, although a customer-controlled VPN would still be dependent upon the ISP for meeting committed service level agreement (SLA) characteristics. This would require a highly trained local IT staff.

▲ **Shared-control VPN** This VPN would usually be owned by the ISP and managed jointly by the customer and the ISP. In this environment, the ISP would most likely be the one that manages the hardware and network equipment, and the customer would manage its own VPN and security policies. This allows both the cost benefits of an ISP-controlled solution and the flexibility and control of the CPE-based solution.

SUMMARY

In this chapter, we have talked about the different VPN architectures and the applications for which they are best suited. We have also talked about the different types of VPN implementations available.

This brings us to our next section of the book, where we will discuss in detail the protocols and protocol combinations that are used to create these VPNs.

PART III

VPN Protocols

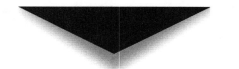

CHAPTER 10

Tunneling Protocols

In Chapter 8, we talked about tunneling. Tunneling has amazing implications for VPNs. For example, you can place a packet that uses a protocol not supported on the Internet (such as NetBEUI) inside an IP packet and send it safely over the Internet. Or, you could put a packet that uses a private (non-routable) IP address inside a packet that uses a globally unique IP address to extend a private network over the Internet. It is important to remember that it is possible to create a tunnel and send the data through the tunnel without encryption. This is not a VPN connection, because the private data is sent across a shared or public network in an unencrypted and easily readable form. So, to review, the *tunnel* is the portion of the connection in which your data is *encapsulated.* The *VPN* is the portion of the connection in which your data is *encrypted.* For secure VPN connections, the data is encrypted and encapsulated along the same portion of the connection.

VPNs and their specific architectures rely on tunneling. In this chapter, we'll discuss the protocols used in the creation of these tunnels and dissect their inner workings. We will start with the Generic Routing Encapsulation (GRE) protocol. It is one of the original tunneling protocols, and many of the other protocols discussed in this chapter were based on it.

GENERIC ROUTING ENCAPSULATION (GRE)

GRE was designed in 1994, in RFCs 1701 and 1702, in a response to earlier attempts to tunnel protocols through the Internet that were too specific. Instead of only tunneling one type of protocol, GRE could encapsulate up to 20 different types of protocols in its protocol type field. (These protocol types are essentially the equivalent of the *Ethertype* values that would be found in Version 2 Ethernet or in the Type field of a SNAP header in an 802.2-compliant protocol like 802.3 or 802.5. These values were discussed earlier in the LAN section of Chapter 2.)

GRE tunneling involves three types of protocols:

▼ **Passenger protocol**   This protocol used on the local area network is encapsulated.

■ **Carrier protocol**   This is the GRE protocol that provides carrier services.

▲ **Transport protocol**   This is the protocol used to transport the two preceding protocols (in the case of the Internet, it is IP) by encapsulating them.

How GRE Works

Here is how GRE works:

1. A source computer on the local LAN using the protocol native to that specific location creates a packet and sends it onto the wire.

2. The local router sees that the computer for which it is destined is not local. That router takes the packet and, using the GRE (carrier) protocol, encapsulates the packet (passenger, which could be IP or any number of LAN protocols) to create GRE tunnels. With GRE tunneling, a router at one site will encapsulate a given type of protocol in an IP header, creating a virtual point-to-point link to a router (or routers) at a site on another end of an IP cloud (such as the Internet), where the IP header is stripped off.

3. Once the IP header is stripped off, it is left with the original packet's protocol header native to the local LAN and, using that information, is delivered to its intended destination.

All of this is completely transparent to the two computers communicating with each other. GRE tunneling allows LAN protocols to take advantage of the enhanced route-selection capabilities of specialized routing protocols. Most LAN protocols, especially NetBEUI, AppleTalk, and Novell IPX, were designed for and optimized for local use. By this we mean they are non-routable (like NetBEUI) or, in Novell's case, have limited route-selection metrics (algorithms to determine optimal path selections) and hop count limitations.

Specific routing protocols allow more flexible route selection and scale better over large internetworks. Figure 10-1 illustrates GRE tunneling across a single IP backbone between sites. Regardless of how many routers and paths may be associated with the IP cloud, the tunnel is seen as a single hop.

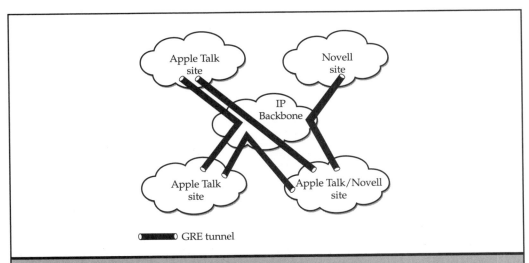

Figure 10-1.   GRE tunneling across a single IP backbone

GRE Strengths

GRE has some unique characteristics that are not found in other encapsulation protocols, such as sequencing and the ability to carry tunneled data at high speeds. Some higher-level protocols require that packets be delivered in correct order. The GRE sequencing option provides this capability. GRE also has an optional key feature that allows you to avoid configuration errors by requiring the same key to be entered at each tunnel endpoint before the tunneled data is processed. You can set policies to decide which types of traffic can use which routes. You could also set different security levels or priorities for certain types of traffic. Capabilities like these are not available in many of the popular proprietary LAN protocols.

Things to Consider When Using GRE

Because encapsulation requires handling of the packets, it is generally faster to route protocols natively than to use tunnels. Tunneling is very CPU intensive, causing the tunneled traffic to be processed at about half the speed.

Performance depends on the passenger protocol characteristics and the bandwidth of the physical link. If not controlled, a lot of administrative traffic (things related to LAN maintenance) could be sent over each tunnel interface and could saturate the physical link with constant routing information and updates if several tunnels are configured over it. Tunneling can also make it difficult to diagnose problems on the physical link if problems occur.

You can control these problems in a number of ways depending on your situation. These updates and other administrative traffic can be filtered to cut down on the amount of traffic being sent over the tunnels to a minimum.

Tunneling can disguise the nature of a link, making it look slower, faster, or more or less costly than it may actually be in reality. This can cause unexpected or undesirable results regarding route selection. Routing protocols that make decisions based only on hop count will usually prefer a tunnel (virtual connection) to an actual connection. This may not always be the best routing decision.

IP networks can consist of many different types of media with many different qualities. The most profound difference in connection quality is that found in WAN and LAN connections. LAN connections are invariably much faster than common WAN connections. For example, Ethernet LANs commonly run at 100 Mbps while many common WAN connections tend to range from (these are approximate) 128 Kbps (Basic Rate ISDN) to 1.5 Mbps (T1, Primary Rate ISDN, and Frame Relay). These subjects were discussed in Chapter 2.

Many protocols make route selection decisions based only on the physical number of hops (router traversals) on the way to the destination. If you were a datagram having to choose between taking ten 100-Mbps links or one 128-Kbps link to your destination, you (as an intelligent human being) would choose the ten high-speed links as opposed to the one slow link. Tunneling interfaces can often appear to be single hops to less-intelligent routing protocols. When using tunneling, you always need to consider the type of media over which virtual tunnel traffic passes and the criteria used by the routing protocols being used to make those decisions. This is not only a factor that can affect performance; it can also cause what are known as routing loops.

Routing Loops   A *routing loop* occurs when the passenger protocol and the transport protocol are identical. The routing loop occurs because the best path to the tunnel destination is the (virtual) tunnel connection itself. Let's take an example of IP over IP. If you follow the steps below you can get a better idea as to how this could happen:

1. The packet is placed in the output queue of the local router's tunnel interface.

2. The tunnel interface adds a GRE header and gives the packet to the transport protocol (IP) for the destination address, which is the router's tunnel interface on the other end of the tunnel.

3. IP looks up the route to the tunnel destination address and learns that the best path to that tunnel endpoint is through the local tunnel connection itself.

4. Once again, the packet is placed in the output queue of the tunnel interface, as described in step 1, and this endless set of instructions is what causes your routing loop.

To avoid routing loops, you can take various measures. Your solution depends on the individual network topologies and protocols unique to your situation. With that in mind, here are a few things you can try:

▼   Use separate routing protocol identifiers.

■   Use more intelligent routing protocols.

▲   Manually assign the local tunnel interface an unattractive metric by lying to (or fooling) the routing protocol being used. That could mean assigning a distance of a million hops for protocols using that criteria or saying the bandwidth is 1 bps (bit per second) for protocols using that criteria.

Firewall Considerations   Two considerations to keep in mind are the location of your firewall and the filtering policies and methods used by that firewall. If a network has sites that use protocol-based packet filters as part of a firewall security scheme, be aware that tunnels encapsulate passenger protocols unchecked. You must establish filtering on the firewall router so that only authorized tunnels are allowed to pass. If tunnels are accepted from unsecured networks, it is a good idea to establish filtering at the tunnel destination or to place the tunnel destination outside the secure area of your network so that the current firewall scheme will remain secure.

Another consideration has to do with a characteristic involving something called a *maximum transmission unit* (MTU). Sometimes when traffic goes through a GRE tunnel, you might experience problems with downloading web pages or transferring files using FTP. A common reason for this problem has to do with the MTU size settings. Before we discuss this further, take a look at Figure 10-2, which shows a typical GRE tunnel setup as it relates to MTU size. (The subtleties of the following example were discussed in Chapter 3.)

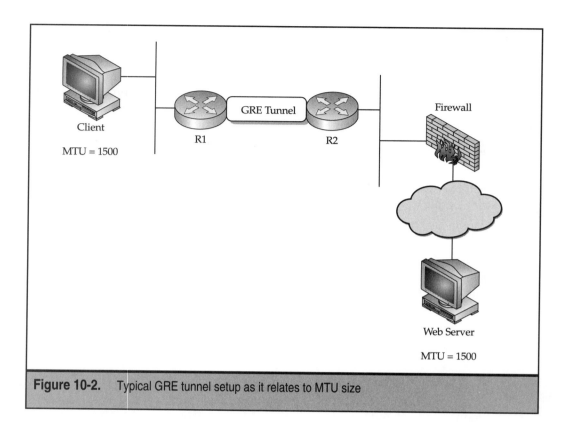

Figure 10-2.   Typical GRE tunnel setup as it relates to MTU size

When a client wants to access a web page, it establishes a TCP session with the web server. During this process, the client and web server exchange information to let each other know the largest TCP segment each can accommodate. The label used for this information is, oddly enough, the *maximum segment size* (MSS). Once these computers know these sizes, they need to agree on the size to be used during the session. This agreement is labeled the *send max segment size* (SMSS) and is essentially the lowest common denominator, or the smaller of the two MSSs.

Let's say the web server in the example determines that it can send packets up to 1500 bytes long. So, as per the agreement, it sends a 1500-byte packet to the client. However, that is the only thing it has agreed to. In the IP header, it has also set the don't fragment (DF) bit.

When the packet arrives at R2, the router tries encapsulating it into the GRE tunnel packet. In the case of GRE encapsulation, the header is 24 bytes. (We will be breaking this down soon in the GRE header section.) So, if we do the math, this means that there is room for only 1476 bytes (1500 original bytes minus 24 bytes). Now you have R2 trying to encapsulate 1500 bytes into a 1476-byte packet. I'm no math professor, but I'm rather certain that this is not possible…not without breaking up the packet. To do this, R2 will have to "fragment" the packet in two, creating one packet of 1476 bytes (the data and IP header) and one packet of 44 bytes (the additional 24 bytes of the GRE header and a new IP header of 20 bytes). But even though the router could easily do this, it can't, because it has been instructed not to do so. Remember, the DF bit was set in the original IP header.

R2 needs to make another agreement. R2 has to instruct the web server to send smaller packets or to allow R2 to fragment them. It does this by sending an Internet Control Message Protocol (ICMP) type 3 code 4 packet. When the web server receives the ICMP message, it will contain the correct maximum transmission unit (MTU) and adjust the packet size accordingly. Great, our problem is solved and now we can all get some sleep, right? No, more like miles to go until we…

A common problem occurs when ICMP messages are blocked along the path to the web server. If this happens, then the ICMP message will never reach the web server, preventing the necessary new agreement we need for this to work.

This is not an insurmountable problem, and we can try the following:

▼ We could try and find out where along the path the ICMP message is blocked and have that device allow fragmentation.

■ Set the Maximum Transmission Unit (MTU) on the client's network interface to 1476 bytes. That would cause the original SMSS to be smaller, obviating the need for the devices to fragment the packets in the first place. This solution would depend on your particular situation and might be very viable if you have a large LAN. If you change the MTU for the client, you should also change the MTU for all devices that share the network with this client.

■ Use a proxy server between R2 and the gateway (firewall), and let the proxy server request all of the web pages.

▲ If the tunnel runs over interfaces that have an MTU greater than 1524 bytes (1500 bytes plus the 24-byte GRE header), you can increase the IP tunnel MTU. Not all equipment is necessarily capable of this. This solution also won't work if any of the links over which the tunnel packets are sent are less than 1500 bytes, because the DF bit of the original packet is copied to the tunnel packet header. The router would be able to encapsulate the original packet, but would not be able to fragment the tunnel packet since the DF bit has to be set.

The bottom line is that GRE tunnels can be very useful in many VPN implementations, but, depending on your particular situation, you will always need to take all of the points we discussed earlier into consideration for your original VPN design or when you make future additions to your networks.

GRE Header (Version 0)

00	01	02	03	04	05	06	07	08	09	10	11	12	13	14	15	16	17	18	19	20	21	22	23	24	25	26	27	28	29	30	31
C	R	K	S	s	Recur			Flags				Version			Protocol															
Checksum																Offset															
Key																															
Sequence Number																															
Routing																															

▼ **C – Checksum Present** This is a 1-bit field. When this bit is set, the Checksum field is present. If either the Checksum Present bit or the Routing Present bit is set, the Checksum and Offset fields are both present.

■ **R– Routing Present** This is a 1-bit field. When this bit is set, the Offset field is present. If either the Checksum Present bit or the Routing Present bit is set, the Checksum and Offset fields are both present.

■ **K – Key Present** This is a 1-bit field. When this bit is set, the Key field is present.

■ **S – Sequence Number Present** This is a 1-bit field. When this bit is set, the Sequence Number field is present.

■ **s – Strict Source Route** This is a 1-bit field. It is recommended that this bit only be set if all of the routing information consists of strict source routes.

■ **Recur – Recursion Control** This is a 3-bit field. It denotes the number of additional encapsulations that are permitted. 0 is the default.

■ **Flags** This is a 5-bit field. These bits are reserved and must be set at 0.

■ **Version** This is a 3-bit field. It denotes the GRE protocol version and must be set to 0.

■ **Protocol** This is a 16-bit field. It denotes the protocol type of the payload packet. In general, the value will be the Ethertype protocol type field for the packet we talked about earlier. Additional values may be defined in other documents.

■ **Checksum** This is a 16-bit field. It contains the IP checksum of the GRE header and the payload packet. (This is an optional field.)

■ **Offset** This is a 16-bit field. It indicates the byte offset from the start of the Routing field to the first byte of the active source route entry to be examined. (This is an optional field.)

■ **Key** This is a 32-bit field. It contains a number that is inserted by the encapsulator and may be used to authenticate the source of the packet. (This is an optional field.)

■ **Sequence Number**   This is a 32-bit field. It contains a number that is inserted by the encapsulator and may be used by the receiver to determine the sequence in which the packets have been transmitted. (This is an optional field.)

▲ **Routing**   This field is a variable-length field. This field, defined next in the SRE packet section, is a list of *source route entries* (SREs). (This is an optional field.)

Source Route Entry (SRE) Packet

00	01	02	03	04	05	06	07	08	09	10	11	12	13	14	15	16	17	18	19	20	21	22	23	24	25	26	27	28	29	30	31
Address Family																SRE Offset								SRE Length							
Routing Information																															

▼ **Address Family**   This is a 16-bit field. This indicates the syntax and semantics of the Routing Information field (defined shortly).

■ **SRE Offset**   This is an 8-bit field. This indicates the byte offset from the start of the Routing Information field to the first byte of the active entry to be examined.

■ **SRE Length**   This is an 8-bit field. It indicates the number of bytes in the SRE. When this field is set to 0, it indicates this is the last SRE in Routing.

▲ **Routing Information**   This is a variable-length field. This contains information to be used in the routing of the packet.

POINT-TO-POINT TUNNELING PROTOCOL (PPTP)

"PPTP" stands for "Point-to-Point Tunneling Protocol." PPTP is a Layer 2 protocol that essentially encapsulates PPP frames inside a modified GRE header and is placed into the IP datagrams for transmission. The protocol was originally designed as an encapsulation mechanism to allow the transport of native LAN protocols (such as IPX and AppleTalk) over public infrastructures like the Internet. The specification itself is fairly generic and allows for a variety of authentication mechanisms and encryption algorithms.

PPTP Background

A group called the PPTP Forum developed PPTP. The PPTP Forum originally was made up of the companies Ascend Communications, Microsoft Corporation, U.S. Robotics, 3Com, Copper Mountain Networks, and ECI Telematics. (Some of those companies have been absorbed by other companies since then.) PPTP was first demonstrated in March 1996 at the NetWorld Expo by U.S. Robotics and Microsoft. The Internet Engineering Task Force (IETF) met in Montreal a few months later. At this meeting, the PPTP Forum submitted a draft of the specifications to the Working Group on Point-to-Point-Protocol

Extensions (PPP-WG). The IETF's PPP-WG is responsible for additions, improvements, or changes to the draft. PPTP was documented in RFC 2637 in July 1999.

Microsoft has by far had the biggest influence on and was the first to implement widely the protocol. PPTP has been available in Windows since 1996. Microsoft uses Password Authentication Protocol (PAP), Challenge Handshake Authentication Protocol (CHAP), and Microsoft CHAP (MS-CHAP) versions 1 and 2 for authentication. Microsoft also uses RC4 and Digital Encryption Standard (DES) for encryption.

PPTP is currently available for Windows 95 and can also be found in Windows 98, Windows NT, Windows 2000, Apple, and Linux servers. With respect to NAS equipment, PPTP is also now available on almost all of the routers used by ISPs today. PPTP has been the most widely used of the access VPN protocols. Over time, the most widely used will end up being the L2TP protocol, which is a fusion of L2F and PPTP. L2F is a VPN protocol that Cisco was developing at the time Microsoft PPTP was announced. This collaborative application is particularly good news. It means that there will be just one industrywide IETF specification for a Virtual Private Networking protocol and that it will consist of the best elements of L2F and PPTP. We will be discussing these protocols in the next chapter.

PPTP Overview

PPTP is a protocol that encapsulates other common LAN protocols such as NetBEUI, IPX/SPX (Novell), and TCP/IP into an IP packet. The nice thing about PPTP is that you can route any of these protocols over an IP-only network (such as the Internet) in a manner that is completely transparent to the user. From the user's and the machine's perspective, though they might be a thousand miles away, it still seems as though they were directly connected to the LAN in their office that might be running IPX, NetBEUI, or any other protocol exclusively.

PPTP is generally viewed as a remote access solution. It was originally designed by Microsoft to allow a user to connect to a corporate Access Server by using the Internet, as opposed to dialing directly into the corporate Access Server over a standard PSTN line. A connection can be accomplished in a couple of ways, depending on whether the ISP being used to access the Internet has hardware that is PPTP aware or not. The variations differ only slightly—should the ISP Network Access Server (NAS) be PPTP aware, then the remote user need only initiate a standard PPP connection to an ISP. At this point, the PPTP-aware NAS at the ISP creates the PPTP tunnel directly to the user's corporate Access Server. This is an example of the compulsory tunnel we discussed in Chapter 8.

If the ISP NAS is not PPTP aware, then after the user has connected to the ISP NAS via a PPP connection, he or she then needs to initiate a second PPTP connection, over the existing connection, straight through to the corporate Access Server. Data sent using this second connection is in the form of IP datagrams that contain PPP packets referred to as *encapsulated PPP packets*. Now the tunnel exists all the way from the corporate Access Server in the office to the remote PC, as opposed to only extending from the office to the ISP NAS the user dialed into initially. This is an example of the voluntary tunnel we discussed in Chapter 8.

At this point, the corporate Access Server can dynamically assign either a unique Global IP Address from the pool it has registered or any number of nonroutable IP address defined in RFC 1597 (10.0.0.0, 172.16.0.0, 192.168.0.0, and so on) that we talked about earlier in this chapter.

The client connects to a Network Access Server (NAS) at the ISP facility. Once connected, the client can send and receive packets over the Internet. The network access server uses the TCP/IP protocol for all traffic to the Internet.

After the client has made the initial PPP connection to the ISP, a second connection is made over the existing PPP connection. Data sent using this second connection is in the form of IP datagrams that contain encapsulated PPP packets. The second call creates the virtual private networking (VPN) connection (the tunnel) to a PPTP server on the private enterprise LAN. A voluntary PPTP tunnel or VPN can be seen in Figure 10-3.

When the PPTP server receives the packet from the Internet connection, it places it onto the private network connection and sends it to the destination computer. The PPTP server does this by unwrapping the PPTP packet to obtain the private network computer name and address information found in the encapsulated PPP packet. That encapsulated PPP packet can consist of almost any type of protocol data, such as AppleTalk, TCP/IP, IPX, or NetBEUI protocols.

Figure 10-4 illustrates the multiprotocol support built into PPTP. A packet sent from the PPTP client to the PPTP server passes through the PPTP tunnel to a destination computer on the private network. PPTP encapsulates the encrypted and compressed PPP packets into IP datagrams for transmission over the Internet. These IP datagrams are routed over the Internet until they reach the PPTP server that is connected to the Internet and the private network. The PPTP server disassembles the IP datagram into a PPP packet and then decrypts the PPP packet using the native network protocol being used on the private network.

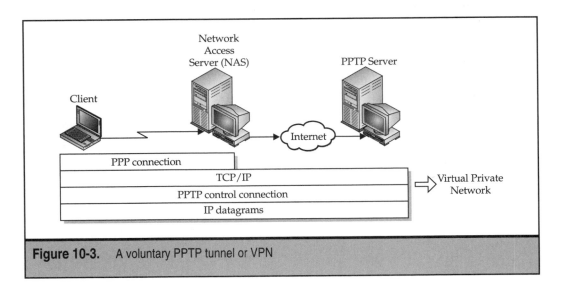

Figure 10-3.   A voluntary PPTP tunnel or VPN

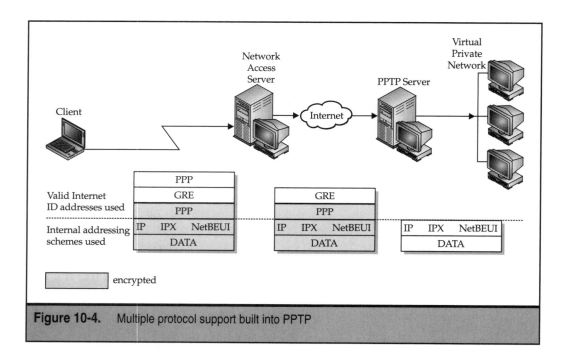

Figure 10-4. Multiple protocol support built into PPTP

WAN Clients and LAN Clients

A PPTP client can connect to a PPTP server in two ways:

▼ **Using an ISP's NAS that supports inbound PPTP connections** PPTP clients that use an ISP's NAS must be configured with a modem and PPTP client software to make the separate connections to the ISP and the PPTP server. First, a connection is made using the PPP protocol over the modem to an ISP. The second connection is a VPN connection using PPTP, over the modem and the ISP connection, to tunnel across the Internet to the PPTP server. The second connection requires the first connection because the tunnel between the VPN devices is established by using the modem and PPP connection to the Internet.

▲ **Using a Local LAN connection to connect to a PPTP server** Instead of a dual connection requirement, avoided by the initial dial-up connection to the ISP, we can use PPTP to create the tunnel between computers that are physically connected to the same private internal LAN. In this case, a PPTP client is already connected to the network and only uses the PPTP client software to create the connection to a PPTP server on the LAN.

PPTP packets from a remote access PPTP client and a local LAN PPTP client are processed differently. A PPTP packet from a remote access PPTP client is placed on the Public Switched Telephone Network (PSTN) physical media, while the PPTP packet from a LAN PPTP client is placed on the LAN physical media over the network interface card (NIC). Figure 10-5 illustrates how PPTP encapsulates PPP packets and then places the outgoing PPTP packet on either a modem, ISDN, or LAN network media.

PPTP Elements and Characteristics

Now we are going to talk about some of the PPTP elements and the characteristics related to this protocol.

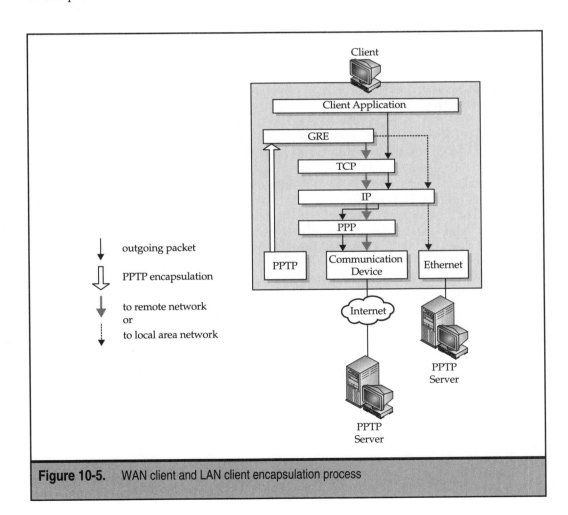

Figure 10-5.   WAN client and LAN client encapsulation process

PPTP Architecture Overview

The secure communication created using the PPTP protocol typically involves three processes, each of which requires successful completion of the previous process. This section explains these three processes and how they work:

▼ **PPP connection** A PPTP client uses PPP to connect to an ISP by using a standard telephone line or ISDN line. This connection uses the PPP protocol to establish the connection and to encrypt data packets.

■ **PPTP control connection** Using the connection to the Internet established by the PPP protocol, the PPTP protocol creates a control connection from the PPTP client to a PPTP server on the Internet. This connection uses TCP to establish the connection and is a called a *PPTP tunnel.*

▲ **PPTP data tunneling** Finally, the PPTP protocol creates IP datagrams containing encrypted PPP packets, which are then sent through the PPTP tunnel to the PPTP server. The PPTP server disassembles the IP datagrams and decrypts the PPP packets, and then routes the decrypted packets to the private network.

PPP Protocol

The PPTP protocol is built upon and used in conjunction with the Point to Point Protocol (PPP). PPP is a remote access protocol used by PPTP to send multiprotocol data across TCP/IP-based networks. PPP supports multiple protocols, authentication, privacy, and compression of data. PPTP allows a PPP session to be tunneled through an existing IP connection, no matter how it was set up. An existing connection can be treated as if it were a telephone line, so a private network can run over a public one.

PPP encapsulates IP, IPX, and NetBEUI packets between PPP frames and sends the encapsulated packets by creating a point-to-point link between the sending and receiving computers. Most PPTP remote access sessions are started by a client connecting to an ISP NAS. (Keep in mind that in some situations, remote clients may have direct access to a TCP/IP network. With a direct IP connection, the initial PPP connection to an ISP is unnecessary. The client can initiate the connection to the PPTP server, without first making a PPP connection to an ISP.)

The PPP protocol is used to create the dial-up connection between the client and the ISP NAS over the PSTN after completion of the following phases. (Phase 3 is optional.)

Phase 1: Link Establishment Phase PPP uses Link Control Protocol (LCP) to establish, maintain, and end the physical connection. During the initial LCP phase, basic communication options are selected. Note that during the link establishment phase (Phase 1), authentication protocols are selected, but they are not actually implemented until the connection authentication phase (Phase 2). Also, during this phase, the LCP will decide if and how the two peers will negotiate the use of compression and/or encryption. The actual choice of compression/encryption algorithms and other details occurs during the Network layer protocol phase (Phase 4).

Phase 2: Authentication Phase   PPTP clients are authenticated using the PPP protocol. Authentication can be done using cleartext or encrypted authentication protocols such as PAP and CHAP. (There are others, but this will give you the basic idea behind cleartext and encrypted protocols.) Most implementations of PPP provide limited authentication methods, typically Password Authentication Protocol (PAP) and Challenge Handshake Authentication Protocol (CHAP).

▼   **Password Authentication Protocol (PAP)**   PAP is a simple, cleartext authentication scheme. The NAS requests the user name and password, and PAP returns them in cleartext (unencrypted). Obviously, this authentication scheme is not secure, because a third party could capture the user's name and password and use it to get subsequent access to the NAS and all of the resources provided by the NAS. PAP provides no protection against replay attacks or remote client impersonation once the user's password is compromised.

■   **Challenge Handshake Authentication Protocol (CHAP)**   CHAP is an encrypted authentication mechanism that avoids transmission of the actual password on the connection. The NAS sends a challenge, which consists of a session ID and an arbitrary challenge string, to the remote client. The remote client must use the MD5 one-way hashing algorithm to return the user name and an encryption of the challenge, session ID, and the client's password. (These concepts and specific algorithms will be discussed in great detail in later chapters.) The user name is sent unhashed. CHAP is an improvement over PAP in that the cleartext password is not sent over the link. Instead, the password is used to create an encrypted hash from the original challenge. The server knows the client's cleartext password and can therefore replicate the operation and compare the result to the password sent in the client's response. CHAP protects against replay attacks by using an arbitrary challenge string for each authentication attempt. CHAP protects against remote client impersonation by unpredictably sending repeated challenges to the remote client throughout the duration of the connection. Secure authentication schemes provide protection against *replay attacks* and *remote client impersonation:*

■   A *replay attack* is when a third party monitors a successful connection and uses captured packets to play back the remote client's response so that it can gain an authenticated connection.

■   *Remote client impersonation* occurs when a third party takes over an authenticated connection. The intruder waits until the connection has been authenticated and then traps the necessary communication parameters, disconnects the authenticated user, and takes control of the authenticated connection.

During this authentication phase (Phase 2), the NAS collects the authentication data and then validates the data against its own user database or against a central dedicated authentication database server.

Phase 3: Callback Control Phase (Optional) Some PPP implementations include an optional callback control phase. This phase uses a Callback Control Protocol right after the authentication phase. If this phase has been implemented, once the user has been authenticated, both the remote client and NAS disconnect. Then the reverse happens. The NAS calls the remote client back at a prespecified phone number. This provides an additional level of security since, even if someone were to capture the passwords during the initial call, they would also have to know and be connected to the prespecified phone number. This way the NAS only allows connections from remote clients that physically reside at specific phone numbers.

Phase 4: Network Control Phase This last phase implements the various Network Control Protocols (NCPs) that were selected during the link establishment phase (Phase 1) to configure protocols used by the remote client. During this phase, addresses can be dynamically assigned to the remote clients computer during dial-in. This is also where the negotiation and implementation of compression control protocols and data encryption protocols take place.

If you look at Figure 10-6, you will get a better idea of what actually takes place during the initial PPP connection from the remote client's computer to the ISP's NAS.

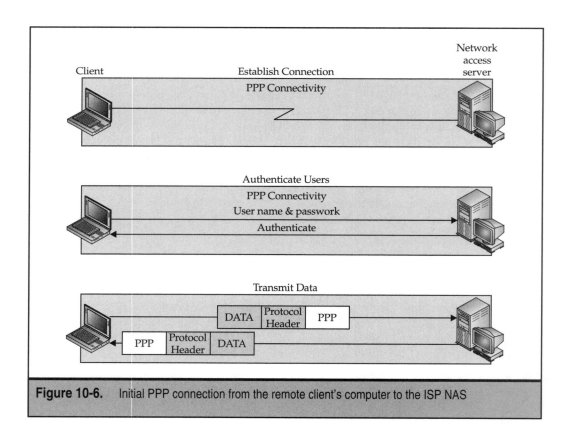

Figure 10-6. Initial PPP connection from the remote client's computer to the ISP NAS

PPTP Control Connection

The control connection is a standard TCP session over which PPTP call control and link management information is passed involving the establishment, release, and maintenance of sessions and of the tunnel itself.

The control channel that is set up then uses a standard TCP connection to the Registered Port number 1723 on the corporate Access Server. (The Well-Known Ports are those from 0 to 1023. The Registered Ports are those from 1024 to 49151. The Dynamic and/or Private Ports are those from 49152 to 65535. All assigned numbers can be found in RFC 1700.)

The PPTP protocol specifies a series of control messages sent between the PPTP-enabled client and the PPTP server. The control messages establish, maintain, and end the PPTP tunnel. Table 10-1 lists the primary control messages used to establish and maintain the PPTP tunnel.

Control messages are transmitted in control packets in a TCP datagram. One TCP connection is created between the PPTP client and the PPTP server. This connection is used to exchange control messages. The control messages are sent in TCP datagrams containing the control messages. A datagram contains a PPP header, a TCP header, a PPTP control message, and appropriate trailers. Take a look at Figure 10-7 to see a PPTP TCP datagram with control messages.

The exchange of messages between the PPTP client and the PPTP server over the TCP connection is used to create and maintain a PPTP tunnel. This entire process can be seen in Figure 10-8.

Message Type	Purpose
PPTP_START_SESSION_REQUEST	Starts session
PPTP_START_SESSION_REPLY	Replies to start session request
PPTP_ECHO_REQUEST	Maintains session
PPTP_ECHO_REPLY	Replies to maintain session request
PPTP_WAN_ERROR_NOTIFY	Reports an error on the PPP connection
PPTP_SET_LINK_INFO	Configures the connection between client and PPTP Server
PPTP_STOP_SESSION_REQUEST	Ends session
PPTP_STOP_SESSION_REPLY	Replies to end session request

Table 10-1. PPTP Control Message Types

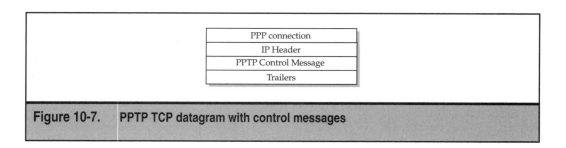

| PPP connection |
| IP Header |
| PPTP Control Message |
| Trailers |

Figure 10-7. PPTP TCP datagram with control messages

In Figure 10-8, the control connection is where the remote access client's computer is the PPTP client. In a case where the remote access client's computer is not PPTP capable, it will connect to a PPTP-aware NAS at the ISP, at which point the PPTP control connection (tunnel) begins at the ISP NAS.

PPTP Data Transmission

After the PPTP tunnel is established, user data is transmitted between the client and PPTP server. Data is transmitted in IP datagrams containing PPP packets. The IP datagrams are created using a modified version of the Internet Generic Routing Encapsulation (GRE version 2) protocol. The IP datagram created by PPTP can be seen in Figure 10-9. The IP delivery header provides the information necessary for the datagram to traverse the Internet. The GRE header is used to encapsulate the PPP packet within the IP datagram. The PPP packet was created by RAS. Note that the PPP packet is just one unintelligible block, because it is encrypted. Even if the IP datagram were intercepted, it would be nearly impossible to decrypt the data.

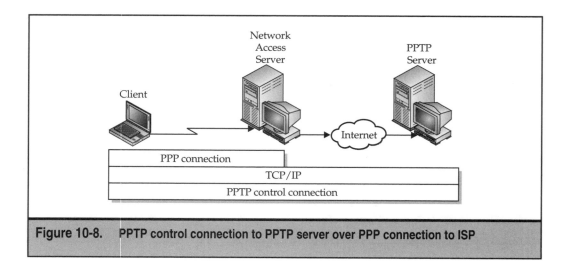

Figure 10-8. PPTP control connection to PPTP server over PPP connection to ISP

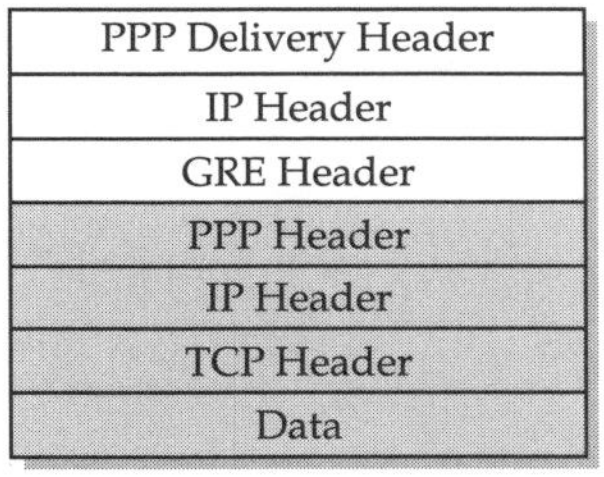

| PPP Delivery Header |
| IP Header |
| GRE Header |
| PPP Header |
| IP Header |
| TCP Header |
| Data |

Figure 10-9.   IP datagram containing encrypted PPP packet as created by PPTP

PPTP Security Issues

There is nothing in the PPTP standard that guarantees a secure architecture. PPTP security is only as good as the specific vendor implementation. Any vendor implementing PPTP can incorporate security services into their product. These services can be built using the underlying TCP transport, the PPP protocol itself, or using any protocol riding on top of the PPP packet.

PPTP control channel messages are neither authenticated nor checked for integrity. Because the PPP negotiations are carried out over the tunnel in the clear, it may be possible for an attacker to eavesdrop and modify those negotiations by hijacking the underlying TCP connection. It is also possible to manufacture false control channel messages and to alter genuine messages in transit without detection because the GRE packets forming the tunnel itself are not cryptographically protected.

The security of user data passed over the tunneled PPP connection is addressed by the PPP protocol, as is the authentication of the PPP peers. Unless the PPP payload data is cryptographically protected, it can be captured, read, or modified. It does not offer protection from substitution attacks or playback attacks, nor does it provide perfect forward secrecy (in other words, protection against reading recorded sessions when provided with session initialization passwords). If protection is not addressed at the underlying transport level, it is possible to hijack the entire session or to insert packets into an existing tunnel.

SUMMARY

In this chapter, we talked about two of the protocols used in the creation of tunnels and VPNs. We started with the Generic Routing Encapsulation (GRE) protocol, which is one of the original tunneling protocols upon which many of the other tunneling protocols were based. We then talked about PPTP and how it worked using GRE. We also discussed in detail how PPTP and GRE were constructed and how they are applied. In the next chapter, we will discuss two other related protocols, L2F and L2TP.

CHAPTER 11

L2F and L2TP

In this chapter, we will discuss two protocols that were designed to facilitate the implementation of Virtual Private Dialup Networks (VPDN) just as PPTP was. VPDNs are synonymous with the Access VPNs that we spoke about in Chapter 9. The two protocols you will learn about are the Layer 2 Forwarding (L2F) and the Layer 2 Tunneling Protocol (L2TP).

ACCESS VPNs OR VPDNs

What makes Access VPNs — or VPDNs — so attractive is that they offer a very cost effective way of providing access to the home or corporate network from anywhere in the world. Sending a user's call over the Internet provides dramatic cost savings for the corporate customer. Instead of connecting directly to the corporate network using the expensive public switched telephone network (PSTN), access VPN users only need to use the PSTN to connect to an ISP's local NAS (also known as the local Point-of-Presence (POP). The ISP then uses the Internet to forward users from the POP to the enterprise customer network.

Access VPNs use Layer 2 tunneling technologies to create a virtual point-to-point connection between remote users and the corporate network. These tunneling technologies provide the same direct connectivity as the expensive PSTN by using the Internet. This means that users anywhere in the world have the same connectivity as they would at the corporate headquarters network.

Access VPNs connect a variety of users from a variety of locations. This could be a single mobile employee to an entire branch office. Once the tunnel is established, the ISP is transparent to the user and the enterprise customer. The tunnel creates a secure connection between the user and the enterprise customer's network over the insecure Internet and is indistinguishable from a point-to-point connection. Figure 11-1 illustrates the methods of connecting to the corporate network using Access VPNs.

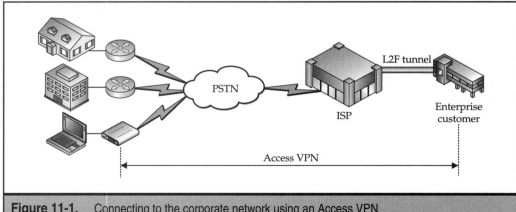

Figure 11-1. Connecting to the corporate network using an Access VPN

Access VPNs are generally created using L2F, L2TP, and PPTP. We will start by discussing L2F because it (along with PPTP) was the basis for the ultimate design of L2TP.

LAYER 2 FORWARDING (L2F) PROTOCOL

L2F is a specification that attempts to make possible the tunneling of the Layer 2 functionality of (in theory) any higher level framing protocol (such as PPP or SLIP) over any protocol that can be made to establish point-to-point connectivity. L2F uses UDP datagrams (Layer 4) to accomplish this.

L2F does not modify the PPP or higher level frames in any way, it simply provides a way to carry the frame on top of any protocol that permits the point-to-point connectivity. Though this is most commonly done over TCP/IP networks (like the Internet), it is not limited to TCP/IP. L2F can be used over other infrastructures like frame relay and X.25 virtual circuits as well. Some of the basic services provided by L2F are

▼ *Tunneling*: *Tunneling* provides the capability of defining virtual tunnels between two points using PPP to encapsulate the underlying traffic.

■ *Authentication*: L2F for the most part relies on PPP's implementation of the PAP and CHAP protocols. The standard encourages the use of shared secret and clear-text authentication during the creation of the tunnel.

■ *Multi Protocol Support*: *Multi Protocol Support* provides the capability of carrying non-IP protocols such as IPX/SPX, AppleTalk, NetBEUI, and any other protocol designed to function at (or above) the Layer 3 of the OSI model. Because L2F can ride over TCP/IP, all these non-IP protocols can be transparently tunneled across TCP/IP based networks (like the Internet).

■ *Data Integrity*: L2F supports basic checksum integrity checks.

▲ *Anti-Spoofing*: Basic anti-spoofing is provided based on key field values calculated at tunnel inception.

How L2F works

In Chapter 9, we discussed the two types of Access VPNs and the two types of tunnels they created. The two types of Access VPNs and their corresponding tunnel types are

▼ Client-initiated VPNs using Voluntary Tunnels

▲ NAS-initiated VPN using Compulsory Tunnels

L2F was designed to create NAS-initiated VPNs over compulsory tunnels. In this case, remote users dial in to the ISP's local NAS. Once connected to the ISP, the NAS then establishes a tunnel to the enterprise's private network. NAS-initiated VPNs scale better and are more robust than client-initiated VPNs in that they allow users to connect to multiple networks by using multiple tunnels, and do not require any VPN client software to be installed on the remote users computer. NAS-initiated VPNs do not extend the tunnel from over the connection between the client and the ISP, but this is not a concern for most enterprise customers because the PSTN is much more secure than the Internet.

NAS-initiated VPNs, unlike client-initiated VPNs, involve the cooperation of the ISP and the corporate enterprise customer. Each entity is responsible for their part of the VPN. In most cases, the ISP is responsible for maintaining the modem pools and network access servers. The corporate customer needs to decide how and who will maintain authentication and internal address assignment for the remote user.

An ISP will configure its NAS to receive calls from the remote users and forward the calls to the corporate customer's network. The ISP only maintains information about the corporate customer's network that is the tunnel endpoint. The customer maintains the corporate network users' IP addresses, routing, and other user database functions. Administration between the ISP and corporate customer's network is reduced to IP connectivity.

Access VPNs use L2F tunnels to tunnel the link layer of high-level protocols (like PPP frames or SLIP frames). By doing this, it is possible to detach the location of the ISP's NAS from the location of the corporate customer's network, where the dial-up protocol connection terminates and access to the corporate customer's network is provided.

Before we get into the specifics of what takes place during the setup and creation of a typical L2F VPN implementation, we'll examine the packet structure and the purpose of the fields within.

L2F Packet Structure

Below is a breakdown of a typical packet. Then the actual L2F header is

Basic Packet Format

L2F header	Payload Packet (PPP/SLIP)	L2F checksum (optional)

L2F packet header

00	01	02	03	04	05	06	07	08	09	10	11	12	13	14	15	16	17	18	19	20	21	22	23	24	25	26	27	28	29	30	31
F	K	P	S	0	0	0	0	0	0	0	0	C	Version			Protocol								Sequence (Optional)							
Multiplex ID (MID)																Client ID (CLID)															
Length																Offset (Optional)															
Key																															
Payload Data (variable)																															
....																															
L2F Checksum (Optional)																															

▼ **F bit:** 1 bit.

■ **K bit:** 1 bit.

■ **P bit:** 1 bit. If the P bit in the L2F header is set, this packet is considered a "priority" packet. When possible, a packet received with the P bit set should be processed in preference to previously received unprocessed packets without the P bit set. For example, it is generally recommended that PPP keepalive traffic be sent with the P bit set.

- **S bit:** 1 bit. This is the sequence valid bit. When this bit is set, it indicates that a sequence number will be used.
- **C bit:** 1 bit. This is the checksum present bit. When this bit is set, it indicates that a 16 bit checksum will follow the encapsulated payload.
- **Version Field:** 3 bits. This is the version of the L2F software creating the packet. It must be set to [001] or the packet will be considered invalid.
- **Protocol Field:** 8 bits. This field specifies the protocol that had been encapsulated within the L2F packet.

Hex Value	Type Abbreviation	Description
0x00	L2F_ILLEGAL	Illegal.
0x01	L2F_PROTO	L2F management packet.
0x02	L2F_PPP	PPP tunneling inside L2F.
0x03	L2F_SLIP	SLIP tunneling inside L2F.

- **Sequence Number Field (Optional):** 8 bits. The Sequence number is present if the S bit in the L2F header is set to 1. This bit must always be set to 1 for all L2F management packets. It may be set to 1 for non-L2F management packets. If a non-L2F management packet is received with the S bit set, then all future L2F packets sent for that Multiplex ID (discussed next) must have the S bit set and must be sent using sequence numbers.
- **Multiplex ID Field (MID):** 16 bits. The Multiplex ID (MID) identifies a particular connection within the tunnel. Each new connection is assigned a MID unique within the tunnel. A MID with a value of 0 is special. This field is used to communicate the state of the tunnel itself, as distinct from any connection within the tunnel.
- **Client ID Field (CLID):** 16 bits. The Client ID (CLID) is used to assist endpoints in de-multiplexing tunnels when the underlying point-to-point transport lacks an efficient or dependable method of doing it on its own.
- **Length Field:** 16 bits. Length is the size in octets of the entire packet, including header, all fields present, and the payload itself. Length does not include the addition of the checksum, if one is present.
- **Offset Field (Optional):** 16 bits. The Offset Field is present when the F bit is set in the header flags. This field specifies the number of bytes past the L2F header at which the payload data is expected to start. If the value is 0, or the F bit is not set, then the first byte following the last byte of L2F header is the first byte of payload data.

■ **Key Field:** 32 bits. The Key field is present when the K bit is set in the L2F header. The Key is based on the authentication response last given to the peer during tunnel creation (the details of tunnel creation are provided in the next section). It serves as a key during the life of a session to resist attacks based on spoofing. If a packet is received whose Key does not match the expected value then the packet must be discarded.

■ **Payload Data:** Variable. This is the actual payload being carried immediately following the L2F header.

▲ **Packet Checksum Field (Optional):** 16 bits. The Checksum is present when the C bit is set in the header flag. This is a 16-bit Cyclical Redundancy Check (CRC) used by PPP/HDLC. This is applied over the entire packet starting with the first byte of L2F flags, through the last byte of payload data. The checksum is then added as two bytes immediately following the last byte of payload data.

L2F management message types

In the *Protocol Field* of the L2F header, we discussed above the hex value of 0x01 indicates an L2F management message. When an L2F packet's Protocol field specifies an L2F management message, the body of the packet is encoded as zero or more options. An option is a single octet "message type," followed by zero or more sub-options. Each sub-option is a single byte sub-option value, and further bytes as appropriate for the sub-option. Refer back to these tables when we cover the event sequences for establishment of an access VPN.

L2F options and sub-options are:

Hex Value	Type Abbreviation	Description
0x00	Invalid	Invalid message
0x01	L2F_CONF	Request configuration
0x02	L2F_CONF_NAME	Name of peer sending L2F_CONF
0x03	L2F_CONF_CHAL	Random number peer challenges with
0x04	L2F_CONF_CLID	Assigned CLID for peer to use
0x02	L2F_OPEN	Accept configuration
0x01	L2F_OPEN_NAME	Name received from client
0x02	L2F_OPEN_CHAL	Challenge client received
0x03	L2F_OPEN_RESP	Challenge response from client
0x04	L2F_ACK_LCP1	LCP CONFACK accepted from client
0x05	L2F_ACK_LCP2	LCP CONFACK sent to client
0x06	L2F_OPEN_TYPE	Type of authentication used
0x07	L2F_OPEN_ID	ID associated with authentication

Hex Value	Type Abbreviation	Description
0x08	L2F_REQ_LCP0	First LCP CONFREQ from client
0x03	L2F_CLOSE	Request disconnect
0x01	L2F_CLOSE_WHY	Reason code for close
0x02	L2F_CLOSE_STR	ASCII string description
0x04	L2F_ECHO	Verify presence of peer
0x05	L2F_ECHO_RESP	Respond to L2F_ECHO

Access VPN Establishment Sequence

Now that you have been given a basic understanding of the L2F protocol and its components, we'll take a look at a typical VPN connection and the sequence of events that takes place as it is being established. This sequence of events takes place in three stages:

▼ Protocol Negotiation Process

■ L2F Tunnel Authentication Process

▲ Three-Way CHAP Authentication Process

Protocol Negotiation Process

This process description will give you an overall view of what takes place to set up the Access VPN from the remote user to the home gateway of the user's corporate network.

You have a remote user on the road somewhere who wants to connect to the corporate home gateway. The user will dial a designated number and establish a standard PPP connection to the ISP's NAS.

The ISP needs to determine whether you're a regular customer requiring normal Internet access, or whether you are a remote user needing VPN access to the corporate home gateway. This can be done by using some form of structured addressing like "sramos@rferguson.com" or by using CLID/DNIS based authentication (CLID and DNIS is defined after the next paragraph). If structured addressing is used, the ISP can identify the user by using PAP or CHAP authentication.

The ISP's NAS will then establish an L2F tunnel with the corporate home gateway by sending a message using UDP on port 1701. Finally, the home gateway authenticates the client's username and password, and establishes the PPP connection with the client. Figure 11-2 illustrates a numbered sequence of protocol negotiation events between the ISP's NAS and the enterprise customer's home gateway followed by a short description of what occurs at each step.

DNIS and CLID

The Network Access Servers (NAS) used by ISP can contain hundreds and thousands of modems each. Obviously, they can't advertise individual phone numbers for each line the modems are attached to. Instead, a small list of numbers is advertised and it is those numbers

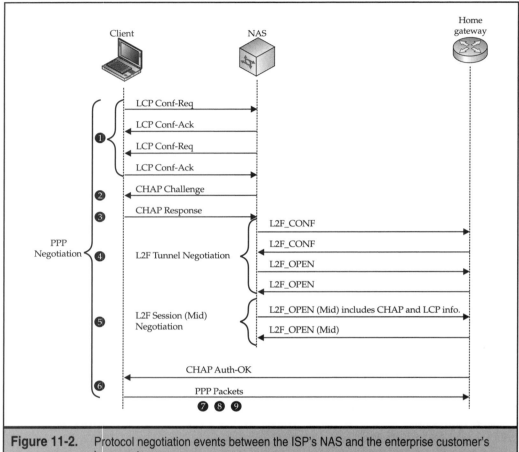

Figure 11-2.   Protocol negotiation events between the ISP's NAS and the enterprise customer's home gateway

that their legions of customers dial into. ISPs will often use both the DNIS and CLID to perform a rudimentary form of authentication and their respective descriptions are as follows:.

▼ *DNIS*: This stands for Dialed Number Identification Service. The "Called" number is determined by its DNI (Dialed Number Identification). DNIS is a service that identifies for the *receiver of a call* the number that the *caller dialed*. It's a common feature of 800 and 900 services. If you pay for the use of multiple 800 or 900 numbers that terminate on the same destination trunks, DNIS can identify the number originally dialed.

▲ *CLID*: This stands for Calling Line Identification. The "Calling" number exists in the telephone network as an ANI (Automatic Number Identification). A code is generated from this ANI that is known as Calling Line ID (CLID) and it is the

CLID code that is passed on to the receiver. The reason for the differentiation of CLID and ANI is for legal and privacy reasons. You can hide your identity by blocking Caller ID if you choose. You cannot, however, hide your identity from phone companies, or the 911 service (some of you troublemakers might want to keep that in mind).

The protocol negotiation event descriptions are as follows:

1. The user's client and the ISP's NAS conduct a standard PPP link control protocol (LCP) negotiation.

2. The ISP's NAS begins PPP authentication by sending a Challenge Handshake Authentication Protocol (CHAP) challenge to the client.

3. The client replies with a CHAP response.

4. When the ISP's NAS receives the CHAP response, it authenticates the user using either the phone number (when using CLID/DNIS-based authentication) or the user's domain name (when using domain name-based authentication). Once authenticated, the ISP's NAS creates a VPN to forward the PPP session to the home gateway by using an L2F tunnel. To do this, the ISP's NAS and the home gateway exchange L2F_CONF packets, which prepares the L2F tunnel. Then they exchange L2F_OPEN packets, which opens the L2F tunnel.

5. Once the L2F tunnel is open, the ISP's NAS and home gateway exchange L2F session packets. The ISP's NAS sends an L2F_OPEN (MID) packet to the home gateway that includes the client's information from the LCP negotiation (from step 1), the CHAP challenge, and the CHAP response. The home gateway uses this information to open a virtual interface it has created for the client and responds to the ISP's NAS with an L2F_OPEN (MID) packet.

6. The home gateway authenticates the CHAP challenge and response (using either local or remote authentication server) and sends a CHAP Auth-OK packet to the client completing the three-way CHAP authentication.

7. Once the client receives the CHAP Auth-OK packet, it can send PPP encapsulated packets through the tunnel to the virtual interface on the home gateway where the L2F tunneling header is stripped off sending the PPP packet on its way to the internal destination.

8. The client and the home gateway can now exchange PPP encapsulated packets as though it was dialed into the home gateway. Now that the L2F tunnel has been established, the ISP's NAS simply acts as a transparent PPP frame forwarder.

9. Any other incoming PPP sessions from any other authorized users (destined for the same home gateway) will not have to go through the earlier L2F tunnel negotiations because the L2F tunnel is already open. They will be forwarded transparently by the ISP's NAS as well.

L2F Tunnel Authentication Process

Now that you have seen the overall process of the Access VPN setup starting from the perspective of the remote user, we can move in a little closer to examine the tunnel authentication process that takes place between the ISPs NAS and the corporate home gateway.

When the NAS receives a call from a client that instructs it to create an L2F tunnel with the home gateway, it first sends a challenge to the home gateway. The home gateway then sends a combined challenge and response to the ISP's NAS. Finally, the ISP's NAS responds to the home gateway's challenge, and the two devices open the L2F tunnel.

Before the NAS and home gateway can authenticate the tunnel, they must have a common "tunnel secret." A tunnel secret is a pair of usernames with the same password that is configured on both the ISP's NAS and the home gateway. By combining the tunnel secret with random value algorithms, which are used to encrypt to the tunnel secret, the ISP's NAS and home gateway authenticate each other and establish the L2F tunnel. Figure 11-3 illustrates a numbered sequence of what takes place during L2F tunnel authentication.

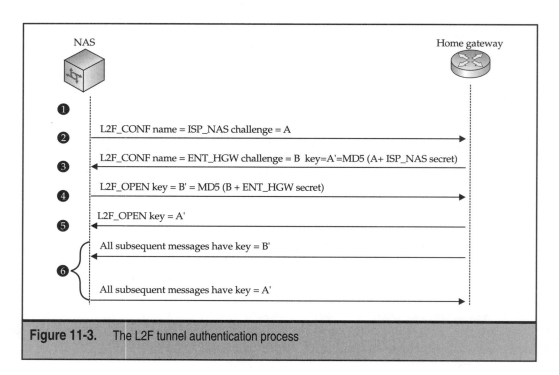

Figure 11-3.   The L2F tunnel authentication process

Here are the L2F tunnel authentication event descriptions:

1. Before the ISP's NAS and home gateway open an L2F tunnel, both devices must have a common tunnel secret in their respective configurations.

2. The ISP's NAS sends an L2F_CONF packet that contains the ISP's NAS name and a random challenge value that we will call A.

3. After the home gateway receives the L2F_CONF packet, it sends an L2F_CONF packet back to the ISP's NAS with the home gateway name and its own random challenge value that we will call B. This message also includes a key containing A', which was derived using an MD5 hash of the ISP's NAS secret and the value A (we will be discussing hash algorithms in Section IV, Secure Communications).

4. When the ISP's NAS receives the L2F_CONF packet, it compares the key A' with the MD5 of the ISP's NAS secret and the value A. If the key and value match, the ISP's NAS sends an L2F_OPEN packet to the home gateway with a key containing B' (the MD5 hash of the home gateway secret and the value B).

5. When the home gateway receives the L2F_OPEN packet, it compares the key B' with the MD5 hash of the home gateway secret and the value B. If the key and value match, the home gateway sends an L2F_OPEN packet to the ISP's NAS with the key A'.

6. All subsequent messages from the ISP's NAS include key=B'; all subsequent messages from the home gateway include key=A'.

Three-Way CHAP Authentication Process

Now that we have discussed how the tunnel itself was authenticated, we'll examine the process used to authenticate the user and the home gateway.

When establishing an access VPN, the client, ISP's NAS, and home gateway use three-way CHAP authentication to authenticate the client's username and password. CHAP is a challenge/response authentication protocol in which the password is sent as a 64-bit signature instead of as plain text. This enables the secure exchange of the user's password between the user's client and the home gateway.

First, the ISP's NAS challenges the client, and the client responds. The ISP's NAS then forwards this CHAP information to the home gateway, which authenticates the client and sends a third CHAP message (either a success or failure message) to the client. This three-way CHAP authentication process can be seen in a numbered sequence event description in Figure 11-4.

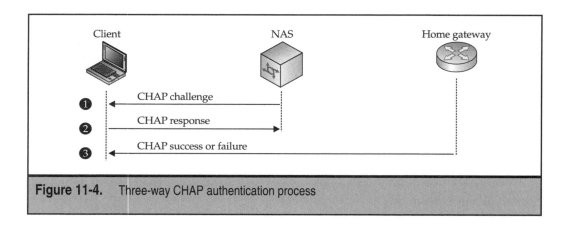

Figure 11-4. Three-way CHAP authentication process

Here are the three-way CHAP authentication event descriptions:

1. When the user initiates a PPP session with the ISP's NAS, the ISP's NAS sends a CHAP challenge to the client.

2. The client sends a CHAP response, which includes a plain text username, to the ISP's NAS. The ISP's NAS uses either the phone number (when using CLID/DNIS-based authentication…see bullets in the previous DNIS CLID section) or the user's domain name (when using domain name-based authentication) to determine the IP tunnel endpoint information. At this point, PPP negotiation is suspended, and the ISP's NAS asks its authentication server for IP tunnel information. The authentication server supplies the information needed to authenticate the tunnel between the ISP's NAS and the home gateway. Next, the ISP's NAS and the home gateway authenticate each other and establish an L2F tunnel. Then, the ISP's NAS forwards the PPP negotiation to the home gateway.

3. The third CHAP event takes place between the home gateway and the client. The home gateway authenticates the client's CHAP response, which was forwarded by the ISP's NAS, and sends a CHAP success or failure to the client.

Once the home gateway authenticates the client, the access VPN is established. The L2F tunnel creates a virtual point-to-point connection between the client and the home gateway. At this point, the ISP's NAS simply acts as a transparent packet forwarder and the final result all ends up looking something like Figure 11-5. Remember, if any subsequent clients dial in to the ISP's NAS to be forwarded to the home gateway, the ISP's NAS and home gateway do not repeat the L2F session negotiation because the L2F tunnel has already been established.

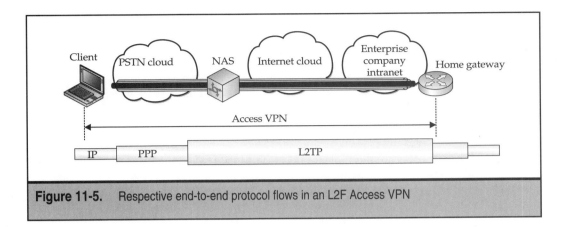

Figure 11-5.   Respective end-to-end protocol flows in an L2F Access VPN

LAYER 2 TUNNELING PROTOCOL (L2TP)

Up to this point, we have been discussing three tunneling protocols. In the sections immediately preceding this, we have been talking about GRE, PPTP, and L2F. In this section we will discuss L2TP. The reason we mention this is because, like us, it too started with those same three protocols. We will be referring to these protocols and comparing their characteristics, PPTP and L2F in particular. This is because L2TP was created using the best features of PPTP and L2F as a direct result of a mandate from the Internet community. Rather than having two incompatible proprietary protocols compete for the Access VPN marketplace, the Internet Engineering Task Force (IETF) merged the two into an open standard protocol defined in RFC 2661.

A Little L2TP History

To better understand how the L2TP protocol came about, it might help if we include a little background behind impetus of its creation. A volatile situation can exist when you consider the potential revenue from the future market for VPNs combined with the sheer might of the two companies that enjoy almost complete monopolies of the two technologies comprising the most significant elements of a VPN; Clients/Servers and a medium over which they can connect.

The differences between L2F and PPTP stem from the types of companies that designed them. L2F was developed by Cisco Systems, a *hardware* company whose routers are used to direct almost 80 percent of all traffic traveling over the entire Internet every day. PPTP was developed (for the most part) by Microsoft Corporation, a *software* company whose operating system is running on the majority of all PCs in the world.

Consider the objectives that Microsoft and Cisco initially had for developing PPTP and L2F in the first place. Both vendors were looking for ways to provide remote access functionality for companies. This could be accomplished by allowing companies to use their cheap local connection to their ISPs and use the Internet infrastructure. This would obviate the need for using the telephone companies' infrastructure thereby saving them from the expensive long distance charges that went along with it. This would also reduce administration and hardware costs because companies would no longer have to maintain modem banks and extra phone lines for remote access.

Both companies came up with a solution to achieve those objectives by creating L2F and PPTP. Both protocols make it possible for remote clients to connect to and have their PPP sessions accepted by the ISP and authenticated by the corporate and/or ISP security databases at which point the PPP session would be tunneled through to their corporate network transparently and using the native LAN protocol of their choice.

The difference was that in order for L2F to work, the ISPs and corporate networks had to be using routers and remote access servers that support L2F. With Microsoft's PPTP solution, the ISP wouldn't have to support tunneling at all. The advantage to that was the only thing necessary to create the tunnels was PPTP software running on the client PCs and the NT servers on the corporate network at home. It shouldn't come as a surprise that, if you remember the historical generosity of Microsoft providing free Internet browsers to the masses, the PPTP protocols would be included at no charge!

However, L2F had its own advantages over PPTP. Unlike PPTP, which required an IP-based infrastructure, L2F was not tied to any specific protocol and could run on any number of Link layer technologies with little overhead. L2F could also provide better security for endpoint authentication by offering options on top of the usual simple user ID and password.

L2TP Similarities and Differences with PPTP and L2F

Cisco and Microsoft have incorporated into L2TP the best characteristics of their respective protocols and thrown in some extras as well. Like L2F and PPTP, L2TP establishes a virtual PPP connection at Layer 2 of the OSI model. Like PPTP, the L2TP specification does not require built-in hardware support, however, equipment that does support L2TP will provide faster performance and greater security than hardware that does not. As with L2F, L2TP is not bound to IP, so it also works well over popular Link layer technologies like ATM and Frame Relay. And like PPTP, flow control was added to L2TP, ensuring that systems aren't fed more data than they can handle.

L2TP differs from L2F in that tunnels can be initiated from the remote user via tunneling software. In this example, the tunnel endpoints become the remote user and the corporate gateway. Both L2TP and L2F support NAS-initiated tunneling also. From an end-user's perspective, L2F and L2TP appear to be the same.

L2TP encapsulates PPP frames to be sent over IP, X.25, Frame Relay, or ATM networks. However, when used over an IP network, L2TP frames are encapsulated inside

UDP. L2TP uses these UDP messages over IP internetworks for both tunnel maintenance *and* tunneled data. That differs in two ways from PPTP. PPTP not only uses TCP instead of UDP for the control channel, it also requires a second channel for data channel communications using the GRE tunneling protocol.

The payloads of encapsulated PPP frames can be encrypted, compressed, or both. L2TP also has the added option of using smaller headers than in PPTP, and this is accomplished through header compression. When compression is enabled, L2TP uses a 4-byte header as opposed to PPTP which uses a 6-byte header. L2TP allows for tunnel authentication, in addition to user authentication. PPTP does not provide tunnel authentication, it only offers user authentication by way of the PPP protocol. These are some of the main differences between PPTP and L2TP. Despite the advantages held by L2TP, both L2TP and PPTP are still quite prevalent in the tunneling protocol market.

Even though L2TP was created from the two existing technologies, it has also added a capability not found in its predecessors. L2TP makes it possible to have multiple simultaneous tunnels opened between end points. This can ensure a certain degree of QOS by allowing administrators to dedicate channels to specific tasks.

L2TP Security Considerations

L2TP is not perfect. There is nothing in L2TP that guarantees traffic confidentiality. Authentication and/or encryption could be added inside of the PPP packet but this solution would leave the connection vulnerable at the transport level. If security is a major consideration in the underlying transport level, then L2TP would have to be coupled with another security protocol like IPSec.

So if IPSec is so great, why would we need to combine it with L2TP at all? The answer is that L2TP addresses at least two requirements that IPSec cannot fulfill by itself:

▼ L2TP permits the tunneling of non-IP protocols such as IPX and Appletalk.

▲ L2TP supports tunneling of IP and non-IP protocols alike across non-IP based network links like ATM, X.25, and Frame Relay.

Many networks today rely on non-IP protocols. In some cases, there may be a need for these protocols to traverse public IP networks. In other cases, a non-IP based ATM or Frame Relay infrastructure may be used to carry many different types of heterogeneous network traffic. L2TP can be used in conjunction with IPSec in any of these network environments to create very robust and secure VPN implementations.

L2TP Protocol Description

Before we get into the mechanics of how L2TP works, we will define some terms that are common to this protocol. In our discussions of the preceding tunneling protocols, we used terms like *NAS* and *Home Gateway* to describe tunnel endpoints, but the Internet standard for L2TP uses a different set of terms to describe these devices. In this part of our discussion, we will limit our definitions to the devices themselves. We will define the terms specific to the header fields and Control messages just after this in the packet format and header descriptions part.

- ▼ **L2TP Access Concentrator (LAC):** A node that acts as one side of an L2TP tunnel endpoint and is a peer to the L2TP Network Server (LNS). The LAC sits between an LNS and a remote system and forwards packets to and from each. Packets sent from the LAC to the LNS requires tunneling with the L2TP protocol. The connection from the LAC to the remote system is either local or a PPP link.

- ■ **L2TP Network Server (LNS):** A node that acts as one side of an L2TP tunnel endpoint and is a peer to the L2TP Access Concentrator (LAC). The LNS is the logical termination point of a PPP session that is being tunneled from the remote system by the LAC. A Home Gateway could also be a LAC.

- ■ **Management Domain (MD):** A network or networks under the control of a single administration, policy, or system. For example, an LNS's Management Domain might be the corporate network it serves. An LAC's Management Domain might be the Internet Service Provider that owns and manages it.

- ■ **Network Access Server (NAS):** A device providing local network access to users across a remote access network such as the PSTN. An NAS may also serve as an LAC, LNS, or both.

- ■ **Peer:** When used in context with L2TP, peer refers to either the LAC or LNS. An LAC's Peer is an LNS and vice versa. When used in context with PPP, a peer is either side of the PPP connection.

- ■ **Remote System:** An end-system or router attached to a remote access network (i.e., a PSTN), which is either the initiator or recipient of a call. Also referred to as a *dial-up* or *virtual dial-up* client.

- ■ **Session:** L2TP is connection-oriented. The LNS and LAC maintain state for each Call that is initiated or answered by an LAC. An L2TP Session is created between the LAC and LNS when an end-to-end PPP connection is established between a Remote System and the LNS. Datagrams related to the PPP connection are sent over the Tunnel between the LAC and LNS. There is a one-to-one relationship between established L2TP Sessions and their associated Calls.

- ■ **Tunnel:** A Tunnel exists between a LAC-LNS pair. The Tunnel consists of a Control Connection and zero or more L2TP Sessions. The Tunnel carries encapsulated PPP datagrams and Control messages between the LAC and the LNS.

- ■ **IPSec Authentication Header (AH):** This used to ensure the integrity, data origin authentication for IP datagrams, and protection against replay. Data integrity is assured by the checksum generated by a message authentication code (for example, MD5); data origin authentication is assured by including a secret shared key in the data to be authenticated; and replay protection is provided by use of a sequence number field within the AH header. The AH authenticates additionally and secures much of the packet's IP header.

▲ **IPSec Encapsulation Security Payload (ESP):** The Encapsulating Security Payload provides data confidentiality (encryption), data integrity, data origin authentication, and protection against replay. ESP always provides data confidentiality, and can also optionally provide data origin authentication, data integrity checking, and replay protection. When comparing ESP to AH, you will find that ESP provides encryption, while either can provide authentication, integrity checking, and replay protection. When ESP is used to provide authentication functions, it uses the same algorithms used by the AH protocol but doesn't cover parts of the IP header like AH does.

Combining IPSec with L2TP

We will be devoting an entire section on IPSec and covering it in great detail. For now, we will only be discussing a solution to a shortcoming of L2TP by using IPSec as a secondary security protocol. Considering that PPTP, L2F, and L2TP are most vulnerable whenever the protocol over which it is transported is left unprotected, when you add protection to a lower layer it extends the implemented protection to all layers riding above it.

PPTP, L2F, and L2TP use Layer 4 protocols when using a TCP/IP based network like the Internet. When combined with IPSec, a Layer 3 protocol, it too inherits that layer's protection status. Why not combine IPSec with the other standards then? Mainly because PPTP and L2F are considered older and less capable than L2TP but other than the additional complexity, there is nothing to stop the combining of IPSec with any of the other PPP tunneling protocols.

We have covered enough of these new terms to be able to use them to describe the process of encapsulation using IPSec with L2TP for encryption.

L2TP Control messages using IP as the underlying transport are sent as UDP datagrams using port 1701. L2TP Control messages sent as UDP datagrams, when using IPSec, are sent as the encrypted payload of the IPSec Encapsulating Security Payload (ESP) header as illustrated in Figure 11-6.

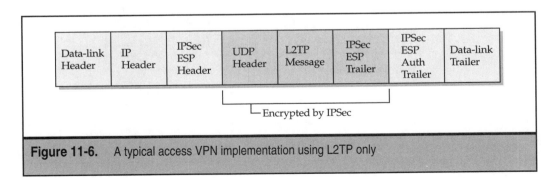

Figure 11-6. A typical access VPN implementation using L2TP only

IPSec may also be used *inside* of the L2TP frame. This implementation would be in situations where the underlying carrier protocol used for the L2TP connection is not TCP/IP like ATM, X.25, or Frame Relay. Under this scenario, the PPP payload carried by L2TP can be protected using encryption, key based authentication, or any of the services provided by IPSec. You can see an example of L2TP data tunneling using IPSec and multiple levels of encapsulation in Figure 11-7.

Now, let's go through the steps of the encapsulation process using L2TP over IPSec Tunneled Data:

1. The initial PPP payload begins by being encapsulated into a packet with a PPP header by the sending L2TP LAC/LNS.

2. The Encapsulation of the initial PPP payload and header into a packet with an L2TP header by the sending L2TP LAC/LNS.

3. The L2TP encapsulated packet itself is now encapsulated by the sending L2TP LAC/LNS into a UDP segment. This UDP segment header includes the source and destination ports set to UDP port 1701.

4. IPSec encapsulation process of the UDP segment is performed at this point by the sending L2TP LAC/LNS where it is encrypted and encapsulated with an IPSec Encapsulating Security Payload (ESP) header.

5. IPSec encapsulation of the ESP encapsulated UDP segment is performed at this point by the sending L2TP LAC/LNS where it is encrypted and encapsulated with an IPSec Authentication (AH) trailer.

6. The IPSec packet is encapsulated by the sending L2TP LAC/LNS into an IP packet. This IP packet's header contains the source and destination IP addresses of the VPN client and VPN server.

7. IP datagram is finally encapsulated by the sending L2TP LAC/LNS with a header and trailer for the Data-Link layer media access of the outgoing physical interface and sent on its way.

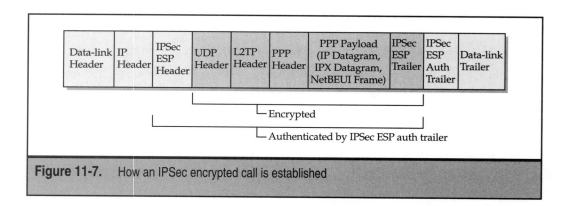

Figure 11-7. How an IPSec encrypted call is established

Now let's follow the de-encapsulation process using L2TP over IPSec Tunneled Data:

1. Upon receipt of the L2TP over IPSec tunneled data, the L2TP LAC client or L2TP LNS server processes and removes the Data-Link header and trailer. The receiving L2TP LAC/LNS then processes and removes the IP header. The receiving L2TP LAC/LNS uses the IPSec ESP Auth trailer to authenticate the IP payload and the IPSec ESP header.

2. The receiving L2TP LAC/LNS uses the IPSec ESP header to decrypt the encrypted portion of the packet.

3. The receiving L2TP LAC/LNS processes the UDP header and sends the L2TP packet to L2TP.

4. The receiving L2TP LAC/LNS uses the Tunnel ID and Call ID in the L2TP header to identify the specific L2TP tunnel containing the initial PPP payload and header.

5. The receiving L2TP LAC/LNS finally uses the PPP header to identify the PPP payload and forwards it to the proper protocol driver for the intended application process.

Now that we have seen how this procedure works at the packet level, let's look at a typical scenario of how it might be implemented in the real world. In this example, we have users at the enterprise remote office that are connected to the peer. When users need to connect to the headquarters office, the peer initiates a VPDN session that uses L2TP to tunnel the user session to the headquarters office. First, the peer establishes a PPP session with the local LAC of the ISP. The LAC retrieves L2TP tunneling information from the LAC server, and then establishes an L2TP tunnel and session with the LNS at the enterprise headquarters. The LNS forwards the user name and password to the LNS server for authentication and authorization. Once the user is authenticated and authorized, the user has an L2TP tunnel in the headquarters network and you end up with something like the situation shown in Figure 11-8.

This network is still vulnerable. This network uses L2TP to tunnel PPP traffic over the Internet. L2TP does not encrypt traffic. L2TP tunnel authentication between the LAC and the LNS occurs at tunnel origination. This method does not protect control channel or data channel traffic on a per-packet basis. PPP authenticates the client to the LNS, but it doesn't have any per-packet authentication, integrity, or replay protection either. This situation leaves both the control and data packets of the L2TP open to the following types of attacks:

▼ Discovery of user identities by sniffing the data packets.

■ Modification or duplication of the data.

■ Hijacking the L2TP tunnel or the PPP connection inside the tunnel.

▲ DoS attacks by terminating the PPP connections or L2TP tunnels.

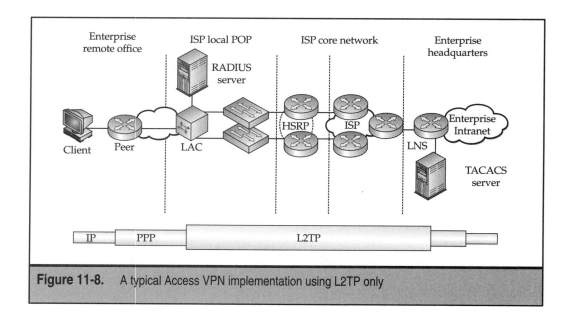

Figure 11-8. A typical Access VPN implementation using L2TP only

If they combine IPSec with L2TP, those attacks can be avoided because IPSec will provide a solution using the following features:

▼ Authentication, integrity, and replay protection for the control channel packets.

■ Integrity and replay protection for the data channel packets.

▲ Encryption of the control channel packet.

Take a look at Figure 11-9 to see how an IPSec encrypted call is established. The following list describes the sequence of events shown in Figure 11-9:

1. A user in the enterprise remote office initiates a call to the headquarters office from a PC connected to the peer. The peer places a call over the Public Switched Telephone Network (PSTN) to forward the user PPP session to the LAC.

2. When the LAC receives the PPP session, it negotiates an L2TP tunnel with the LNS at the enterprise headquarters.

3. After the LAC and LNS establish an L2TP tunnel, the LAC forwards the user information to the LNS. The LNS authenticates the user and establishes an L2TP session for the user.

4. Once the L2TP tunnel and session are established, the remote user has connectivity to the enterprise headquarters. The peer then initiates IPSec negotiation directly with the LNS (the LAC is not involved in IPSec).

Now that we have combined IPSec with L2TP in our new network, the peer, LAC, and LNS establish L2TP tunnels and sessions exactly as they did in the original network. But

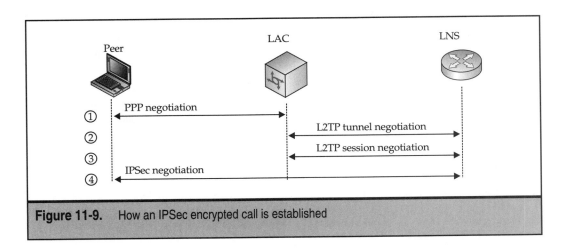

Figure 11-9. How an IPSec encrypted call is established

now, once the L2TP tunnel and session are established, the peer initiates IPSec negotiation with the LNS (the LAC is not involved in this process). Once the peer and LNS complete IPSec negotiation, the connection is secure from the peer in the remote office to the LNS at the headquarters office and you are left with something like the situation shown in Figure 11-10.

L2TP Packet Structure

This section provides a breakdown of a typical packet. Then we examine the L2TP header as well as the individual fields and their purposes.

There is an important distinction you must learn about tunnel maintenance using L2TP. Recall that PPTP requires two separate connections using two different protocols. This is what is referred to as an "out-of-band" control method. L2TP tunnel maintenance and data transmission travel the same channel. This is an "in-band" control method. L2TP packets for the control channel and data channel share the same header format. This is made possible by using Attribute Value Pairs (AVP) to identify the characteristics of the Control message. Before data is sent, Control messages are passed back and forth to set up the tunnel. These Control messages are distinguished from Data messages by setting the very first bit in the packet to 1 or on. This is called the T-bit or Message Type bit.

Also, in each case where a field is marked (optional), its space does not exist in the message if the field is marked *not present*. Note that while optional on Data messages, the Length, Ns, and Nr fields marked as optional below, are required to be present on all Control messages.

We will now define some of the key terms you will encounter when examining L2TP packets. Note that some of the terms are actually associated with the PPP protocol considering its close relation to L2TP and its place as a passenger.

▼ **Attribute Value Pair (AVP):** The variable length concatenation of a unique Attribute represented by an integer and a Value identified by the attribute.

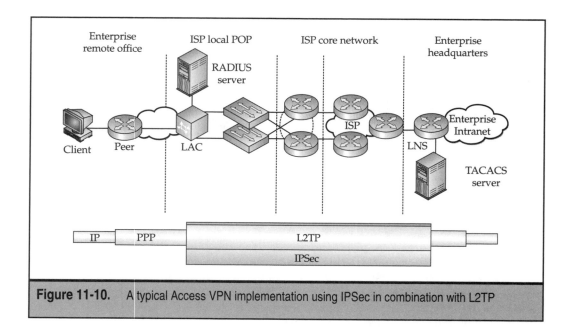

Figure 11-10. A typical Access VPN implementation using IPSec in combination with L2TP

Multiple AVPs make up Control messages that are used in the establishment, maintenance, and teardown of tunnels.

■ **Call:** A connection, or attempted connection, between a Remote System and a LAC. A Call (Incoming or Outgoing) from a PSTN that is successfully established between a Remote System and LAC will result in an L2TP Session within a previously established Tunnel between the LAC and LNS.

■ **Called Number:** An indication to the receiver of a call as to what telephone number the caller used to reach it (recall our discussion of CLID and DNIS based authentication in the Access VPN Establishment Sequence section earlier in this chapter).

■ **Calling Number:** An indication to the receiver of a call as to the telephone number of the caller (recall our discussion of CLID and DNIS based authentication in the Access VPN Establishment Sequence section earlier in this chapter).

■ **Control Connection:** A control connection operates "in-band" over a tunnel to control the establishment, release, and maintenance of sessions and of the tunnel itself. Control messages are exchanged between LAC and LNS pairs, operating "in-band" within the tunnel protocol. Control messages govern aspects of the tunnel and sessions within the tunnel.

- **Digital Channel:** A circuit-switched communication path intended to carry digital information in each direction.

- **Incoming Call:** A Call received at an LAC that will be tunneled to an LNS.

- **Link Control Protocol (LCP):** This is a PPP protocol used to establish communications over a PPP link. LCP packets are sent by PPP peers on each end of a link to configure and test the integrity of Data Link connection. Once a link has been established, both peers can be verified using an authentication method. It is at this point that the Network Control Protocol (NCP) is used to establish and configure one or more Network Layer protocols that will be used over the link. When the connection is completed successfully, the datagrams from the Network Layer protocols can then be sent over the link connection. The link will remain open for communication while either protocol (LCP or NCP) sends a frame to close the link.

- **Network Control Protocol (NCP):** Network Control Protocol (NCP) is a PPP protocol used to establish and configure one or more Network Layer protocols that will be used over the link. Before this happens, the LCP protocol must be used to establish communications over a PPP link. LCP packets are sent by PPP peers on each end of a link to configure and test the integrity of Data Link connection. Once a link has been established, both peers can be verified using an authentication method.

- ▲ **Zero-Length Body (ZLB) Message:** A control packet with only an L2TP header. ZLB messages are used for explicitly acknowledging packets on the reliable control channel.

Now that you have an understanding of these terms, let's go ahead and take a look at the L2TP packets in both forms and see how this protocol operates.

Basic Packet Format (Control Channel and Data Channel)

MAC header	IP header	UDP header	L2TP header	Data

L2TP header (Control Channel and Data Channel)

00	01	02	03	04	05	06	07	08	09	10	11	12	13	14	15	16	17	18	19	20	21	22	23	24	25	26	27	28	29	30	31
T	L	0		S	0	0	P				0					Version				Length (Optional)											
Tunnel ID																Session ID															
Ns (Optional)																Nr (Optional)															
Offset Size (Optional)																Offset pad (Optional) variable...															
Data (Optional) variable...																															

▼ **T, Message Type bit:** 1 bit. This bit indicates whether the message is a date or Control message.

T   **Description**

0   Data Message

1   Conrol Message

■ **L, Length present bit:** 1 bit. Control messages must have this bit set.

■ **S, Sequence present bit:** 1 bit. If set, the **Ns** and **Nr** fields are present. Control messages must have this bit set.

■ **O, Offset present bit:** 1 bit. If set, the **Offset Size** field is present. Control messages must have this bit set.

■ **P, Priority bit:** 1 bit. If set, this Data message should receive preferential treatment during the queue and transmission process. Also, anything like PPP's Link Control Protocol (LCP) echo requests being used as a keepalive for the link should normally have this bit set. Without it, a temporary interval of local congestion could result in interference with keepalive messages and unnecessary loss of the link. This feature is only for use with Data messages. Control messages must have this bit set.

■ **Version Field:** 4 bits. This field indicates the L2TP protocol version and must be set to 2. The value 1 is reserved to allow the detection of L2F packets should they arrive intermixed with L2TP packets. Obviously, L2F will always use version 1. Packets received with an unknown value must be discarded.

■ **Length Field (Optional):** 16 bits. This field is the total length of the message in bytes. This field exists only if the **L** bit is set.

■ **Tunnel ID Field:** 16 bits. This field indicates the identifier for the control connection. L2TP Tunnel IDs have local significance only. This just means that the very same tunnel will be given different Tunnel IDs by the LAC/LNS at each end of the tunnel. This is similar in concept to the DLCIs in Frame Relay that you learned about in Chapter 2. The Tunnel ID in each message is that of the intended *recipient*, not the *sender*. Tunnel IDs are selected and exchanged using Assigned Tunnel ID AVPs during the creation of a tunnel.

■ **Session ID Field:** 16 bits. This field indicates the Session ID for the tunnel. L2TP Session IDs are also locally significant. Just as in the case of the Tunnel ID, each end of the session will be given different Session IDs. The Session ID in each message is that of the intended *recipient*, not the *sender*. Session IDs are selected and exchanged as Assigned Session ID AVPs during the creation of a session.

■ **Ns, sequence number Field (Optional):** 16 bits. This optional field indicates the sequence number for this data or Control message, beginning at zero and incrementing by one for each message sent.

■ **Nr, sequence number expected Field (Optional):** 16 bits. This optional field indicates the sequence number expected in the next Control message to be received. Thus, **Nr** is set to the **Ns** of the last in-order message received plus one. In Data messages, **Nr** is reserved and, if present (as indicated by the **S bit**), must be ignored upon receipt.

■ **Offset Size Field (Optional):** 16 bits. This optional field specifies the number of bytes past the L2TP header at which the payload data is expected to start. Actual data within the offset padding is undefined. If the offset field is present, the L2TP header ends after the last byte of the offset padding. This field exists if the **O** bit is set.

■ **Offset Pad Field (Optional):** Variable length. (Optional).

▲ **Data Field (Optional):** Variable length. (Optional).

AVP Format

00	01	02	03	04	05	06	07	08	09	10	11	12	13	14	15	16	17	18	19	20	21	22	23	24	25	26	27	28	29	30	31
M	H	Reserved				Length										Vendor ID															
Attribute Type																Attribute Value (Variable...defined In Length Field)															
Attribute Value continues until value in the Length Field is reached (maximum of 1023 octets)																															

The AVP Format is used for Control message data. The first six bits are a bit mask that describe the general attributes of the AVP. Only the first two bits are used as the rest are reserved for future extensions. **Reserved** bits must be set to 0. An AVP received with a **Reserved** bit set to 1 must be treated as an unrecognized AVP. The bit descriptions are as follows:

▼ **Mandatory M, bit:** Controls the behavior required of an implementation which receives an AVP which it does not recognize. If the **M bit** is set on an unrecognized AVP within a message associated with a particular session, the session associated with this message must be terminated. If the **M bit** is set on an unrecognized AVP within a message associated with the overall tunnel, the entire tunnel (and all sessions within) must be terminated. If the **M bit** is not set, an unrecognized AVP must be ignored. The Control message must then continue to be processed as if the AVP had not been present.

■ **Hidden H, bit:** Identifies the hiding of data in the Attribute Value field of an AVP. This capability can be used to avoid the passing of sensitive data, such as user passwords, as cleartext in an AVP.

- **Length Field:** Encodes the number of octets (including the Overall Length and bit mask fields) contained in this AVP. The Length is calculated as 6 plus the Length of the Attribute Value field in octets. The field itself is 10 bits, permitting a maximum of 1023 octets of data in a single AVP. The minimum Length of an AVP is 6. If the Length is 6, then the Attribute Value field is absent.

- **Vendor ID Field:** A value dictated by the Internet assigned number authority (IANA) assigned "SMI Network Management Private Enterprise Codes" covered RFC 1700 value. The value 0, corresponding to IETF adopted attribute values, is used for all AVPs defined within this document. Any vendor wishing to implement their own L2TP extensions can use their own Vendor ID along with private Attribute values, guaranteeing that they will not collide with any other vendor's extensions or with future IETF extensions. There are 16 bits allocated for the Vendor ID, limiting this feature to the first 65,535 enterprises.

- **Attribute Type Field:** A 2 octet value with a unique interpretation across all AVPs defined under a given Vendor ID.

▲ **Attribute Value Field:** This is the actual value as indicated by the Vendor ID and Attribute Type. It follows immediately after the Attribute Type field, and runs for the remaining octets indicated in the Length Field (Length minus 6 octets of header). This field is absent if the Length is 6.

L2TP Connection Control

By now, you should have a pretty good understanding of L2TP at the packet level. We have broken down the types of messages used by the protocol and how those messages are assembled. Now let's talk about the connection controls defined using the AVPs we have just learned about.

Before we talk about the specific messages used, let's examine the title of this section. "Connection Control" ... that sounds like it is a connection-oriented protocol. But how is that done? L2TP uses UDP, a connectionless Layer 4 protocol. L2TP has a lower level reliable transport method for its Control messages. The upper level functions of L2TP don't care about the order or retransmission of the Control messages.

Since a TCP connection is not used, L2TP use message sequencing to ensure reliable delivery of L2TP messages. Within the L2TP Control message, the Next-Received (**Nr**) field and the Next-Sent (**Ns**) field to maintain the sequence of Control messages. This is similar to the TCP Acknowledgement field and TCP Sequence Number fields, respectively. The Next-Sent (**Ns**) and Next-Received (**Nr**) fields can also be used for sequenced delivery and flow control for tunneled data using a sliding window mechanism (also like TCP). The receiving peer is responsible for making sure that Control messages are delivered in order and without duplication to the upper level. Messages arriving out of order may be stored and delivered when the missing messages are received, or they can discard the message and ask for a retransmission by the sending peer.

You may have noticed that there is an option to use sequence numbers in Data messages, too. However, unlike the control channel, the data channel does not use sequence numbers to *retransmit* lost Data messages. It can use sequence numbers to *detect* lost packets and restore the original sequence of packets that arrive out of order.

L2TP supports multiple calls for each tunnel. In the L2TP, Control message and the L2TP header for tunneled data is a Tunnel ID that identifies the tunnel and a Call ID that identifies a call within the tunnel. The purpose of the L2TP Control messages are very similar to the ones defined in PPTP. However, unlike PPTP, each of these messages may include additional information in the format of Attribute Value Pairs (AVPs). AVPs are very useful because they maximize extensibility of the L2TP protocol and provide a mechanism which lets any company using these attributes ensure that their implementation will work with other vendors' implementations. AVPs provide a uniform method for encoding message types and bodies used throughout L2TP. The following control connection messages are used to establish, clear, and maintain L2TP tunnels.

Control Message	Message Code	Message Definition
Start-Control-Connection-Request	1 (SCCRQ)	Used to initialize a tunnel between an LNS and an LAC.
Start-Control-Connection-Reply	2 (SCCRP)	Sent in reply to a received SCCRQ message. SCCRP is used to indicate that the SCCRQ was accepted and establishment of the tunnel should continue.
Start-Control-Connection-Connected	3 (SCCCN)	Sent in reply to an SCCRP. SCCCN completes the tunnel establishment process.
Stop-Control-Connection-Notification	4 (StopCCN)	Sent by either the LAC or the LNS to inform its peer that the tunnel is being shutdown and that the control connection should be closed. In addition, all active sessions are implicitly cleared.
Reserved	5 (reserved)	
Hello	6 (HELLO)	Sent by either peer of a LAC-LNS control connection. This control is used as a "keepalive" for the tunnel.
Outgoing-Call-Request	7 (OCRQ)	Sent by the LNS to the LAC to indicate that an outbound call from the LAC is to be established. It is the first message sent in a three-way handshake used for establishing a session within an L2TP tunnel.

Control Message	Message Code	Message Definition
Outgoing-Call-Reply	8 (OCRP)	Sent by the LAC to the LNS in response to a received OCRQ message. This is the second message sent in a three-way handshake used for establishing a session within an L2TP tunnel.
Outgoing-Call-Connected	9 (OCCN)	Sent by the LAC to the LNS following the OCRP and after the outgoing call has been completed. It is the final message in a three-way handshake used for establishing a session within an L2TP tunnel.
Incoming-Call-Request	10 (ICRQ)	Sent by the LAC to the LNS when an incoming call is detected. It is the first message in a three-way handshake used for establishing a session within an L2TP tunnel.
Incoming-Call-Reply	11 (ICRP)	Sent by the LNS to the LAC in response to a received ICRQ message. It is the second in the three-way handshake used when establishing sessions within an L2TP tunnel.
Incoming-Call-Connected	12 (ICCN)	Sent by the LAC to the LNS in response to a received ICRP message. It is the third and final message of the three-way handshake used for establishing sessions within an L2TP tunnel.
Reserved	13 (reserved)	Reserved
Call-Disconnect-Notify	14 (CDN)	Sent by either the LAC or the LNS to request disconnection of a specific call within the tunnel. Its purpose is to inform the peer of the disconnection and the reason why the disconnection occurred.
WAN-Error-Notify	15 (WEN)	Sent by the LAC to the LNS to indicate WAN error conditions (conditions that occur on the interface supporting PPP).
Set-Link-Info	16 (SLI)	Sent by the LNS to the LAC to set PPP-negotiated options. Because these options can change at any time during the life of the call, the LAC must be able to update its internal call information dynamically and perform PPP negotiation on an active session.

Control Connection States

Okay, so we have discussed what the actual Control messages used L2TP are. Now, let's talk about the way these Control messages are sent to and from a LAC and LNS.

L2TP Control messages are exchanged using a set of *state tables*. There are tables used for the placement of incoming calls and outgoing calls. There are also tables used during the initiation of the tunnel itself. In this section, we are going to cover all of the L2TP Control Connection messages that are used to support and exchange all of the Control messages we learned about in the previous section.

Whenever an invalid or damaged Control message is received, it should be logged and a Stop-Control-Connection-Notification (StopCCN) sent to have the control connection cleared to recover and get back in to a known state. Then, the control connection can be started again by the initiator of the call. An invalid Control message is one that has a Message Type that is set to mandatory and is not recognized by the machine or Control messages that arrive in the wrong order.

Connection Control messages are not determined by whether or not they are from the LNS or LAC. They are determined by which device was the originator and which one was the recipient. The originating peer is the one that first initiates establishment of the tunnel. Since either an LAC or LNS can be the originator of a connection, a collision can occur creating a "tie breaker" situation that is settled using Tie Breaker AVPs. When this happens, it is the originator that is the winner of the tie.

Connection Establishment or Tunnel States

In this situation, either side can originate a tunnel and have the following states in common.

States used by both LAC and LNS   The states associated with the LNS or LAC for control connection establishment are

Idle	Both originator and recipient start from this state. The originator transmits a Start-Control-Connection-Request (SCCRQ), while the recipient remains in the idle state until receiving an SCCRQ.
Wait-ctl-reply	The originator checks to see if another connection has been requested from the same peer, and if so, handles the collision using Tie Breaker AVPs. When an Start-Control-Connection-Reply (SCCRP) Control message is received, it is examined for version compatibility. If the version of the reply is lower than the version sent in the request, then the older (lower) version should be used, provided it is supported and moved to the established state. If the version is not supported, a StopCCN Control message must be sent to the peer and the originator cleans up and terminates the tunnel.

wait-ctl-conn	This is where an Start-Control-Connection (SCCCN) Control message is awaited. Once received, the challenge response is checked. If successful, the tunnel is established. It is taken down if an authorization failure is detected.
Established	An established connection may be terminated by either a local condition or the receipt of a Stop-Control-Connection-Notification (StopCCN) Control message. In the event of a local termination, the originator must send a StopCCN Control message and clean up the tunnel.

Call Control or Session States

In this situation, sessions are governed by the LNS. The LNS can be asked by the LAC to set up a call session but not told to. The LNS, on the other hand, can instruct the LAC to make a call for the session to be established. The following explains the differences in states you will be reading about next.

Incoming calls   Incoming calls are initiated by an LAC and sent to the LNS. Three Control messages are used for incoming calls:

▼   Incoming-Call-Request (ICRQ)

■   Incoming-Call-Reply (ICRP)

▲   Incoming-Call-Connected (ICCN)

An Incoming-Call-Request (ICRQ) Control message is generated by the LAC when an incoming call is detected from the phone network by a remote user's computer. At this point, it assigns a Session ID and serial number and sends it to the LNS and waits for a response. This is the first step of a three-way handshake process. When the LNS receives the ICRQ, it responds with an Incoming-Call-Reply (ICRP) Control message. When the LAC receives the ICRP, it sends a Incoming-Call-Connected (ICCN) Control message to the LNS. This indicates that the call states for both the LAC and the LNS should enter the established state. Now the three-way handshake process is complete. When the dialed-in remote user hangs up, the call is cleared normally and the LAC sends a Call-Disconnect-Notify (CDN) Control message. If the LNS wishes to clear a call, it sends a CDN Control message and cleans up its session.

LAC Incoming Call States   The states associated with the LAC for incoming calls are

Idle	The LAC detects an incoming call on one of its interfaces. Typically, this means an analog line is ringing or an ISDN TE has detected an incoming Q.931 SETUP message. The LAC initiates its tunnel establishment state machine, and moves to a state waiting for confirmation of the existence of a tunnel.

wait-tunnel	In this state, the session is waiting for either the control connection to be opened or for verification that the tunnel is already open. Once an indication that the tunnel is opened, session Control messages may be exchanged with the first being an Incoming-Call-Request (ICRQ).
wait-reply	The LAC receives either a Call-Disconnect-Notify (CDN) Control message from the LNS, indicating it will not accept the call and moves back into the idle state, or an Incoming-Call-Reply (ICRP) Control message indicating the call is accepted. Then, the LAC sends an Incoming-Call-Connected (ICCN) Control message and enters the established state.
Established	Data is exchanged over the tunnel.

LNS Incoming Call States

Idle	An Incoming-Call-Request (ICRQ) Control message is received. If the request is not acceptable, a Call-Disconnect-Notify (CDN) Control message is sent back to the LAC and the LNS remains in the idle state. If the ICRQ Control message is acceptable, an Incoming-Call-Reply (ICRP) Control message is sent to the LAC. The session moves to the wait-connect state.
wait-connect	If the session is still connected on the LAC, the LAC sends an Incoming-Call-Connected (ICCN) Control message to the LNS, which then goes into established state. The LAC may send a Call-Disconnect-Notify (CDN) Control message to indicate that the incoming caller could not be connected, like when someone accidentally places a standard voice call to an LAC.
Established	The session is terminated when a Call-Disconnect-Notify (CDN) Control message is received from the LAC or by the LNS sending its own CDN.

Outgoing Calls

Outgoing calls are initiated by the LNS and instruct an LAC to place a call. Three Control messages are used for outgoing calls:

▼ Outgoing-Call-Request (OCRQ)

■ Outgoing-Call-Reply (OCRP)

▲ Outgoing-Call-Connected (OCCN)

The LNS sends an OCRQ specifying the phone number it wishes the LAC to call. Then, once the LAC has determined it can place the call, the LAC must respond to the OCRQ Control message with an OCRP Control message. Once the outbound call is connected, the LAC sends an OCCN Control message to the LNS indicating the final result of the call attempt.

LAC Outgoing Call States The states associated with the LAC for outgoing calls are

Idle	If Outgoing-Call-Request (OCRQ) is received from the LNS in error, respond with a Call-Disconnect-Notify (CDN) Control message. Otherwise, allocate a physical channel for the call to be placed on and send an Outgoing-Call-Reply (OCRP) Control message to the LNS. Place the outbound call and move to the wait-cs-answer state.
wait-cs-answer	If the call is not completed or a timer expires waiting for the call to complete, send a Call-Disconnect-Notify (CDN) Control message with the appropriate error conditions set and go to idle state. If a connection is established, send an Outgoing-Call-Connected (OCCN) Control message indicating success and go to established state.
established	If a Call-Disconnect-Notify (CDN) Control message is received by the LAC, the call placed must be released and the session cleaned up. If the call is disconnected by the remote user's computer, a CDN Control message must be sent to the LNS. The sender of the CDN Control message returns to the idle state after sending the message is complete.

LNS Outgoing Call States The states associated with the LNS for outgoing calls are

idle, wait-tunnel	When an outgoing call is initiated, a tunnel is first created, much as the idle and wait-tunnel states for an LAC incoming call. Once a tunnel is established, an Outgoing-Call-Request message is sent to the LAC and the session moves into the wait-reply state.
wait-reply	If a Call-Disconnect-Notify (CDN) is received, an error occurred, and the session is cleaned up and returns to idle. If an Outgoing-Call-Reply (OCRP) is received, the call is in progress and the session moves to the wait-connect state.
wait-connect	If a Call-Disconnect-Notify (CDN) is received, the call failed and the session is cleaned up and returns to idle. If an Outgoing-Call-Connected (OCCN) is received, the call has succeeded and the session may now exchange data.

| Established | If a Call-Disconnect-Notify (CDN) Control message is received, the call has been terminated for the reason indicated in the Result and Cause Codes; the session moves back to the idle state. If the LNS chooses to terminate the session, it sends a CDN Control message to the LAC and then cleans up and idles its session. |

Tunnel Disconnection

The disconnection of a tunnel can be initiated by an LAC or LNS by sending Stop-Control-Connection-Notification (SCCCN) Control message. The sender waits a finite period of time for the acknowledgement of this message before releasing the control information associated with the tunnel. The receiver sends an acknowledgement and then releases the associated control information.

Tunnel Creation and Authentication

Okay, now it is time to discuss how L2TP uses all of those Control messages and States. Tunnel Creation and Authentication begins with a series of communication processes between the LAC and LNS. The entire L2TP packet uses the well-known UDP port 1701 version 2 (remember UDP port 1701 version 1 is used by L2F) and is encapsulated in a UDP datagram. Take a look at Figure 11-11 to see how the tunnel creation and authentication process happens (the I and O in the drawings represent Incoming and Outgoing messages).

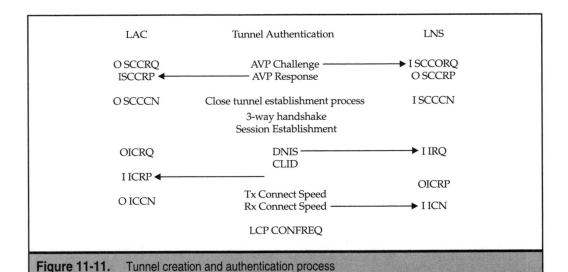

Figure 11-11.   Tunnel creation and authentication process

Tunnel Creation and Authentication Control Message Exchange and States Using Control Connection Messages

To be able to see the Control messages used and the States maintained by the Control Connection messages, we need to look at the process from the perspective of the LAC and the LNS. When you are looking at a sequence of events taking place at either side of a connection between an LAC and an LNS, always remember that Tunnel IDs and Session IDs are locally significant!

Let's start by taking a look at the tunnel creation and tunnel authentication process from the perspective of the LAC:

▼ LAC: idle
■ LAC: SCCRQ sent to LNS
■ LAC: Tunnel state change from idle to wait-ctl-reply
■ LAC: wait-ctl-reply
■ LAC: SCCRP received from LNS
■ LAC: Got a challenge from remote peer, LNS
■ LAC: Got a response from remote peer, LNS
■ LAC: Tunnel Authentication success
■ LAC: Tunnel state change from wait-ctl-reply to established
■ LAC: established
■ LAC: SCCCN sent to LNS Tunnel ID 5
▲ LAC: established

Here is the initial tunnel creation from tunnel authentication process from the perspective of the LNS:

▼ LNS: idle
■ LNS: SCCRQ received from LAC
■ LNS: New tunnel created for remote LAC
■ LNS: Got a challenge in SCCRQ, LAC
■ LNS: SCCRP sent to LAC Tunnel ID 1
■ LNS: Tunnel state change from idle to wait-ctl-reply
■ LNS: wait-ctl-reply
■ LNS: SCCCN received from LAC over Tunnel 1
■ LNS: Got a Challenge Response in SCCCN from LAC
▲ LNS: Tunnel Authentication success, wait-ctl-reply

Three-way Session Establishment

Refer to Figure 11-11 again. The LAC and LNS negotiate session creation. This is the start of the three-way handshake beginning with the Incoming-Call-Request (ICRQ) Control message from the LAC. The session is created and an Incoming-Call-Reply (ICRP) Control message is sent back to the LAC. The Incoming-Call-Connect (ICCN) Control message is the last step in the three-way handshake process, and the session is established. After the session is established, the PPP session can begin using the PPP protocol's LCP CONFREQ. Let's take a look at this process from the perspective of the LNS:

▼ LNS: Tunnel state change from wait-ctl-reply to established

■ LNS: established

■ LNS: ICRQ received from LAC over Tunnel 1

■ LNS: Session enabled, idle

■ LNS: Session state change from idle to wait-for-tunnel

■ LNS: wait-for-tunnel, new session created

■ LNS: ICRP sent to LAC over Tunnel 1

■ LNS: Session state change from wait-for-tunnel to wait-connect

■ LNS: ICCN received from LAC over Tunnel 1

■ LNS: Session state change from wait-connect to established

▲ LNS: established

Forwarding PPP Frames

Now our tunnel has been established. At this point, all incoming PPP frames from the remote user's computer are received at the LAC, are encapsulated using L2TP, and are sent over the appropriate tunnel. The LNS receives the L2TP packet, and processes the encapsulated PPP frame just as though it had been received on a local PPP interface.

The sender of a message associated with a particular session and tunnel uses the Session ID and Tunnel ID values that were specified by its peer and places them inside the Session ID and Tunnel ID header for all outgoing messages. This way, all of its PPP frames can be multiplexed and de-multiplexed over a single tunnel between the LNS-LAC pair. L2TP does not limit you to just one tunnel. Multiple tunnels can exist between an LNS-LAC pair, just as multiple sessions can exist within a tunnel.

It is important to note that the value of 0 for Session ID and Tunnel ID is *special* and must *never* be used as an Assigned Session ID or Assigned Tunnel ID. When a situation exists where a Session ID has not yet been assigned by the peer, like during establishment of a new session or tunnel, the Session ID field needs to be sent as 0, and the Assigned Session ID AVP within the message must be used to identify the session. The same holds true for cases where the Tunnel ID has not yet been assigned from the peer; the Tunnel ID must be sent as 0 and Assigned Tunnel ID AVP used to identify the tunnel.

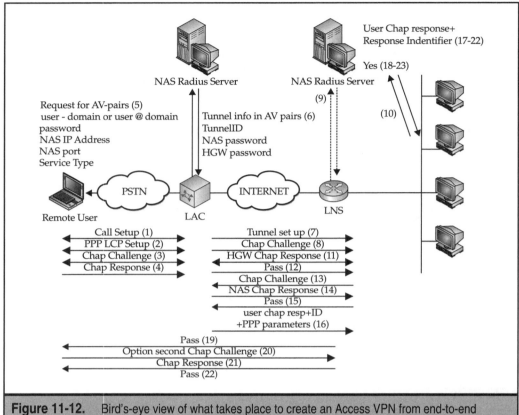

Figure 11-12. Bird's-eye view of what takes place to create an Access VPN from end-to-end

High-Level View of the L2TP Access VPN

We have covered quite a bit of material in this chapter and you have been given a lot of information to digest. So to finish off, let's take a look at Figure 11-12 to get a bird's-eye view of everything that takes place to create an L2TP Access VPN from end-to-end. Each step of the process can be seen using the numbers in parentheses next to the descriptions.

SUMMARY

In this chapter, we discussed L2F and the newer L2TP protocols. These, along with PPTP and GRE, are the protocols used most commonly in Access VPNs. In the next section, "Secure Com," we will discuss cryptag, the fundamental foundation of VPN.

PART IV

Secure Communication

CHAPTER 12

Cryptography

Modern data communication (and voice communication, for that matter) makes cryptography increasingly important in everyday life, especially with regard to the public Internet infrastructure. If you plan to use the Internet, you'll need to use cryptography for encoding personal or secret material.

The structure of the Internet allows almost anyone with a little know-how to read e-mails on their passage across some server. A simple example would be the payment for something you bought on the Net: You have to mail your credit-card details to the respective shop. Surely you wouldn't want anyone else to be able to read this information. It is true that millions of e-mails pass through the Internet each day and nobody could ever hope to find credit-card information by actually reading others' e-mails. But 16-digit credit-card numbers accompanied by a 4-digit expiration date can easily be spotted by any simple robot looking for such information.

The underlying technology that makes VPNs a viable solution in today's public networks, including but by no means limited to the Internet, is cryptography. Cryptography is not only the single most important factor in secure communications of any kind, but is also very interesting once you get beneath the surface to see what is taking place. Thus, in this chapter we will explore the history, methodologies, and concepts of cryptography. We will also discuss the many different subcategories and their histories, methodologies, and concepts.

CRYPTOGRAPHY

Here are some basic definitions that are used in this whole concept. *Cryptography* is the art or science of keeping messages secret. This "art or science" allows two entities—they can be machines or humans—to communicate in a secure fashion. A number of terms are used when talking about cryptography and secure communications. The best way to learn these terms is by means of a simplified example. Suppose someone wants to send a message to a receiver and wants to be sure nobody else can read the message. However, there is always the possibility that someone will find the message or intercept it during its transfer from one location to another.

Using cryptographic terminology, the message is referred to as *plaintext* or *cleartext*. Encoding (altering) the contents of the message in such a way that hides its contents from outsiders is called *encryption*. The steps taken to alter or encrypt a message are known as an *algorithm*. The encrypted result of that message is the *ciphertext*. The process of retrieving the plaintext from the ciphertext is called *decryption*. Encryption and decryption generally require the use of a *key* of some sort. To be able to encrypt and/or decrypt the message requires knowledge of what that key is. *Cryptanalysis* is the art and science of figuring out what the plaintext is without knowing the proper key. People who invent ways of encrypting messages are *cryptographers*, and finally, those who make it their business to undo the fruits of such labor are called *cryptanalysts*. *Cryptology* is the branch of mathematics that studies the mathematical foundations of cryptographic methods.

Some other definitions we should cover are the commonly used names of people in describing the mechanics of secure communications. They are invariably:

▼ **Alice (A)**   This is the person who generally initiates the communication.

■ **Bob (B)**   This is the person on the receiving end of such an initiation.

▲ **Mullet (M)**   This is the malicious man-in-the-middle intent on discovering what the message contents of that communication are. Why "Mullet"? Your guess is as good as mine. Maybe because nobody ever names their children Mullet and don't want them associated with malicious busybodies. That's not to say it's not a nice name! (Honest, Mullet, it is…)

Cryptography deals with all aspects of secure communication. That includes messaging, authentication, digital signatures, data integrity, and the like. Now, let's define what secure communication is. Let's say that Alice wants to communicate "securely" with Bob. That creates some dilemmas. Obviously, Alice wants only Bob to be able to understand the message that she has sent, even though they are communicating over an "insecure" medium where Mullet, the malicious man-in-the-middle, may intercept, read, and perform any other unspeakable acts upon its contents as it travels from Alice to Bob. Bob also wants to be sure that the message that he receives from Alice was indeed sent by Alice, and Alice wants to make sure that the person with whom she is communicating is indeed Bob. Alice and Bob also want to make sure that the contents of Alice's message have not been altered in transit. Given these needs, we can identify the following properties required in a secure communication. (This is where many people get confused when they hear about cryptographic algorithms.)

▼ **Secrecy**   Only the sender and intended receiver should be able to understand the contents of the transmitted message. Because Mullet can intercept the message, this requires that the message be encrypted so that, even if Mullet were to intercept the message, he wouldn't be able to understand it. This can be accomplished by encrypting the contents of the message. It is important to note that secrecy of contents is only one of the ingredients of secure communication. For example, Alice might also want the mere fact that she is communicating with Bob kept secret. This can be accomplished using cryptographic encryption algorithms.

■ **Authentication**   Both the sender and receiver need to be certain of the identity of the other party involved in the communication—a way to be sure that the other party is indeed who or what they claim to be. Once the message is sent over a medium where they cannot see the message as it travels, see the other party, or see who else can see the message or the other party, it becomes very important that some method of authentication must be used. If someone were to call you on the telephone claiming to be from your bank and asking for your account number and ATM card PIN (personal ID number), would you give that information out over the phone? (If you said yes to that, could you e-mail me…oh, and could you please include your phone number?) You get the idea—it's important to know with whom we are dealing, and authentication provides a way of doing that. This can be done using a combination of cryptographic encryption and hashing algorithms.

▲ **Message integrity**   You may have found a way to ensure that your message can't be understood and may have even discovered a great way of authenticating the identity of each other. That doesn't mean that you can be certain the contents haven't been tampered with. You also need a way of ensuring that the contents of that communication are not altered, maliciously or by accident, during its transmission. This can be accomplished by using cryptographic hashing algorithms.

So now we know what is necessary to communicate securely, and we know that cryptography is the key to that. We will be going over all of these concepts and their cryptographic algorithms in great detail in the next few chapters. Before we do, let's talk about some of the other important subtleties of cryptography.

History of Cryptography

We know what cryptography is, but how did it come to be? To understand how it came to be, we need to go back to about 100 B.C., when a certain Roman emperor named Julius Caesar began to distrust the messengers who carried sensitive items to those he actually trusted. This first recorded *cipher* was an algorithm that has come to be known as Caesar's Cipher. This cipher would take the message and substitute characters in the Roman alphabet with other characters of the same alphabet. History clearly shows, however, that the messengers would end up being the least of Caesar's problems as far as trust was concerned.

Have you ever heard of the "Enigma Machine"? In World War II, the allies became aware of Enigma, and it is a good thing they did; otherwise, this could very well have been written in German. It was because of this device that most governments now classify cryptographic methods as weapons in the most literal sense.

The algorithm used by Enigma was broken, in 1941, by what many people regard as the first computer, the Colossus. The power of using automatic machines for cryptography was revealed with the breaking of Enigma. The incredible growth rate in use and power of computers became the impetus for frequent and enormous advances in cryptography during the 1970s. In 1977, the U.S. National Bureau of Standards implemented a cipher using a symmetric encryption technique called the *Data Encryption Standard* (DES). In 1980, the American National Standards Institute adopted the same DES cryptographic algorithm for commercial use.

This development gave rise to the implementation and advancement of other, stronger ciphers using asymmetric cryptographic methods. The strength of these new asymmetric ciphers and their hybrids was seen as a threat to national security. This has led to some very complex laws concerning public use here and abroad. This is the reason you may have heard the terms "128-bit domestic versions" and "40-bit export versions" used with certain types of software. We will talk about symmetric and asymmetric algorithms in the section "Cryptographic Algorithm Methods" later in this chapter.

Ciphers

A number of different ciphers have been devised since the beginning of cryptography as we know it. Let's take a look at two of the types that were commonly used.

Substitution Ciphers

The first recorded *cipher*, Caesar's Cipher, is an example of a *substitution cipher*, also known as a *monoalphabetic cipher*. Caesar's Cipher was an algorithm that replaced each character in the Roman alphabet with a character three positions ahead of it. This algorithm is still used commonly even to this day. Perhaps you may have run across an e-mail or newsgroup message that has a bunch of random characters with "<rot13>" at the top of this message. The "rot" stands for "rotate," and the "13" indicates by how many. The "13" in this case is lucky because you don't need to specify which direction, considering the number of letters in the English alphabet (stop counting, it's 26).

Transposition Ciphers

In 1586, the Vigenere cipher was born. This is an example of a *transposition cipher*, also known as a *polyalphabetic cipher*. Transposition ciphers rearrange the position of characters in the plaintext. This avoids the problem of frequency analysis, but can provide clues to a cryptanalyst. Instead of substituting each letter, you rearrange them. Let's say you have some graph paper that has ten rows and ten columns. You could place the letters of the message in each box left to right on one paper and on another paper transpose them in a top to bottom order, backwards, diagonally, or any way you choose. A cipher is something that protects a message by modifying or altering the actual components of that message. The components in these cases would be the letters, themselves. This gives you the ability to encrypt virtually any message regardless of content.

Combination Ciphers

A *combination cipher* is any cipher that is the product of two or more cryptographic algorithms. The Germans used this in the 1920s when they created their infamous Enigma machine. The Enigma employed components of both transposition and substitution ciphers. These ciphers were automated in rotor machines that used a modified version of the Vigenere transposition cipher and involved the substitution of a given combination of these rotor wheels.

Codes

Now let's talk about a cousin of the cipher, the *code*. On the surface they might seem the same, but the difference lies in the method. While ciphers alter the contents of a message to encrypt it, a code alters the meaning. I'm sure we have all heard of "codebooks" used in clandestine operations throughout history. For each party to discover the original contents of a coded message, each would need a *codebook*, as opposed to a key.

Let's say I want to send my brother in England a message that only we would be able to understand. I could tell him to rotate the letters by 13 (a cipher), but that is easy to break, considering the consistency and repetitive nature of written language. If you have ever watched the endgame in *Wheel of Fortune*, you may have noticed that contestants are asked to pick five letters to help them solve the puzzle. They will (if they're not idiots, and I am aware that there is no shortage of those on that show) always say, "OK, Pat, I'll take R, S, T, L, N." That is because these are the most common consonants in the English language. The letter E is by far the most common letter of all. So a pattern could quickly be observed and eventually figured out with some ease.

A codebook is a different story. You might say, "Oh swell, all I have to do is make a codebook and send it to my brother." Not so—we could speak on the phone, decide to buy a 1985 third edition Webster's paperback dictionary, and for every *word* we use, substitute the word located 13 definitions in front of it. Now you would need to know the method *and* have the codebook. The ability to find a pattern here is infinitely harder than that of a 13-letter rotation cipher. Patterns are a cryptographer's nightmare and a cryptanalyst's dream.

Randomness

This brings us to the subject of *randomness*. The most effective weapon against patterns is randomness. Having to perform the task of encrypting and decrypting messages manually would certainly become quite tedious and unproductive. That is why machines are perfect for these repetitive and complex tasks. One of the earliest, and easily the most famous example of implementing this idea, was the three-rotor Enigma machine.

No machine on Earth is more capable of doing repetitive and complex tasks than computers. However, even though it is a fast and powerful tool for implementing all of the cryptographic algorithms being used today, use of computers poses an interesting problem. When executing an algorithm to generate a cryptographic key, the single most import element in this procedure is *randomness*. The key that is produced by any cryptographic protocol depends completely on the unpredictability of this key. If you can predict a key's value, or even narrow down the number of keys that need to be tried, that key can be broken with much less effort than if a completely random number, known as a *seed value*, had been used in the algorithm to create that key. Consequently, it is of the utmost importance that this key be generated from the seed value of a truly unpredictable random number source.

The problem that exists is that Computers by their very nature are extremely predictable and deterministic. It wouldn't do us much good to have a calculator that managed to produce a completely random number every time we entered the equation $2 + 2$. We depend on the fact that a computer will operate exactly the same way every single time it runs any given algorithm. So, by definition, it is impossible for a computer to generate a *truly* random number of any kind. The only source of true randomness exists in nature. For example, if we were to take some graph paper that was 20 feet square, place it on the ground while it was raining, and use the coordinates of each raindrop that fell at 30-second intervals, *that* would be truly random. Truly random and truly a major pain in the…oh, let's say neck.

The best a computer can hope to do is what is known as a *Pseudo-Random Number Generator* (PNRG). A PNRG can use any available source of randomness for its initial seed value. Then it repeatedly scrambles this seed. Usually, the seed is a short, random number that the PRNG expands into a longer, random-looking number. Some PNRGs use a multiple of this seed with the number of seconds that have passed from a given date. Another method is for the computer to take a random measurement of time between the keystrokes of a user at any given moment. (This one is actually quite good.)

STEGANOGRAPHY

So, what is steganography? Remember way back, somewhere between the middle of the Jurassic period up through the late Cretaceous period, when brave photographers would wonder out into the wilderness and take those beautiful photographs of Stegosaurs you see in all of the encyclopedias? The word used to define that activity was…well, it isn't a word really. It's more like placing yourself in unnecessary danger and at a time when you had no business being there.

Steganography is formed from two Greek words, *stego-* meaning "covered" and *graphy* meaning "writing." The word translates literally to "covered writing"; however, in practice, it's more like secret writing. The art and science of steganography is the hiding of secret messages within otherwise innocent material. This is not only clever but formidable as well. A simplified example of steganography that children find popular is "Where's Waldo?" If you have never seen a "Where's Waldo?" picture, it is generally a drawing of hundreds of similar looking characters. If the question "Where's Waldo?" weren't posted, you would never even begin to look for this single "Waldo" character.

One problem inherent in purely cryptographic methods is that should a message be discovered, the output is so nonsensical that it is painfully obvious that someone is trying to hide something. This triggers the human nature in us to automatically want to find out why someone would want to hide this item and, even more, now want to know what it is. If you were to combine steganography with cryptography, you would have yourself one formidable and rather sneaky cryptographic solution.

History of Steganography

Sending hidden messages predates even the use of ciphers. The Egyptians concealed secret messages within hieroglyphics on their monuments. Often there was so much information hidden in a story that the meaning of the story became lost to all but the writer. The Chinese used a system whereby secret messages were written on a thin piece of silk. The silk would be rolled into a ball, given a wax coating, and swallowed. The message could be retrieved a day or two later with relative ease. How is that you ask? Well, what you do is, you see….you're just gonna have to figure that out on your own.

Steganography was often used together with ciphers to provide an added layer of protection. In the 16th century, Girolamo Cardano created a device called the Cardano Grille that did just that. It was a rigid sheet with letter- or word-sized holes cut into it at random intervals, not unlike that you may have seen when teachers would grade certain kinds of tests. To create the message, you would lay the grille over an empty page and write the secret message through the holes in the grille. The grille was then removed and the blank spaces on the page filled in to produce an innocent-looking cover message. To retrieve the secret message, you would need a duplicate of the Cardano Grille used. The problem with this method is that the language of the cover message would generally be awkward and stilted. The 16th century brought with it the invention of invisible ink. Invisible inks were used widely during the fight for American independence. Invisible inks were also used in both world wars.

Steganographic Methods

There are many methods of steganography. One method used during World War II was to mark key letters in certain inconspicuous documents being mailed all over the world every day. Postal chess games, crossword puzzles, newspaper clippings, children's drawings, and report cards were all good prospects. Another method used by Nazi spies would be to split the paper, write an invisible message on the inside, and then re-join the paper.

The U.S. and British governments knew the threat steganography posed. It even became illegal to send cables ordering that specific types of flowers be delivered on a specific date, and eventually all international flower orders were banned between these two countries.

Microdots

Microdots were developed by the Germans during World War II. They were constructed using circular photographs of messages reduced in size to 0.05 inch across. The photograph would then be inserted in the periods of an innocent message. The message would be sent, and the dots would be removed and enlarged to see the original photograph.

Spread Spectrum

Spread spectrum techniques were originally developed for radio transmissions and are also used by the military for radar transmissions and to counter electronic counter measures (ECM). These methods are commonly used today for things like wireless networking.

This is done by shifting a carrier frequency over some predetermined range or set of frequencies for the duration of the transmission. To intercept the message, an eavesdropper would need to know what frequencies were being used as well as how often and for what duration the channels would be changed during the transmission.

Wrapping It in "Noise"

Electrical signals contain an element of noise or interference. In analog forms of communication, noise can never be completely eliminated. When an analog signal is converted to a digital value for communication (like modems do), it is done by putting discrete limits on the measurements of the signal so that the noise component of the signal being sampled is rejected, or "cropped" if you will, before assigning a value.

You can use the noise component in an analog signal and maybe exploit it, too, by "wrapping" the secret message inside of that noise. In analog systems, noise stems from inaccuracies in the components of the system and from electromagnetic interference.

That doesn't make Digital communication immune to such methods, however. Many types of digital data contain intrinsic "noise" suitable for use as a wrapper in secret messages. Multimedia data such as digital photographs, motion video, and sound will contain an element of sampling error that is usually below the perceivable threshold. When transmitting multimedia across expensive long-distance networks, it is desirable to transmit only the data that is necessary. Several groups on the Internet have been assembled to produce ways of sending that multimedia efficiently, most notably the Joint Photographic Expert's Group, who produced the JPEG standard for image compression. When you do that in very high color graphic images, the low-order bits that describe each pixel, when altered, do not make a distinguishable difference in the original quality of the image, making that a perfect steganographic medium since you can alter the low-order bits to carry whatever you like in that space.

Substitution and Selective Wrapping

Changing the wrapper to embed a secret is called *substitution*. In a noisy wrapper, the noise can be altered to carry the secret. This will change the noise profile of the wrapper, but is relatively simple to implement. This is the method that is used by almost all of the currently available steganography programs for message hiding.

With *selective* steganography, a wrapper could be chosen that exactly matches the secret. This method could be cracked, but not that easily. If the wrapper to be used is a scanned image, then the image could be repeatedly rescanned until its noise component is identical to the secret message. This would be very time-consuming, and it could not be guaranteed to work for all secret messages.

CRYPTOGRAPHIC ALGORITHM METHODS

We will be building on your understanding of cryptography in more detail as you learn about specific protocols and algorithms in the chapters ahead. There are two basic kinds of algorithm methods: those you want to be able to undo and those you don't ever want undone. Both methods have their respective purposes and are often used together. The kinds you don't want undone are called *one-way functions*, also known as *hash algorithms* or *message digests*. The kinds you want to undo are *two-way functions* and use public keys, private keys, or both.

Two-Way Functions

There are three classes of two-way encryption algorithms used presently:

▼ Symmetric or private-key algorithms

■ Asymmetric or public-key algorithms

▲ Hybrid algorithms

Symmetric Algorithms

Symmetric algorithms use the same key for encryption and decryption. They are also referred to as *private-key algorithms* since it is essential to keep the key value private or "secret." Symmetric algorithms use two types of ciphers:

▼ Stream ciphers

▲ Block ciphers

Stream ciphers can encrypt plaintext one bit at a time. *Block ciphers* take a number of bits called a *block* and encrypt them as a single entity. Symmetric algorithms are much faster to execute on a computer than asymmetric ones, because the computations used in the process are less complex.

Asymmetric Algorithms

Asymmetric algorithms use one key for encryption and one for decryption. Asymmetric algorithms are such that the decryption key cannot be derived from the encryption key. Asymmetric ciphers are also referred to as *public-key algorithms* and are designed to allow, even encourage, that the encryption key value become publicly available to anyone. This method requires that a pair of keys be used. The key used to encrypt the message is called the *public key*, and the key used to decrypt the message is called the *private key* or *secret key*.

Using this method, you would encrypt a message using the public key of the person you are sending the message to. However, only the person for whom the message was intended can decrypt it using his or her own private key, which must always be kept secret.

Asymmetric algorithms are more secure than symmetric algorithms, but as we stated earlier, they are slower and are much more CPU intensive.

Hybrid Algorithms

Hybrid algorithms use a combination of symmetric and asymmetric algorithms. When these are used together, a public-key algorithm is used to encrypt a randomly generated encryption key to be used in a symmetric algorithm. The way you would send a message to someone would be by using that person's public key to create a standard symmetric key and then sending that key to the other person. Then, you would use the symmetric key you just created to encrypt the message and send it to the intended recipient.

One-Way Functions

Algorithms using two-way functions work fine. In fact, they are the most secure forms of encrypting a document, sending it over a public medium, and having it decrypted. The only flaws with using two-way functions exclusively are

- ▼ The signature is *as long as the original document,* which means an enormous waste of disk space and network bandwidth.
- ▲ You have to *encrypt the whole document*—remember, public-key encryption algorithms are comparatively slow and CPU intensive.

It would certainly be better if we only had to encrypt some short but unique identification of the whole document. In other words, find a way or mathematical function that takes an *arbitrarily long* document and produces a *short* output of *fixed length.* Such a function is generally known to computer scientists as a *hash function.*

However, care must be taken. We can't just use an arbitrary hash function to get a secure digital signature scheme. Since the hash value calculated for the unique identification or "signature" of our document is normally much shorter than the original document, there will be infinitely many documents whose identification is identical.

It is therefore necessary that *there must not be any known method to invert the hash function.* That is, given a signed document *A*, you must not be able to generate a document *B* such that the identifications of *A* and *B* are equal. Otherwise, the signature of document *A* would also be valid for document *B*. You could "recycle" other people's signatures without knowing their secret encryption keys.

A hash function that meets this condition is called a *hash algorithm* or a *message digest code* (MDC), and the identification or signature of a message created using that code is called its *message digest.*

CRYPTANALYSIS

Cryptanalysis is, as we have already discussed, the art and science of figuring out the plaintext encrypted by cryptographers. There are different ways of approaching cryptanalysis, and the best method depends on the methods used to encrypt the message.

When you are trying to figure out how something is done, you need to assess items you have available to do it. You need to decide what you are going to analyze in this group of items.

A good example of this would be something like when you were a kid. Chances are that when you were taken out to a restaurant some nice waitress gave you a placemat full of activities and a handful of crayons. Invariably, one of those activities was to trace a path through a drawing of a maze of some kind. Now, you may have tried to start from the beginning a number of times to get to the center, without success. Then it would dawn on you that if you just started from the center and worked your way backward, you would find the solution in no time. You were provided with all of the same items from the very beginning, but it was your approach in the analysis that was the key to the solution.

Cryptanalysis is no different. It all depends upon what it is you have to work with. In some cases, you combine information you have about the key with encrypted information that you have managed to sniff from a given network.

Maybe you have a way of getting possession of plaintext that you know is going to be encrypted and can use that to discover the key. Knowing what you have to work with and the easiest way of approaching the discovery of the algorithm used is the key to cryptanalysis.

Any discovery of even the simplest of faults in a cryptographic system being used can also help in cryptanalysis and eventual discovery of the key. It was the interest in cryptographic hardware devices that led to this discovery. It turned out that some algorithms behaved very badly upon introduction of small faults in the internal computational values. Private-key operations are very susceptible to fault attacks. It has been shown that causing one bit of error at a suitable point can give enough information to determine the value of the key.

Cryptographic Attacks

Cryptanalysis is the art of deciphering encrypted communications without knowing the proper keys. There are many cryptanalytic techniques. Let's talk about some of the ways a cryptographic algorithm can be attacked.

As important as it is to know how to defend yourself from an enemy, it is also important that you find out the methods they will use to attack you. Keep one thing in mind as well: the faster and more powerful computers become to help us implement newer and better cryptographic algorithms, the faster and more powerful they become to crack those algorithms. Cryptographers develop encryption algorithms. Cryptanalysts find ways to break them—well, cryptanalysts and a few pale, pimply teenagers. The attacks are broken down into types of attacks because, like in any investigation, it all depends on how much evidence you have to start with. There are two main types of categories that you should be aware of: cryptanalysis and protocol subversion. Think of *cryptanalysis* as a kind of encryption forensics and *protocol subversion* as the equivalent of what a con artist does—but to data instead of people. We'll start with cryptanalysis methods first.

Ciphertext-Only Attacks

In this type of attack, the cryptanalyst has no idea what the message contains; all he or she has is garbage the algorithm produces. In this method, the cryptanalyst tries to gather as much of the ciphertext as possible in order to look for patterns. The ultimate goal is to figure out the key being used so that it may be used to decipher future messages. This is the most difficult attack, because the cryptanalyst has little to work with. This is known as the *brute force* method, because it involves trying every possible key value. This is not as "out of the question" as it might seem. An RSA algorithm using a 140-digit key (that is a big number) was cracked in just 8.9 CPU years. Well, that 8.9 years took only 11 weeks. By using many computers at once or, in cases where you don't own that many computers, using the spare CPU cycles of someone else's computers, you could accomplish the same thing without them even knowing your doing it. There are several known instances of this happening since the Internet came into being.

Known-Plaintext Attack (Also Known as Linear Cryptanalysis)

What happens here is that the cryptanalyst has samples of both the ciphertext and the plaintext. He or she either has some of the plaintext or tries to guess what that plaintext might be. Many documents begin in a very predictable way. In business letters you may have your "to whom it may concern" and "sincerely" consistencies to work with. If you know the name of the company sending the document, you could even use the letterhead as a constant in your search for the key. The more you have to work with, the better your chances of using these items to discover the key, thereby enabling you to decipher any future correspondence.

Chosen-Plaintext Attack (Also Known as Differential Cryptanalysis)

In this case, the cryptanalyst has not only the ciphertext and plaintext. He or she also has the ability to send any other text he or she likes through the algorithm to produce more ciphertext to work with. He or she can now encrypt text with the unknown key. Each time he or she does this, he or she creates more clues to revealing the key. This is a popular method of attacking many of the RSA algorithms.

Chosen-Ciphertext Attack

This is a situation where the cryptanalyst might actually have access to whatever means were used to decrypt the ciphertext, but be unable to extract the actual decryption key.

Man-in-the-Middle Attack

This attack is used in cryptographic communication and key exchange. The idea is that when two parties are exchanging keys for secure communications, Mullet is located between the parties on the communication line. Mullet then performs a separate key exchange with each party. The parties will end up using a different key, each of which is known to the adversary. The adversary will then decrypt any communications with the proper key and encrypt them with the other key for sending to the other party. The parties will think that they are communicating securely, but the adversary is hearing everything.

Protocol Subversion

Protocol subversion is another method that can cause problems for your network. Obviously, it is very important to use a good encryption technique, but without using the proper combination, you are vulnerable to subversion methods. This is a man-in-the-middle attack where Mullet intercepts packets off of the wire and starts substituting his own packets in place of theirs. Another clever way of protocol subversion would be in the form of running a denial of service attack to knock out the computer that is thought to be a server, and assigning that address to Mullet's computer before the next communication with a client takes place.

Attack Against or Using the Underlying Hardware

In the last few years, numerous small, mobile, cryptographic hardware devices have come into widespread use. A new category of attacks has become relevant that aims to break the cryptographic implementation on the hardware level instead of breaking the mathematical algorithm.

The attacks use the data from measurements taken on one of the cryptographic hardware devices doing encryption and compute the key information from these measurements. Then, the same analysis used in the other kinds of attacks is simply applied, like guessing some key bits and attempting to verify the correctness of the guess by studying output of information created using the device.

FUTURE CRYPTOGRAPHIC METHODS

We will just add these as food for thought. These are some suggestions that have been proposed but not yet implemented since we are not far along enough technologically.

Quantum Cryptography

This is based on an idea presented by Peter Shor to use quantum computers in future cryptography. Quantum computing is a recent field of research that uses quantum mechanics to build computers that are, in theory, more powerful than the modern serial computers.

The power is derived from the inherent parallelism of the quantum computer. So instead of doing the tasks one at a time, as serial machines do, it just performs them all simultaneously, making quantum computers exponentially faster than serial computers. Obviously, if we can manage to do this for cryptography, think of all the other problems we could solve using quantum computers that are infeasible with serial machines. These results suggest that if quantum computers could be implemented effectively, then most of the public-key cryptography used today could become history.

Quantum computation may also be a source for new ways of data hiding and secure communication. As quantum computers are fundamentally different from current serial computers, new applications of quantum computation to cryptography are likely to be found.

Current state-of-the-art of quantum cryptography does not appear alarming, as only very small machines have been implemented. It is not yet known whether quantum computers are truly more powerful than serial computers in practice. The theory of quantum computation gives much promise for better performance; however, whether it will be realized in practice is an open question.

DNA Cryptography

Leonard Adleman is the "A" in "RSA." You will be seeing those three letters a lot in this book. Adleman came up with the idea of using DNA molecules as computers. DNA molecules could be viewed as a very large computer capable of parallel execution. This parallel nature could give DNA computers exponential advantages in speed when compared with the modern serial computers used most commonly today. One of the problems with DNA computers, such as the exponential speed-up, also requires exponential growth in the volume of the material needed. Thus, in practice, DNA computers would have limits on their performance. Also, it is not very easy to build one.

SUMMARY

In this chapter, we talked about the very heart of VPNs and secure communication, and that is the art and science of cryptography. We learned about the history, methodologies, and concepts of cryptography and their many different subcategories.

We learned about the different terms such as plaintext and cleartext, encryption and decryption, and cryptology and cryptanalysis. We learned about codes, ciphers, and steganography and the importance that randomness plays in their success or failure.

In the remaining chapters of this section of the book, we will see exactly how all of those things are applied, exploited, and used together to make VPNs and secure communication possible.

CHAPTER 13

Cryptographic Algorithms

Before we get into some of the specific algorithms, we should first discuss the three main categories that these algorithms fall into. The categories we will discuss in this chapter are

- ▼ Hash algorithms
- ■ Private-key (or symmetric) algorithms
- ▲ Public-key (or asymmetric) algorithms

At certain points throughout this section of the book ("Secure Communications"), I will be describing some algorithms in a series of numbered steps. To do this, I will be referring to the individuals we talked about in Chapter 12. Note that I have added a name to the cast, Ted. Please refer back to this list for the cast of characters, because their function in these steps will remain consistent.

- ▼ **Alice**   This is the first participant in all of the algorithms (in keeping with cryptographic tradition).
- ■ **Bob**   This is the second participant in all of the algorithms (also in keeping with cryptographic tradition).
- ■ **Mullet**   This is the one that is malicious or *in the middle*.
- ▲ **Ted**   This is a trusted third party (because you can't spell "trus*ted*" without Ted).

HASH ALGORITHMS

One of the more interesting and impressive cryptographic subjects is the hash algorithm. The reason it is so elegant and impressive is that the product of a hash algorithm is so *easy* to calculate in one direction and just plain *awful* to try and calculate in the reverse direction. Along with being a very powerful tool, it is very compact. Hash algorithms are often referred to as *one-way hash functions*. You will also hear terms like *message digests* (MD4, MD5, and so on), *fingerprints*, or even *trapdoors*, but for the most part they all do the same thing. The main characteristic of a hash is that it's easy to compute, ridiculously hard to reverse, and extremely unlikely to repeat. To fully appreciate this, we will give you a basic idea of what happens in a hash function through the following scenario:

Enter Spy 1—let's call him Yuri. He has been ordered by his boss to send sensitive messages to the mother country via the post office and told that the contents of these messages can never be revealed to their capitalist enemy. Spy 2 we will call John. Through diligent intelligence work, John has discovered the actual post office that Yuri's country uses to mail these messages. John's boss says, "Good work, John! I want you to apply for a job at this post office, intercept these messages, and inform me of the contents therein." This should be a cakewalk for John. He just needs to retrieve the message from the outgoing mailbag, copy the message, and bring it to his boss.

Now, Yuri has an idea. He decides that he is going to take this message, convert the words of this message into a numeric value, divide that value into the numeric equivalent of that day's date, and write down the number that occupies the second position (or digit) of the remainder. He calls his comrade in the mother country and informs him of this exact procedure. Let's just say the words in the message calculate into the number 512. (How he did this is not important.) It is October 31, 2000, so, written numerically, this works out to be (10312000). He takes out a pencil and a piece of scrap paper and divides 512 into 10312000. As it turns out, the second digit of the remainder is 2. This little procedure takes him maybe a minute and a half. Yuri then takes out a nice new sheet of stationery and writes the 2 on it. He then places this in an envelope and drops it into the mail slot at his local post office.

Night has fallen and John painstakingly sifts through hundreds of envelopes and Bingo! He finds the envelope. With great enthusiasm he returns to his boss, opens the envelope, and plunks down a piece of paper with the number "2" written on it. John's boss replies, "So, you have intercepted a 2. Well, that's just swell. Here is my pencil and there is a calculator over there on the desk. Just go ahead and reverse whatever steps they took to arrive at that number, and get back to me with the original message."

I can tell you this for free. I wouldn't want to apply for that job! Even though this is a simplistic view of the one-way hash process, it is not all that different from what actually happens. What is actually taking place in a hash function is what is known as *modular arithmetic* or *modulo math*. Here is how it works and why it is virtually impossible to undo: Let's look at an example. When we say "5 mod 3 = 2," the way we arrive at 2 is quite simple. "5 mod 3 = x" means "take 5, divide it by 3, and whatever the remainder is will be x." So, 5 divided by 3 is 1 with a remainder of x, which is 2. The number that we use to divide by (3, in this example) is called the *modulus*. The reason this is virtually impossible to undo is that 17 mod 3 = 2, as does 8 mod 3 = 2, and 10 mod 8, and so on. So, if like our friend, Spy 2 (John), you discover this 2, how can you possibly determine which mod was performed on which set of numbers? I'll tell you how. *You can't*, that's how! Computers use extremely large prime and almost-prime numbers that will always, regardless of their original size, mod to 0, 1, 2,...you get the idea.

Where hashes excel is when you want to check whether or not a message is valid or if any kind of tampering has occurred. When a one-way hash function is used to compute a hash value, any change to the message produces a different hash value when the same hash function is used after the message is received. Another common use is to check a password. This is the method used by Windows NT. Instead of maintaining a plaintext list of the passwords being used by all authorized users, Windows NT saves a one-way hashed value of the password.

The benefits are twofold. For one, if someone were to obtain the "password" list file, all they would have on their hands would be a bunch of very small numbers. And two, the list becomes very compact, because instead of saving hundreds or thousands of ASCII characters, it saves a very small numeric hash value. There are two types of hash functions. One is called a *Message Authentication Code* (MAC) hash, which uses a secret key (sometimes called a *keyed hash*). The other is known as a *Manipulation Detection Code*

(MDC) hash, and this one does not involve a key. Their uses are pretty self-explanatory. With a MAC hash it is impossible to compute the hash value without knowledge of the secret key. An MDC hash need only be a one-way function whose purpose is making sure that the data has not been "manipulated" or tampered with. Hash functions can be used to protect the authenticity of large quantities of data with a short secret-key MAC hash, or to protect the integrity of a short string MDC hash. Sometimes an MDC hash is used in conjunction with encryption and can guarantee both confidentiality and authenticity. Here are some of the more common hash algorithms being used today. The reason for the list is that you are very likely to run across many of these when working with VPN's and you should be familiar with them when you do.

MD5, MD4 (Message Digests)

These are secure hash algorithms developed by Ron Rivest from MIT and RSA Security, Inc. What the MD5 algorithm does is take a message of any given length and produce a 128-bit *fingerprint* or *message digest.* It is virtually impossible to produce two messages that will have the same message digest. It is also impossible to produce any message having a prespecified target hash value. The MD5 algorithm is generally used for digital signature applications, where a large file must be "compressed" in a secure manner before being encrypted with a private key before being signed with an RSA public-key type of cryptosystem. You can read an in-depth description in RFC 1321 for MD5 and RFC 1320 for MD4. While MD4 may be a bit quicker, it is less secure and generally not recommended. A better choice might be the RIPE Message Digest mentioned shortly.

SHA (Secure Hash Algorithm)

Also know as the Secure Hash Standard (SHS), this is a cryptographic hash algorithm published by the U.S. government. It produces a 160-bit hash value from an arbitrary length string. This and the RIPE 160-bit Message Digest are considered to be better than the MDx series.

RIPEMD-160 (RACE Integrity Primitives Evaluation Message Digest)

This hash algorithm is relatively new. It is a 160-bit cryptographic hash function that was designed to replace the MDx series of message digests. It produces a digest of 20 bytes that can supposedly run at up to 40 MBps on just a 90-MHz Pentium computer and has been placed in the public domain by its designers.

PRIVATE-KEY (SYMMETRIC) ALGORITHMS

Private-key algorithms are very simple in principle. You use the same key to encrypt a message as you do to decrypt a message. This is why it is called a *symmetric key* algorithm.

The benefit of using private-key or symmetric algorithms is that it doesn't require nearly as much computing power as public-key or asymmetric algorithms—in some cases, a thousand times less than public-key algorithms. When sending vast amounts of information, this becomes a huge factor, especially when using WAN links, whose bandwidth is limited anyway. The drawback comes from finding a way to exchange this key over a public network without it being disclosed. One way to overcome this dilemma is to use a hybrid system (to be explained in the "Diffie-Hellman Algorithm" section). That is, use a public-key algorithm to encrypt the actual symmetric keys first, and then use those keys to encrypt the rest of the data.

The most important factor in symmetric key encryption is key length. The longer the key, the harder it is to crack. The same principle applies to passwords. A chain is only as strong as its weakest link. So if you use the password "dog" and encrypt it with even a 128-bit key, you weaken its strength profoundly. This means that if we were to take the letters *d*, *o*, and *g* and scramble them, chances are good that you can rearrange them back into the word "dog" without much effort. Even if you were not sure of whether it was "god" or "dog," you wouldn't have to waste a very big portion of your day to try them both. The private-key approach is well suited to situations where one location requires secure interaction with relatively few other applications or users. Let's go over some of the more common symmetric key algorithms. Again, these are being listed because you are very likely to run across many of them, and it will be better that you have at least some understanding of what they are when you do. Take a look at Figure 13-1 to see what takes place when using symmetric or "shared" key algorithms.

Data Encryption Standard (DES)

DES is an algorithm that was developed in the 1970s in an arrangement with IBM and the National Security Agency (NSA). It was made a standard by the U.S. government and has also been adopted by several other governments worldwide. Its use is extremely common, especially in the financial industry. Virtually all ATMs use DES to protect the data that is transmitted from the ATM to the transaction processing location.

DES is a symmetric block cipher that uses 64-bit block size. A *block cipher* operates on blocks of data where several bytes are put together prior to encryption. A *stream cipher* encrypts the bytes or bits individually. Block ciphers are the more common of the two. DES uses 56-bit keys; this is because the last eight bits of the 64 are used for parity checking. DES was designed to be resistant to differential cryptanalysis, but it is still somewhat susceptible to linear cryptanalysis. DES, for the most part, is strong enough to keep most hackers and individuals out, but it can easily be cracked with today's computers.

Triple DES (3DES)

A derivative of DES, Triple DES (3DES) is an *iterative block cipher*. An iterative block cipher encrypts the input multiple times. Triple DES (3DES) uses the same key length as the standard DES key, 56 bits, but now it encrypts the plaintext three times with three different

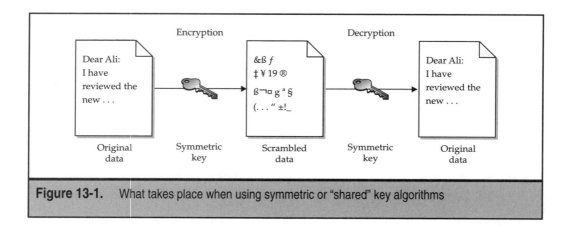

Figure 13-1. What takes place when using symmetric or "shared" key algorithms

keys. There are different ways of encrypting the data with each key. The most common methods are

▼ **DES-EEE3** This is Triple DES encryption using three completely different keys.

■ **DES-EDE3** These are Triple DES operations using an encrypt-decrypt-encrypt method with three different keys.

▲ **DES-EEE2 and DES-EDE2** These are Triple DES operations using an encrypt-decrypt-encrypt method also, except now the same key is used during the first and third operations.

The "EEE" and "EDE" that were appended to the "DES" refer to the method—"encrypt-encrypt-encrypt" and "encrypt-decrypt-encrypt," respectively. The most secure form is Triple DES using the three different keys. Triple DES is considered to be much safer than using just plain DES.

International Data Encryption Algorithm (IDEA)

The International Data Encryption Algorithm (IDEA) is a symmetric block cipher algorithm with a 64-bit block length and a 128-bit key, which is twice as long as that of DES. IDEA is immune to not only differential cryptanalysis, but also linear cryptanalysis. It has patents in the United States as well as Europe.

Blowfish Algorithm

Blowfish is a 64-bit symmetric block cipher that was invented by Bruce Schneier back in 1993. It can be used as a replacement for DES or IDEA, which are both patented. The data is encrypted and decrypted in 64-bit chunks. It can have a key length that varies from 32 to 448 bits. The algorithm uses 16 rounds, or iterations, of the main algorithm. The number of rounds is exponentially proportional to the amount of time required to find a key using a brute force attack. What this means is that as the number of rounds increases, the security of the algorithm will increase exponentially. It, too, is immune from both differential and linear cryptanalysis. Blowfish was placed in the public domain so that there would be no restrictions on the use or distribution of the algorithm. Blowfish has been extensively analyzed publicly, and no real weaknesses have been found. It has yet to be broken.

RCx Algorithms (RC2, RC4, and RC5)

The "RC" is sometimes referred to as "Rivest Cipher" or "Ron's Code." These ciphers were developed by Ron Rivest of MIT and RSA Security. RC2 is a 64-bit *block cipher*. It uses a variable key size and can be made more or less secure than DES by varying the key size. It is two to three times faster than DES. RC4 is a *stream cipher* that also uses a variable key length size. RC5 is a *block cipher* that is an algorithm that not only has a variable block size, but also has a variable key size and a variable number of rounds. (I guess you can consider this cipher to be variable.) The block size will be 32, 64, or 128 bits long. The key size can be anywhere from 0 (not very big) all the way up to 2048 bits. (You won't be breaking this cipher with an abacus!) The number of rounds can range from 0 to 255.

Skipjack Algorithm

This was developed by the National Security Agency (NSA). Skipjack was first called Clipper and then Capstone. It requires the use of a hardware-based chip scheme that incorporates the Skipjack algorithm (which is classified) using a 64-bit block size and an 80-bit key. This key is used in combination with an *escrow key* (or keys) used by the government or some other kind of escrow agency. Escrow keys were designed to enable law enforcement agencies to decrypt messages if they had reason to believe laws were being broken. The fact that the algorithm is classified also means that it is not subject to public scrutiny, therefore making it impossible to test its security fully. This one is not very popular for obvious reasons.

PUBLIC-KEY (ASYMMETRIC) ALGORITHMS

Public-key algorithms not only solve the problem of how two people can exchange a secret key over a public medium, but they also, by their very nature, provide some powerful services other than encryption. These protocols also allow for digital signatures, secure private-key exchange, authentication, and for some, all of the above. You might ask, "If they are so good, why bother using anything else?" The answer lies in speed. These protocols require a lot more processing power and can take, as was mentioned before, a thousand times longer to encrypt the same amount of data than does a symmetric protocol. To get a basic idea of what takes place using asymmetric or public-key algorithms, take a look at Figure 13-2.

Before we start talking about how some of the individual protocols work, let's discuss the basics of how a public-key protocol works.

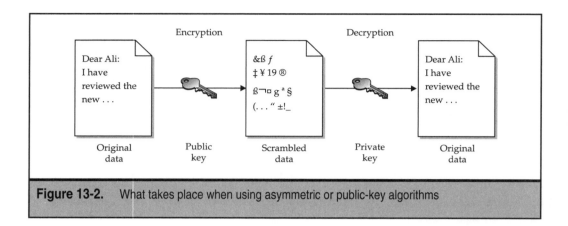

Encryption   Decryption

Dear Ali: &ß ƒ   Dear Ali:
I have   ‡ ¥ 19 ®   I have
reviewed the   ß¬¤ g ª §   reviewed the
new . . .   (. . . " ±!_   new . . .

Original   Public   Scrambled   Private   Original
data   key   data   key   data

Figure 13-2.   What takes place when using asymmetric or public-key algorithms

Most everyone has been to a post office at one time or another and seen the many rows of P.O. boxes with locks and numbers on them. For our example, let's use the same P.O. boxes, but arrange them a little differently. Picture our post office with all of the P.O. boxes filling up one whole side of the room. Instead of each person getting one box, imagine each person having a column of mailboxes from floor to ceiling with each name printed at the top. At the foot of the column sits a container of keys identical to each other but different from those of the other columns, with the boxholder's name inscribed. These keys are open to the *public* and free for the taking. Every individual P.O box has two locks side by side. The lock on the left will *lock* the mailbox, and the lock on the right will *unlock* the mailbox. The keys in the container have one thing in common—they can only *lock* the P.O. boxes.

One and *only* one key has been hand-delivered to the owners of the P.O. boxes. Prior to being issued this key, fingerprints, driver's licenses, and birth certificates had to be verified by the company that made the keys. The owners were instructed to keep this key *privately* secured.

Our post office is open 24 hours a day. We will assume that everyone using this post office knows one another. There may even be friends and enemies among this group. You want to invite four of these friends to a dinner party, but don't want all of your other friends to find out, because it might hurt their feelings. You take invitations to put in the P.O. boxes of these friends: let's call them Rey, Sam, Rio, and Raul. You go to the Rey column, reach down, and pick up one of his *public* keys, place the invitation in one of his open P.O. boxes, turn the key, and lock it shut. You then toss this *public* key back into the container at the bottom. You go to the other columns and repeat the same procedure.

You notice that some of your other friends and even enemies are at the post office. They have all witnessed your actions. This doesn't worry you in the least, because you know that the invitations have been locked away and that only Rey, Sam, Rio, and Raul can retrieve them with the respective *private* keys that were issued to them.

This is exactly how public-key encryption works over the Internet, with the only difference being that the *public* key is used to *encrypt* the invitation instead of to lock the P.O. box. The *private* key is used to *decrypt* the invitation rather than to unlock the P.O. box. Now, what if someone at the post office who didn't like you somehow discovered that you were throwing this party and would like nothing better than to ruin it. Well, it is a public, 24-hour post office, right? What would stop him or her from going into the post office late at night and using the exact same procedure, putting "sorry, party cancelled" notices, ostensibly from you, into each of these four friends' other open P.O. boxes? What could be done to make sure that not only could they not pretend to send something from you, but also to provide a way to let your four friends be absolutely sure the original invitations were from you in the first place? Let's go back to our post office example and find out.

You get a call from the company that originally gave you the *private* key to your P.O. box. They inform you of a new item they are planning to distribute to everyone in this post office. Once again, your fingerprints, driver's license, and birth certificate have to be verified. They deliver a big case of lockboxes with your name printed on them that are just small enough to fit inside the existing P.O. boxes in the post office. You notice that they are very similar to your P.O. boxes. They each have two locks on them side by side. The difference is that now your *private* key is used to *lock* this box. Any of your *public* keys in the post office can be used to *unlock* this box.

It dawns on you that a person who doesn't like you has just seen you put invitations into your four friends P.O. boxes and that this person is just the type to put "party cancelled" notices in their other P.O. boxes. You make up four new invitations, put one in each of your new lockboxes, lock them with your *private* key, and go to the post office. You then place these lockboxes into each of your four friends' remaining open P.O. boxes. You reach down and take a *public* key out of their container and lock their respective P.O. boxes shut with them.

The next day, your friends arrive at the post office, open their respective P.O. boxes with their *private* key and extract the locked box, with your name on it, which is inside. They each walk over to your column of P.O boxes, reach down into the container and use one of your *public* keys to unlock this box and retrieve their invitation.

This is exactly how public-key algorithms are used to digitally sign a message and guarantee its integrity. To take this a step further, you can consider the manufacturer of these boxes and keys as the *issuing authority* or a *trusted third party,* since they make the lockboxes, check identities, and fill up the containers in the post office with public keys whenever they run out. Later you will see the significance of the issuing authority when we talk about *certificates.* But more importantly, this arrangement has now left you sure that your invitation is encrypted, digitally signed, and positive that the integrity of its contents is tamper free. You also take comfort in knowing that your friends are just as sure the invitation is legitimate and from you as well. Pretty neat, yes? That should give you a pretty good feel for what is happening when public-key algorithms are used. Now we can talk about the specific protocols that use this public-key method so that when you run across them while designing VPNs, they will make some sense.

Diffie-Hellman Algorithm

In 1976, Martin Hellman, a professor at Stanford University, and two graduate students who worked with him at the time, Ralph Merkle and Whitfield Diffie, came up with a solution that would allow two people in different locations to safely communicate a secret key over a public medium. This was the birth of public-key cryptography. The Diffie-Hellman algorithm also helps us with another problem. Public-key encryption methods are processor intensive and extremely slow when compared with private-key methods. Why not combine their strengths and reap the benefits of both? This is where Diffie-Hellman shines. Diffie-Hellman is not used to encrypt data per se. This algorithm is designed to exchange symmetric (private) keys securely and very quickly. Once the symmetric keys have been safely exchanged, encryption can begin with a thousand times less overhead. This is also what is referred to as a *hybrid* method. Here is an example of how this hybrid would work:

1. Alice sends Bob her public key unprotected over the Internet. (Bob could have the key already or even get it from a trusted third party.)

2. Bob decides on a secret key to be used with a symmetric algorithm (like DES or IDEA), encrypts this symmetric key using Alice's public key, and sends this key back to Alice.

3. Alice now decrypts, with her private key, the symmetric key that will be used for the duration of the encrypted communication session. (This is where you may hear the term *session key.*)

4. Alice and Bob now communicate with confidence, knowing that all subsequent data is fully encrypted.

There is one drawback that Diffie-Hellman is susceptible to, and that is what is known as a "man-in-the-middle" attack. It is *extremely* hard to pull off in practice, but in theory the danger exists. We will go over these subjects a bit more in later chapters. Here is an example of how this unlikely event would occur:

1. Alice sends Bob her public key unprotected over the Internet. (Bob could have the key already or even get it from a trusted third party.)

2. Mullet (who is *in the middle*) intercepts Alice's public key and sends Bob a copy of his (Mullet's) public key disguised as Alice's.

3. Bob decides on a secret key to be used with a symmetric algorithm (like DES or IDEA), encrypts this symmetric key using Mullet's public key (thinking it's Alice's public key), and sends this symmetric key back to Alice.

4. Mullet intercepts Bob's message containing this secret symmetric key. Mullet now takes this secret key, encrypts it with Alice's public key, and sends it to Alice.

5. Alice now decrypts this secret symmetric key using her private key.

6. Mullet can now create the illusion that Bob and Alice are communicating with each other. In fact, it is really Mullet intercepting the messages midway, subverting the messages, and resending them to each party.

One way around this subversion using a public-key system like Diffie-Hellman, is by first sending messages *already* encrypted with the symmetric key system, and then *afterward* sending that symmetric key necessary to read the message using the normal Diffie-Hellman method. It would work something like this:

1. Alice sends Bob her public key unprotected over the Internet. (Bob could have the key already or even get it from a trusted third party.)

2. Bob sends Alice his public key unprotected over the Internet. (Alice could have the key already or even get it from a trusted third party.)

3. Alice decides on a secret key to be used with a symmetric algorithm (like DES or IDEA), encrypts the message first using this symmetric algorithm, and sends it off to Bob. Bob cannot read the message (yet), but more importantly, neither can Mullet.

4. Bob decides on a secret key to be used with a symmetric algorithm (like DES or IDEA), encrypts the message first using this symmetric algorithm, and sends it off to Alice. Alice cannot read the message (yet) either, but again, neither can Mullet.

5. Alice encrypts her secret symmetric key using Bob's public key and then sends it off to Bob. *Now* Bob can read the original message.

6. Bob encrypts his secret symmetric key using Alice's public key and then sends it off to Alice. *Now* Alice can read the original message.

7. Using the secret key that they now both have in their possession, Bob and Alice can continue to communicate.

This won't necessarily stop Mullet from reading the message eventually, but it will remove his ability to subvert the message and to fool them into thinking they are communicating with each other unmolested. By using *digital signatures* and/or public-key *certificates*, Mullet cannot do either. We will talk about how that works shortly.

RSA Algorithms

About a year after our friends at Stanford came up with this new public-key concept, three professors from MIT, Ron Rivest, Adi Shamir, and Len Adleman, used this underlying technology to come up with RSA, the most popular asymmetric cipher in use today. The way their algorithm works is really quite clever. RSA is a *block cipher* that encrypts data in large blocks using the product of two large prime numbers. Their method is also extremely secure; you will see why shortly.

The algorithm performs what is called a *one-way trapdoor* function. It simply multiplies two prime numbers together to get a product that is used in the creation of the *key*. The whole basis of their method revolves around how computationally easy it is to multiply two *prime* numbers (factors) to get a product and how immensely difficult it is, computationally, to find the two factors (numbers multiplied by each other to get a product) of a *very* large number, consisting of only two numbers in its equation, both of which are *prime* (numbers that, when they are divided, are only the number itself and 1). When using any cryptographic algorithm, the longer the key is, the harder it is to crack. I'll show you what I mean. Let's say I multiplied 7 × 11 to get 77 (a 2-digit key), and the only information I gave you was that I used two prime numbers to calculate this. You would probably be able to figure out that I used 7 and 11 to arrive at this without much trouble. If I then gave you the number 7387 (a 4-digit key), you almost certainly would be there a while longer; go ahead and try, I'll wait. A popular key size for RSA (which is variable) uses a 140-digit key. It took almost 200 computers and a team of cryptographic experts 11 solid weeks to crack this key. This included about 125 Silicon Graphics and Sun workstations running at 175 MHz on average and 60 Pentium class PCs running at 300 MHz on average. The total amount of CPU time spent on the process equaled 8.9 CPU years. (I think it is safe to say we all have friends of this caliber and access to resources like these.)

Let's say you were paranoid and decided to use a 617-digit key. That would involve using the most efficient numeric sieve known to man (something that creates what amounts to a very large spreadsheet) creating a matrix with 10^{150} cells. You might be asking yourself, "So what, is that a big number?" Well, current estimates of the *total* number of atoms in the entire *universe* range between 10^{78} and 10^{100} atoms. Both estimates pale in comparison to our 617-digit key derived matrix. Hopefully, now you have some idea of just how hard a procedure this is. (If you don't, then "Boy, are you hard to impress!")

So without using a hybrid method, a pure public-key example would be as follows:

1. Alice sends Bob her public key unprotected over the Internet. (Bob could have the key already or even get it from a trusted third party.)

2. Bob sends Alice his public key unprotected over the Internet. (Alice could also have the key already or get it from a trusted third party.)

3. Alice uses Bob's public key to encrypt all of the plaintext in her message that will be used during this communication session and sends it to Bob.

4. Bob decrypts Alice's message using his private key and now reads the message in plaintext form.

5. Bob uses Alice's public key to encrypt all of the plaintext in his message that will be used during this communication session and sends it to Alice.

6. Alice decrypts Bob's message using her private key and now reads the message in plaintext form.

This is an excellent way of communicating securely, but it is theoretically possible, although very difficult, for a man-in-the-middle attack. But, fear not, because there is a way to stop this hypothetical malfeasance. This is the hybrid method we spoke of in Chapter 12. The method used to avoid this activity is called the *Interlock Protocol.*

Interlock Protocol

The Interlock Protocol was invented by "R" Ron Rivest and "S" Adi Shamir. (I can only assume that "A" was sick that day.) This is their method of defending against the man-in-the-middle attack. To get an idea of what takes place when using a hybrid protocol like Interlock algorithms, take a look at Figure 13-3.

First, both parties exchange their respective public keys. Then both send, in turn, the first half of an already encrypted message. Next, both send, in turn, the second half of their already encrypted message. Here is how:

1. Alice sends Bob her public key.

2. Bob sends Alice his public key.

3. Alice encrypts her message using Bob's public key. She sends the first half of the resulting encrypted text to Bob.

4. Bob encrypts his message to Alice with Alice's public key. He now sends the first half of the resulting encrypted text to Alice.

5. Alice sends Bob the second half of her encrypted text.

6. Bob puts the two halves together and recovers the plaintext of Alice's message.

7. Bob sends Alice the second half of his encrypted text.

8. Alice puts the two halves together and recovers the plaintext of Bob's message.

Here's why this protocol causes a problem for Mullet. Let's say that Mullet is again sitting in the middle of this conversation and is trying to replace the original messages with subverted ones. He will again replace the two public keys with his own public key. Then he receives the first half of Alice's message to Bob. He has no way to read the contents of this first half of the message without the second half. He has to send Bob something; otherwise, the communication stops. Mullet has to make up a message, completely from scratch, encrypt it with Bob's public key, and send it to Bob. True to the form of the Interlock Protocol, Bob sends Alice the first half of his message. Mullet is in the same boat.

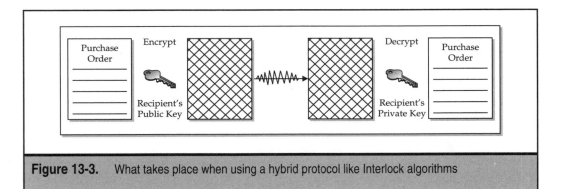

Figure 13-3.   What takes place when using a hybrid protocol like Interlock algorithms

He has to make up a completely new message and send its first encrypted half to Alice. Alice sends out the second half of her message to Bob, which is intercepted by Mullet. Now Mullet can read the contents of the Alice's message, but he cannot adapt the message to Bob's anymore, because he has sent Bob the first half already. Well isn't that a shame. (I never liked that Mullet anyway.)

For this protocol to work, the first half of the encrypted message must not be able to be decrypted without the second half of that same message. This is very easy to achieve using an algorithm that encrypts blocks of text. Since the RSA algorithm is a block cipher and encrypts the data in large blocks, nothing can be done with just half of the digits from a given block of a completely encrypted message. Again, this protocol does not protect Alice and Bob from Mullet reading the message, but it protects them from receiving subverted messages. All of this could be taken care of using *digital signatures* and/or public-key *certificates*, which we will be learning about in the next two chapters.

SUMMARY

In this chapter, we discussed the three main categories that most cryptographic algorithms fall into. These were

▼ Hash algorithms

■ Private-key (or symmetric) algorithms

▲ Public-key (or asymmetric) algorithms

We also talked about some specific algorithms used in each of these categories. We will be talking about certificates in the next chapter, and you will see how the algorithms you have just learned about are used to create them.

CHAPTER 14

Certificates

Think of the number of times you've signed your name to a piece of paper throughout your life. You sign checks, credit-card statements, legal documents, mail delivery slips, and more. Your signature is often used to acknowledge that you have received something. It is also used in a much more serious and binding way, on a contract. A *digital signature* is a cryptographic technique for achieving these same goals.

Like a handwritten signature, a digital signature should verify identity, be impossible to forge, and be incapable of being repudiated. That is, it must be possible to prove that a document signed by an individual was indeed signed by that individual and that *only* that individual could have signed the document. The signature cannot be forged, and a signer cannot later repudiate or deny having signed the document. This can all be accomplished using the concepts and algorithms we have been talking about in the last two chapters. These techniques can also be used in the successful completion of any kind of business transaction using completely electronic methods. This chapter will show the methods being used to achieve these goals.

All of the elements required for such a transaction are listed next.

▼ **Authentication** Authentication refers to the means used to verify the identity of the origin of an electronically transmitted message being sent. The recipient of this message can be confident that the message's origin and author are authentic.

■ **Data integrity** This is the means used to verify that the contents of the message being sent have not been altered or tampered with in any way during transport as well as an acknowledgement of its delivery by the recipient.

■ **Data confidentiality** This is the means used to ensure that data is protected from being observed or understood during transport and/or storage. This generally refers to the method by which the data is encrypted.

■ **Chain of custody** This refers to where the data was kept and by whom, along with the path taken on its way to a recipient.

■ **Timestamping** This applies to documents that are "time sensitive"—for example, the way a document (in all of its forms) is proven to have been created based on the particular time and/or date put on the document along with the signature. This also includes when the document was created or accessed.

■ **Data archiving** In normal paper transactions, you maintain files or folders in which all pertinent documents are stored as long as necessary. In the case of e-documents also, it is imperative to keep a long-term store of all electronic transmissions and to facilitate the retrieval of this electronically transmitted data.

▲ **Audit capabilities** It is necessary for an organization to keep track of all its correspondence. When, who, and how the data was accessed and/or altered are just a few important criteria by which an organization does its auditing.

DIGITAL SIGNATURES

Digital signatures are methods used to verify that the actual contents of any message being sent have not been altered or tampered with in any way during its transport and to ensure that the sender has delivery acknowledged by the recipient.

A number of questions need answering when dealing with data in digital form. "Digital form" could be anything from a hard disk to the wires and airwaves used during transmission. These questions include:

▼ How do you know beyond a doubt that the document you're looking at is identical to the document that was originally transmitted?

■ If you send a database file to someone via e-mail or the U.S. mail using a floppy disk, how does the recipient know that the records were not altered by someone who had access to it anywhere along the way?

■ If you send a credit-card number over the Internet, how does the merchant know that it really was you who placed the order?

▲ How do you know if the person you are sending your credit-card number to is in fact the merchant with whom you are dealing?

The answers to these questions all involve the use of some form of cryptography. The term *cryptography* implies encryption, but cryptographic techniques can be used for things other than just keeping the file's contents scrambled. It can also be used for digital authenticity and integrity. The methods required to do this depend on the level of certainty the situations dictates. This section will talk about some of the methods that can be used in maintaining the integrity of data:

▼ Checksums

■ Cyclic redundancy checks

▲ Hash algorithms or message digests

Checksums

The simplest method of verifying the integrity of digitally transmitted data is by using *checksums.* We talked about modular math in Chapter 13. We talk about it a bit more here. In modular math, a *modulus* is the number by which two given numbers can be divided and produce the same remainder. A *checksum* is a value derived by calculating the sum of all the numbers in the input data. If the sum of all the numbers exceeds the highest value that a checksum can hold, the checksum equals the modulus of the total—that is, the remainder that's left over when the total is divided by the checksum's maximum possible value plus 1. In mathematical terms, a checksum is computed with the equation:

Checksum = Total / (MaxValue + 1)

where the *Total* equals the sum of the input data and *MaxValue* is the maximum checksum value you will allow.

Let's say the document whose contents you wish to verify is the following stream of values:

36 211 163 4 109 192 58 247 47 92

If the checksum is also a 1-byte (8-bit) value, then it can't hold a number greater than 255. The sum of the values in the document is 1159, so the 8-bit checksum is the remainder left when 1159 is divided by 256. That remainder would be 135. If the person who sent the document calculated a checksum of, say, 246, and you get a checksum of 135, then you know the data was altered. A checksum is the simplest form of checking the integrity of a file.

The problem with checksums is that although conflicting checksums are proof positive that a document has been altered, matching checksums doesn't necessarily prove that the data was *not* altered. All you are doing is taking a sum of the numbers. That means that you could put the numbers in any order and maintain the same sum. When you use an 8-bit checksum, there is still a 1 in 256 chance that any two streams of data will have the same checksums. You could expand the checksum length to 16 or 32 bits to decrease the odds of coincidental matches, but checksums are still too susceptible to error to provide a high degree of confidence in the data that they represent.

That's not to imply that they don't have their place, though. They are very efficient and command a low overhead since the algorithm is a simple sum. That makes it OK for things like file-transfer integrity that aren't likely to be subject to serious security risk.

Cyclic Redundancy Checks

A better way is to use a cyclic redundancy check (CRC). CRCs are commonly used in many devices, including NIC, hard-disk controllers, and telephones, to verify the integrity of data traveling through them. CRCs use *polynomial division*, where each bit in a chunk of data represents one coefficient in a large polynomial. A polynomial is a mathematical expression like this one:

$f(x) = 4x3 + x2 + 7$

The *coefficient* is the number or symbol that is multiplied with the variable or unknown quantity in an equation. That would make *f* the coefficient in our expression. For the purpose of CRC calculations, the polynomial that corresponds to the byte value of 85 (whose 8-bit binary equivalent is 01010101) is

$0x7 + 1x6 + 0x5 + 1x4 + 0x3 + 1x2 + 0x1 + 1x0$

which leaves

$x6 + x4 + x2 + 1$

The key to CRC calculations is that polynomials can be multiplied and divided just like ordinary numbers. Dividing a "magic" polynomial (one whose coefficients are dictated by the CRC algorithm you're using) into the polynomial generated from a data stream yields a quotient polynomial and a remainder polynomial. It is the remainder polynomial that is used in a CRC value.

Like checksums, CRCs are very small, usually 16 or 32 bits in length, making them very fast and very efficient. The difference is that they are far more reliable when it comes to detecting minor changes in the input data. If just one bit in a large block of data changes, there's a 100 percent chance that the CRC will change, too. If two bits change, there's better than a 99.99 percent chance that a 16-bit CRC will catch the error. Changing the order of the bits or adding 1 to one and subtracting 1 from another won't fool a CRC as it will a checksum. This makes CRCs very popular for things like checking file transfers.

Hash Algorithms and Message Digests

A one-way hash algorithm is used to take an arbitrarily long message (variable) and compute a shorter fixed value called a *message digest*. Our definition of a message digest may seem quite similar to the definition of a checksum or a more powerful error-detection code such as a cyclic redundancy check. Is it really any different? Checksums, cyclic redundancy checks, and message digests are all examples of hash functions. The problem with even a 32-bit CRC value is that although it's pretty good at detecting inadvertent changes to the input data (such as those introduced by transmission errors), it doesn't stand up very well to intentional attacks. Today's personal computers have considerable processing power, and many programs designed to compromise CRCs are distributed freely and are widely available on the Internet. Thus, using a 32-bit CRC on a file to generate a given value makes it relatively easy for someone to generate a completely different file that produces the same CRC value.

One-way hash functions, however, are an entirely different story, as you may recall from Chapters 12 and 13. One-way hash algorithms produce hash values that are computationally impossible to undo. The values that they generate are unique and so difficult to duplicate that not even someone using several supercomputers and several hundred sleepless years could crack them. Most of today's hash values are at least 128 bits in length. The greater the length, the more difficult the hash value is to crack. We talked about many of the popular hash algorithms being used in Chapter 13.

The MD5 message digest algorithm by Ron Rivest is in wide use today. It computes a 128-bit message digest in a four-step process:

1. *The padding step.* This involves adding a 1 followed by enough zeros that the length of the message satisfies the length conditions.

2. *The append step.* This appends a 64-bit representation of the message length before it was padded.

3. *The accumulator initialization step.* This is where the initialization of an accumulator takes place.

4. *The final looping step.* The message's 16-word blocks are put through four rounds of processing.

Another major message-digest algorithm in use today is SHA-1, the Secure Hash Algorithm. This algorithm is a U.S. federal standard that is based on principles similar to those used in the design of MD4, the predecessor to MD5, and produces a 160-bit message digest. The use of SHA-1 is required whenever a secure message digest algorithm is required for federal applications.

Protection Against Accidental Corruption

Protecting against accidental corruption is relatively easy as data integrity goes. The most primitive step would be to store multiple copies. The likelihood that all of them would get corrupted is slim. However, with only a few copies, you might get into a situation where they do not match each other. How would you decide which copy is the true original?

Hash functions are an ideal choice here. If you recall, we simply hash the file to get its value. This is our message digest. We have mentioned the MDx family of algorithms and the SHA-1 algorithm. These are both fine for this kind of application.

There is also the RIPEMD-160 algorithm. Many complex multimedia documents can be quite large. The reliability of a hash function declines with the size and complexity of a message. RIPEMD-160 seems to have a definite advantage.

An excellent choice of hashing algorithm to digest very large documents is RIPEMD-160. It is reasonably robust and the performance is acceptable on low-end computers.

Protection Against Damage or Tampering

To prove that a given document has not been altered, we need two things:

▼ A message digest of that document, as discussed in the previous section

▲ Some sort of proof as to who generated the digest

Proving who generated the digest can be done by digitally signing the digest. You could use a digital signature algorithm (DSA) to create a message digest and a signature. A common DSA algorithm is used for generating a message digest SHA-1 algorithm or RIPEMD-160 algorithm.

Generating Digital Signatures

You have seen how digital signatures can be used for data integrity. Let's see what other ways message digests can be used. Remember our post office example? What good is keeping something secret if you can't be sure of who sent it to you?

Hash algorithms when combined with asymmetric or public-key algorithms can be used to produce digital signatures that guarantee the authenticity of data the same way a written signature verifies the authenticity of a printed document. Recall that a public-key algorithm is one whereby the method of encrypting and decrypting information relies on a pair of keys. A public key is freely distributed and available to everyone, and its mate, a private key, is known only to its holder. Normally, we would be using someone's public key to encrypt a document. Let's consider how we can do a reverse process to guarantee someone's identity.

We know a one-way hash function is used to compute a hash value. We know that any change to the message produces a different hash value when using the same hash function to recompute the contents of the same message. Using this quality in combination with public-key encryption, we can make a digital signature by performing the following steps:

1. Alice runs a one-way hash function over the purchase order she is signing.

2. Alice encrypts the hash value using her secret key.

3. She appends this encrypted hash value to the end of her purchase order.

Now, anybody that has or obtains her public key can verify that it was Alice who created the purchase order. Using Alice's public key, Bob is now able to recover the original hash value. He then uses the same hash algorithm on the purchase order he receives and compares that value to the hash value that was appended to the end of the purchase order. If the two checksums are identical, he knows for sure that Alice's public key matches the private key that was used to create the digital signature for this purchase order, proving it was Alice. You can see how message digests are used to create a digital signature.

With the goal of accelerating the migration to a "paperless" electronic society, the National Institute of Standards and Technology (NIST) in 1994 adopted the Digital Signature Standard (DSS), also referred to as FIPS-186. (FIPS is an acronym for Federal Information Processing Standard.) There are two major standards of digital signature algorithms (DSAs) currently being used for this application. The first is DSS and the other is the RSA version of a DSA. RSA is generally considered to be the best.

We have seen how public-key encryption technology can be used with a message digest to create a digital signature. One concern with signing data by encryption, however, is that encryption and decryption are computationally intensive. Many network devices and processes need to constantly send routing table updates to their appropriate neighbors. This data may not need to be encrypted, but that is not to say it can be done without concern. If the routers on the Internet were to allow anyone to send any routing table updates anywhere they choose, we would have a real mess. When you do such activities that involve communication of this nature, you want to be sure of two things:

▼ The sender of the data is as claimed; that is, the sender has signed the data and this signature can be checked.

▲ The transmitted data has not been changed since the sender created and signed the data.

Given the overhead involved, signing data using complete encryption and decryption methods may be way more than you need. A more efficient approach that can accomplish these two goals without full message encryption, would be to use a one-way hash function or message digest. So now you can see where such functions can be used to accomplish many things and in a very secure and efficient manner.

CERTIFICATES

We know that to verify a digital signature, the recipient must have access to the signer's public key and have assurance that it corresponds to the signer's private-key pair. This is the only way to decrypt the encrypted hash and authenticate the signature. We need some way to bind that pair and to be absolutely certain that they are authentic.

If there are only two parties to a transaction, then they can communicate this information through reliable and secure means such as a telephone or courier company. (Doing it this way would at least require that the person trying to intercept or misrepresent themselves actually know that the transaction is taking place, know the key and physically be at the receiving end of the transaction itself.) This is no small task, especially if the parties are located far apart. Another point to note is that entities that normally use the Internet as a business communication medium are not individual people but companies or similar artificial entities.

As data communication increasingly moves from a one-to-one to the many-on-many architecture of the Internet and as transactions occur between unknown entities with no prior connection, the more it becomes necessary to provide a means of authentication. Also needed is a method to prevent an entity from later repudiating a transaction.

The solution lies in the use of one or more trusted third parties to associate an identified signer with a specific public key. That trusted third party is referred to as a certification authority, certificate authority, or CA.

Digital Signatures and Certificates

A *certificate* is a digitally signed statement that provides independent confirmation of a person's public key. Certificates can be issued by anyone with a certificate server, but the most common and most reliable certificates are issued by trusted third parties such as governments, financial institutions, or CA companies that specialize in certificate services. The certificate generally includes, at minimum, the following items: (It can contain much more information and in more detail depending on the X.509 class rating.)

▼ The name of the certificate authority

■ The name and attributes of the person that the public key belongs to

■ The actual public key of the person represented by the certificate authority

▲ A digital signature of the certificate authority at the end of the certificate itself

Recall earlier in this chapter when we mentioned *issuing authorities,* we were talking about CAs. In the post office example in Chapter 13, these were the companies that dis tributed and replenished public keys to anyone that needed them and verified the identities of the individuals beforehand. Figure 14-1 shows what takes place when a sender generates a digitally signed purchase order with a certificate prior to transmission.

Now take a look at Figure 14-2 to see what takes place when a sender receives a digitally signed purchase order with a certificate after the transmission.

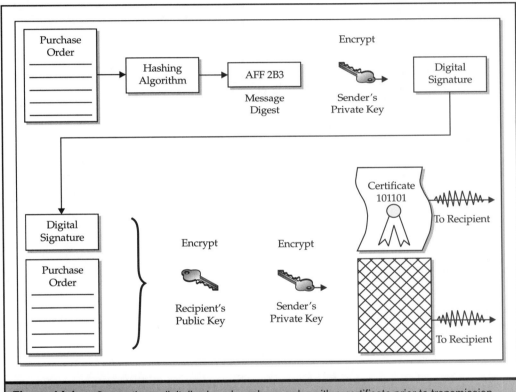

Figure 14-1. Generating a digitally signed purchase order with a certificate prior to transmission

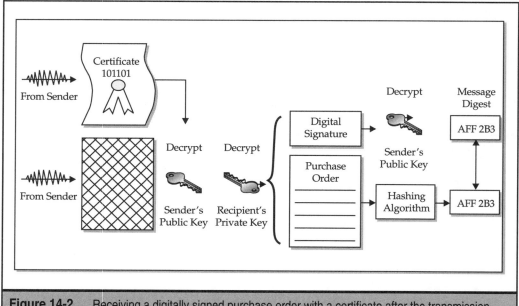

Figure 14-2. Receiving a digitally signed purchase order with a certificate after the transmission

As far as man-in-the-middle attacks go, this would eliminate Mullet's ability to intercept and forge public keys, because Alice and Bob would be getting these public keys from a trusted third party, a certificate authority. These certificates would be digitally signed by the CA as well. The middle with regard to key exchange has been removed from the picture. We will pretend for a moment that there is a guy named Ted who is the world's most famous certificate authority and our trusted third party. Let's see how Mullet deals with this example:

1. Alice e-mails Bob asking for his public key and certificate.
2. Bob sends Alice his public key and certificate and digitally signs it.
3. Alice e-mails Ted asking for Bob's public key and certificate.
4. Ted sends Alice Bob's public key and certificate and digitally signs it.
5. Alice compares Ted's version of the certificate with Bob's version and they match perfectly.
6. Alice encrypts a secret symmetric key using Bob's public key and sends it along with her public key and certificate digitally signed by her.
7. Bob e-mails Ted asking for Alice's public key and certificate.

8. Ted sends Bob Alice's public key and certificate and digitally signs it.

9. Bob compares Ted's version of the certificate with Alice's version and they match perfectly.

10. Alice and Bob communicate very securely.

11. Mullet sees that his future in being a cryptographic middleman is starting to fade and now sets his sights on becoming an attorney, where his future as a middleman is very bright indeed.

We will now discuss the two most common protocols used by certificate authorities. By far the most common and widely recognized certificate in the world is the X.509 version 3 as put forth by the International Standards Organization (ISO). The other is Lightweight Directory Access Protocol (LDAP).

X.509 Certificate Standard

In 1988, the International Telecommunications Union (ITU) recommended a standards specification to be used with cryptographic certificates known as X.509 that currently supports three versions:

▼ **X.509 Version 1** This is the original version and has been available since 1988. This version is still widely deployed and is the most generic certificate.

■ **X.509 Version 2** This version introduced the concept of subject and issuer unique identifiers to handle the possibility of reuse of subject and/or issuer names over time. Most certificate profile documents strongly recommend that names not be reused and that certificates should not make use of unique identifiers. X.509v2 certificates are not widely used.

▲ **X.509 Version 3** This is the most recent (1996) version and supports the notion of extensions, whereby anyone can define an extension and include it in the certificate. Some common extensions limit the use of the keys to particular purposes such as "signing only," allowing other identities to also be associated with this public key (DNS names, e-mail addresses, IP addresses, and so on). Extensions can be marked "critical" to indicate that the extension should be checked and enforced.

All the data in a certificate is encoded using two related standards called ASN.1/DER. Abstract Syntax Notation 1 (ASN.1) describes data. The Definite Encoding Rules (DER) describe a single way to store and transfer that data.

Distinguished Names

An X.509v3 certificate binds a distinguished name (DN) to a public key. A DN is a series of name-value pairs derived from the X.500 naming conventions that are used to uniquely identify an entity. This is called the *certificate subject.* Let's use an example that might be used as a typical DN for an employee of McGraw-Hill Publishing:

▼ uid = doe
■ e = elias@mcgraw-hill.com
■ cn = John Doe
■ o = McGraw-Hill Publishing Inc.
▲ c = US

The abbreviations before each equal sign in this example have these meanings:

▼ **uid** User ID
■ **e** E-mail address
■ **cn** User's common name
■ Organization
▲ **c** Country

DNs may include a variety of other name-value pairs. They are used to identify both certificate subjects and entries in directories that support the Lightweight Directory Access Protocol (LDAP).

A Typical Certificate

Every X.509 certificate consists of two sections: a data section and a signature section. The *data* section includes the following information:

▼ The version number of the X.509 standard supported by the certificate.

■ The certificate's serial number. Every certificate issued by a CA has a serial number that is unique among the certificates issued by that CA.

■ Information.

■ Information about the user's public key, including the algorithm used and a representation of the key itself.

■ The DN of the CA that issued the certificate.

■ The period during which the certificate is valid (for example, between 12:00 P.M. on April 7, 2000, and 12:00 P.M. April 7, 2001).

■ The DN of the certificate subject, also called the *subject name.*

▲ Optional certificate extensions, which may provide additional data used by the client or server. For example, the certificate type extension indicates the type of certificate, such as whether it is a client certificate, server certificate, a certificate for signing e-mail, and more.

The *signature* section includes the following information:

▼ The cryptographic algorithm, or cipher, used by the issuing CA to create its own digital signature

▲ The CA's digital signature, obtained by hashing all of the data in the certificate together and encrypting it with the CA's private key

Lightweight Directory Access Protocol (LDAP) Standard

This is pretty similar to the X.509 in that it uses the X.500 naming convention, but is smaller and easier to implement. Generally, the way it works is when a server gets a request from a client, it asks for the client's certificate before proceeding. The client then sends its certificate to the server. The server then takes the CA listed in the certificate and tries to match it to a trusted CA that is listed on that server. If there isn't a match, some servers end the connection, and some perform a different operation based on the failed match. After the server has determined that the certificate's CA is trusted, the server performs the following steps to map the certificate to an LDAP entry:

1. The server maps the user's DN from the user's certificate to a branch point in the LDAP directory.

2. It searches the LDAP directory for an entry that matches the information about the user of the certificate.

3. It then verifies the user's certificate with one in the LDAP entry that maps to the distinguished name (DN).

After the server finds a matching entry and certificate in the LDAP directory, it uses that information to process whatever transaction is being requested.

Certificate Protocols

We will talk next about two of the most common protocols used for secure communication over the Internet that take advantage of using certificates.

Secure/Multipurpose Internet Mail Extensions (S/MIME) Protocol

Secure/Multipurpose Internet Mail Extensions (S/MIME) protocol uses cryptography to protect against snooping, tampering, and forgery. It protects against snooping by encrypting the message with a *symmetric* algorithm (using the same key for encryption and decryption). This requires that the symmetric key be transmitted with the message, or the recipient won't be able to read it. The key itself must be encoded, or everyone will be able to read it. The symmetric key is encrypted using an *asymmetric* algorithm (using a public- and private-key pair), and every person using S/MIME must have one. The sender uses the public key of the recipient to encrypt the symmetric key, and the recipient uses the private key to decrypt it.

S/MIME uses digital signatures (encryption with a private key) to protect against both tampering and forgery. First, the message text is condensed, using a one-way hash algorithm, into a message digest. That message digest is then encrypted with the sender's private key, and this signature is sent with the message.

The final piece of S/MIME is the trust hierarchy, and it is this method that is based on certificates for verifying that people really are who they say they are. Because of the importance of the trust hierarchy, most implementations of S/MIME include a certificate management system that allows you to set trust levels, but not all do.

SSL Protocol

The Secure Sockets Layer (SSL) protocol, which was originally developed by Netscape, is a set of rules governing server authentication, client authentication, and encrypted communication between servers and clients. SSL is widely used on the Internet, especially for interactions that involve exchanging confidential information such as credit-card numbers.

SSL requires a server SSL certificate, at a minimum. As part of the initial "handshake" process, the server presents its certificate to the client to authenticate the server's identity. The authentication process uses public-key encryption and digital signatures to confirm that the server is the server it claims to be. Once the server has been authenticated, the client and server use techniques of symmetric-key encryption, which is very fast, to encrypt all the information they exchange for the remainder of the session and to detect any tampering that may have occurred.

Servers may optionally be configured to require client authentication as well as server authentication. In this case, after server authentication is successfully completed, the client must also present its certificate to the server to authenticate the client's identity before the encrypted SSL session can be established.

Certificate Types

Different types of certificates are used to accomplish authentication and integrity verification:

▼ **Client SSL certificates**   These are used to identify clients to servers, also known as *client authentication*. Usually, the identity of the client is assumed to be the same as the identity of a human being, such as an employee in an enterprise. Client SSL certificates can also be used for form signing and as part of a single sign-on solution. A bank gives a customer a client SSL certificate that allows the bank's servers to identify that customer and to authorize access to the customer's accounts. A company might give a new employee a client SSL certificate that allows the company's servers to identify that employee and to authorize access to the company's servers.

■ **Server SSL certificates**   These can be used to identify servers to clients, also known as *server authentication*. Server authentication may be used with or without client authentication. Server authentication is a requirement for an encrypted SSL session. Internet sites that engage in electronic commerce will usually support certificate-based server authentication, at a minimum, to establish an encrypted SSL session and to assure customers that they are dealing with a web site identified with a particular company. The encrypted SSL session ensures that personal information sent over the network, such as credit-card numbers, cannot easily be intercepted.

- **S/MIME certificates**   These are used for signed and encrypted e-mail. As with client SSL certificates, the identity of the client is typically assumed to be the same as the identity of a human being, such as an employee in an enterprise. A single certificate may be used as both an S/MIME certificate and an SSL certificate. S/MIME certificates can also be used for form signing and as part of a single sign-on solution. A company can deploy combined S/MIME and SSL certificates solely for the purpose of authenticating employee identities, thus permitting signed e-mail and client SSL authentication but not encrypted e-mail. Another company issues S/MIME certificates solely for the purpose of both signing and encrypting e-mail that deals with sensitive financial or legal matters.

- **Object-signing certificates**   These certificates are used to identify signers of data files or things like Java code, JavaScript scripts, and other signed files. A software company signs software distributed over the Internet to provide users with some assurance that the software is a legitimate product of that company. Using certificates and digital signatures in this manner can also make it possible for users to identify and control the kind of access downloaded software has to their computers.

- ▲ **CA certificates**   These are used to identify CAs. Client and server software use CA certificates to determine what other certificates can be trusted.

Using Certificates

Certificates have a number of applications that can take advantage of the security that certificates were designed to provide. This section talks about a few of those applications.

Signed and Encrypted E-Mail

Some e-mail programs support digitally signed and encrypted e-mail using a widely accepted protocol known as Secure/Multipurpose Internet Mail Extension (S/MIME). Using S/MIME to sign or encrypt e-mail messages requires the sender of the message to have an S/MIME certificate.

An e-mail message that includes a digital signature provides some assurance that it was sent by the person whose name appears in the message header, thus providing the authentication of the sender.

The digital signature is unique to the message it accompanies. If the message received differs in *any* way from the message that was sent, the digital signature cannot be validated. Therefore, signed e-mail also provides some assurance that the e-mail has not been tampered with. As discussed at the beginning of this document, this kind of assurance is known as *nonrepudiation*. In other words, signed e-mail makes it very difficult for the sender to deny having sent the message. This is important for many forms of business communication.

S/MIME also makes it possible to encrypt e-mail messages. This is also important for some business users. However, using encryption for e-mail requires careful planning. If the recipient of encrypted e-mail messages loses his or her private key and does not have access to a backup copy of the key, for example, the encrypted messages can never be decrypted.

Single Sign-On

Network users are frequently required to remember multiple passwords for the various services they use. For example, a user might have to type a different password to log into the network, collect e-mail, and access various servers. Multiple passwords are a big inconvenience for both the user and the system administrator.

When users have to keep track of many different passwords, they will usually either choose easily breakable ones or write them down in obvious places.

A system administrator must keep track of a separate password database on each server and deal with associating rights and permissions as well as the potential security problems related to passwords being sent over the network routinely and frequently. This makes successful sniffing of the password more likely.

Solving this problem requires some way for a user to log in once using a single password and to get authenticated access to all network resources that user is authorized to use without sending any passwords over the network. This is what is known as *single sign-on* capability.

Both client SSL certificates and S/MIME certificates can play a significant role in a comprehensive single sign-on solution. For example, one form of single sign-on relies on SSL client authentication. A user can log in once using a single password to the local client's private-key database and get authenticated access to all SSL-enabled servers that user is authorized to use—without sending any passwords over the network. This approach simplifies access for users, because they don't need to enter passwords for each new server. It also simplifies network management, since administrators can control access by controlling lists of certificate authorities (CAs) rather than much longer lists of users and passwords.

In addition to using certificates, a complete single sign-on solution must address the need to interoperate with enterprise systems, such as the underlying operating system, that rely on passwords or other forms of authentication.

Form Signing

Many kinds of e-commerce require the ability to provide persistent proof that someone has authorized a transaction. SSL doesn't have this capability. You can use S/MIME to provide persistent authentication for e-mail, but e-commerce often involves filling in a form on a web page rather than sending an e-mail.

However, simple form signing is a process that allows you to associate a digital signature with web-based data generated as the result of a transaction, such as a purchase order or other financial document. The private key associated with either a client SSL certificate or an S/MIME certificate may be used for this purpose.

Whenever you click the Submit button on a web-based form that supports form signing, a dialog box generally appears that displays the exact text to be signed. The form designer can either specify the certificate that should be used or allow the user to select a certificate from among the client SSL and S/MIME certificates that are installed in your browser software. When the user clicks OK, the text is signed, and both the text and the digital signature are submitted to the server. The server can then validate the digital signature.

Object Signing

Most Internet browser software products support a set of tools and technologies called *object signing*. Object signing uses standard techniques of public-key cryptography to let users get reliable information about the data they download in much the same way they can get reliable information using the paper certificates you might find in new, shrink-wrapped software. Most importantly, object signing helps you decide whether you trust the data you are about to download based on the level of trust you have for the name represented on the certificate.

DIGITAL SIGNATURES, CERTIFICATES, AND PUBLIC KEY INFRASTRUCTURE (PKI)

As part of the applications of the DSS, the U.S. government is also planning a network of certification authorities (CAs) to create what is called the Public Key Infrastructure (PKI). Digital signatures combined with digital certificates will eventually be used for securing all electronic transactions for the government and are considered the enabling technology for conducting secure electronic commerce and verifying the integrity of software and other forms of data.

What Is a PKI?

A simple PKI starts with a certificate authority (CA), a software package operated in a high-security area by a trusted third party to issue X.509v3 digital certificates. A digital certificate is used to bind the public-key values to information identifying the listed entity (people or devices) that is digitally signed by the issuing CA. You will recall in Chapter 13 when we talked about public-key algorithms that they actually involve a pair of keys. Everyone has their own pair, one that is public and one that is private or secret. The certificate assures any relying party (RP) using an entity's public key that the private key associated with that pair does in fact belong to that entity.

In the high-security area of the certificate authority, there will also be at minimum an X.509v3-compatible database. The CA operator issues the digital certificate to the end entity (EE) and stores a copy of that certificate in its database. The language used to query the X.509v3 database for any reason is Lightweight Data Access Protocol v3 (LDAP v3).

Handling of Invalid Certifications

Even if a certificate has not expired, it may be considered invalid or unusable for several reasons:

▼ The owner may no longer need it.

■ The certificate might have been compromised or stolen.

▲ The owner may have been issued a newer certificate that takes precedence over the existing one.

Should the certificate be considered invalid, the CA may do one of two things:

▼ The certificate may be listed on a Certificate Revocation List (CRL X.509v2) and published at a given interval. The CRL stayed at version 2 because it was agreed that there were no changes on it when the other components went to version 3.

▲ The certificate revocation is published using Online Certificate Status Protocol (OCSP) on an online server, which could be the X.509 v3 database server, providing that service.

Each time an IPSec endpoint checks the validity of a certificate presented to it for authentication, it checks its latest cached CRL or uses OCSP to see if that certificate is listed. If it is listed, that means the certificate is no longer valid, and the IPSec endpoint will reject it.

How a PKI Works

A PKI can utilize multiple CAs with a *root CA*. The root CA holds a self-signed certificate and issues certificates to the subordinate CAs, which in turn can issue certificates to *registration authorities* (RA) or *local registry authorities* (LRA). In operation, the RA or LRA takes the initial request for a certificate from the requesting party and passes the authenticated request to its CA, which issues the certificate. The hierarchy of CAs resembles a tree, which is why the initial CA is identified as the root CA. To see what a CA hierarchy or "certificate chain" looks like, take a look at Figure 14-4.

At this point, a chain of trust has been established between all of the EEs (in this case, IPSec endpoints) from all of the subordinate CAs. But how does EE-1, the *relying party* (RP) whose certificate was issued by CA-1, know that a certificate EE-5 issued by CA-2 is trustworthy? Here are the steps showing how it works:

1. EE-1 either checks the CRLs or uses OCSP to see if EE-5 is a valid certificate.
2. EE-1 checks who signed the EE-5 certificate and finds that CA-2 is the authorizing party.
3. CA-2 is unknown to EE-1, so it checks to see who signed certificate CA-2.
4. EE-1 finds that it was CA-0, the root certificate, who also signed certificate CA-1.
5. CA-1 issued and signed the certificate EE-1.

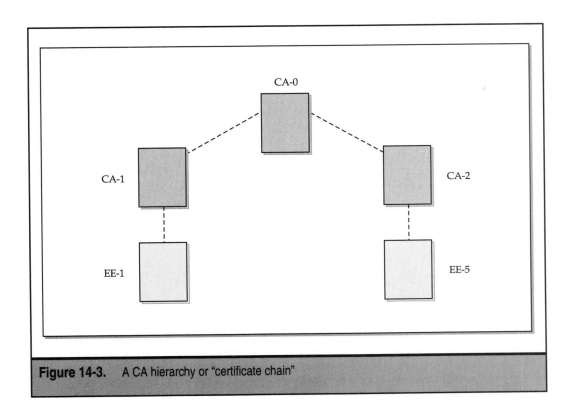

Figure 14-3.   A CA hierarchy or "certificate chain"

This information proves the verification of EE-5 because it is within the same PKI. This procedure is often referred to as "walking the chain of trust."

Certificate Policies

The *certificate policy* (CP) lays out the requirements for authentication to receive a certificate from the CA and the level of authority for the certificate, such as "this certificate allows signature authority for $10,000."

In the case of an IPSec endpoint, the CP defines what information must be submitted to the CA for authentication prior to the issuance of a certificate for that endpoint. It also details what information the individual certificate will contain and how it will look. The CP specifies the CRL update or the requirements for posting the revoked certificate notification to the OCSP server. The CP may also specify the physical security requirements that the CA must meet.

SUMMARY

In this chapter, we learned about digital signatures and the different ways we can create them using tools like message digests, CRCs, and hash functions.

We saw how we could use those digital signatures to create certificates to allow organizations and individuals a way to authenticate each other and discussed the different certificate protocols available to create these certificates in a consistent manner.

We then learned about PKI and the different infrastructures and hierarchies put in place to help distribute those certificates on a global scale.

In the next chapter, we will learn about authentication methods, protocols, and systems used to do this.

CHAPTER 15

Authentication

There is a distinct difference between *authentication* and *encryption*. There tends to be a lot of confusion between these two concepts, perhaps because they both involve the use of cryptography. Authentication and authorization also have very specific meanings, though the two processes are often confounded and in practice are often not clearly distinguished. This chapter is titled "Authentication," and this term is often used to describe the concept of "access management" or *access control systems* (ACS). It often combines the definitions of authentication, authorization, and, in some cases, implementations accounting. This is why, along with *ACS*, you might hear *authentication* used interchangeably with *AAA* or "triple A" services. Suffice it to say that authentication servers generally include all or some combination of authentication, authorization, and accounting (monitoring) of what goes into and out of a corporate network.

We have covered cryptography in some detail, so we will talk in this chapter about authentication and access management. Authentication is not limited to people. Devices, protocols, and locations can also be authenticated in many different ways.

HUMAN IDENTIFICATION

One of the problems when connecting with a system remotely is that the computer needs a way to make sure you are who you say you are. Once a user has identified him- or herself to the computer, the computer has to decide whether that user is "authentic," or who they say they are. This process is known as *authentication*.

There are essentially three ways you can accomplish this with a computer. To prove to someone or something that you are who you say you are, you can provide one or more of the following:

- ▼ Something you know.
- ■ Something you have.
- ▲ Something you are.

Something You Know

By far the most common form of authentication is by providing "something you know." We are all familiar with the almost daily exercise of entering passwords to log onto a computer or a PIN (personal identification number) to use an automated teller or to retrieve messages from a voicemail system. This is generally referred to as *password* or *username/ password authentication*. This is the method used most commonly on the Internet.

To authenticate yourself, you simply provide the site you are visiting with "something you know," such as a username and/or a password. You may even have experienced the extension of the "something you know" method. If you forget your password, you may be asked to answer a whole series of questions that were agreed upon when you signed up to use the site, like your mother's maiden name, the name of a pet, and so on. Once you answer any or all of these questions, you are either permitted to create a new password or to have your original password e-mailed to you.

Unfortunately, this form of authentication is considered to be the least secure, because a lot of this information is predictable and is information that another person could find out. Because of this drawback, another authentication variation that is growing in popularity is that of allowing registrants to make up a couple of questions and answers of their own. For example, you might choose to provide the name of your first dog, the name of your first girlfriend's brother, or maybe the name of your math teacher in high school.

While making up your own questions and answers may virtually eliminate the likelihood of anyone else discovering the "something you know," the problem that remains is the nature of the password itself. These passwords are usually words that exist in any standard dictionary. The average human knows about 8000 words. The average vocabulary of college graduates is about 15,000. To a computer, trying all variations of these common English words is child's play. So, what if you pepper your password with a number or symbol here and there? You might even juxtapose a word or two. Oh, no! Now the poor computer has to worry about hundreds of thousands or even hundreds of millions of permutations. You may have heard the acronym MIPS used before. It means millions of instructions per second and is often used to describe a computer's performance rating. Virtually any personal computer can have most of these passwords deciphered in no time.

Something You Have

Now we move on to "something you have." Almost all of us are familiar with the concept of something you have because most of us have ATM cards. The card is useless without the PIN, and the PIN is useless without the card.

On the Internet, there are a number of "things" you can have, and they are usually supplied by an organization with which you will be interacting. These things are generally called *tokens*. There are *soft tokens* and *hard tokens*. This technology, used in combination with a password or PIN of some kind, is used to create what is known as a *one-time password* (OTP).

Soft Tokens

A *soft token* is a piece of software that emulates a hard token that a given organization has you install on your home computer. The software creates a secure channel between your computer and the organization by using the OTP to authenticate you. While soft tokens provide fairly good authentication, the problem is that software can, if it falls into the wrong hands, be replicated. Another limitation of soft token authentication is that it isn't very mobile. Often, the software that you are sent can only be installed on one computer or, if you try to install it on your computer at work, it might not function through your network firewall. This problem reduces mobility because you can only perform transactions from one computer.

Hard Tokens

The other type of token is a *hard token*. A hard token could be a smart card that you run through a computer with a card reader, or it might be a USB device or PCMCIA card that you plug into a given port on a computer. A hard token authentication system is more secure, because it requires users to present a physical element (something you have) in addition to an ID and password (something you know) to provide an added layer of security against hackers and intruders.

A good example of a hard token is the key-fob hard token device. This piece of hardware, which looks like a tiny calculator display face, works by using time-synchronized number combinations that are synchronized between the server being accessed and the key fob itself. The numbers on the display change every 30 seconds or so, and at the moment that you wish to log in, you key in the number that appears on the key-fob display, along with a password and/or PIN. This kind of hard token provides a pretty high degree of authentication, mainly because the numbers change constantly and provide excellent mobility.

Something You Are

This brings us to "something you are." This method involves the use of *biometrics*. This "something you are" is considered to be the most secure type of authentication. It can include thumbprints, fingerprints, voiceprints, retinal scans, handprints, and even a person's signature. Although these means of authentication are very mobile, very secure, and almost impossible to duplicate, they present a problem and that is change.

Personal traits can change dramatically. A woman's retina changes when she is pregnant. Retinas can also change when cholesterol levels rise or lower dramatically. Thumbprints and fingerprints are a problem because all you have to do is cut your thumb, and it no longer matches the print on the file. Your voice can be affected by illness and aging, as can your signature. As important as it is to keep impersonators from accessing a computer, it is equally important that legitimate users be allowed access to these computers.

Obviously, you need to decide how much you are willing to spend, how important the data being accessed is, and how much people will put up with to maintain any level of productivity. The best solution is using a combination of these methods. Remember when you were in school and a friend would say, "I'll flip you for it," at which point you agreed to flip a coin, only to find it didn't fall your way? You would then immediately suggest "OK, best two out of three" or "best four out of five" (and so on, until *you* would win). Since biometrics are subject to change and passwords can easily be beaten, using all of them in a best three out four or other combination might work best.

ENTITY AUTHENTICATION

Sometimes it is just as important that the computers themselves be authenticated as it is that users be authenticated to computers. Have you ever downloaded software from a web site? You may have done so intentionally or entirely without your knowledge, because some software products update themselves without any intervention from you. Chances are that you have downloaded updates for your operating system from Microsoft.

Computer authentication is important in these instances. We are all aware of the damage, on an unimaginable scale, that viruses, worms, and Trojan horses can inflict. This is where digital signatures and certificates come in. Digital signatures can be used by themselves for data integrity, but by using them to create certificates, you can check the authenticity of a person, organization, or machine with a certificate of authority (CA). This is the most secure method for authenticating a device.

Policy-Based Filtering

This is an access method used for things like processes, services, protocols, source addresses, and the like. Policy-based filtering is not a "mandatory" requirement for implementing a VPN solution, but it significantly increases the security of the network itself as well as the integrity of the data being transported across the Internet. This increased security is accomplished through the use of a firewall or router. You can use either a dedicated firewall or a perimeter router with a firewall feature set used in parallel with a VPN device. You can also use VPN-capable firewalls by themselves.

We talked about firewalls in Chapter 7, but to recap, the main idea of policy-based filtering is to give you the ability to enforce the flow of certain types of network traffic on a per-packet basis. This allows you to control who can and who can't use specific network resources. Filtering can be used in a several different ways to block connections to or from specific hosts and/or networks. You can even block connections to specific ports. You may want to block connections from certain addresses, such as from hosts or networks that you consider hostile or untrustworthy. You might even choose to block connections from *all* addresses external to your network with the exception of a specific few.

Adding TCP or UDP port filtering to address filtering can give you even more flexibility. Services such as Telnet reside (unless specifically changed) at specific ports called *well-known ports*. For example, Telnet resides at well-known port 23. If you set policies to block TCP or UDP connections to or from specific ports, then you can implement policies that call for certain types of connections to be made to some hosts, but not others. You may wish to allow only specific services, such as SMTP for one system and Telnet or FTP connections to another system. With filtering on TCP or UDP ports, this policy can be easily implemented in a firewall, in a packet-filtering router, or by specific hosts with packet-filtering capabilities. Just how granular you can set the level of access to your network depends on the requirements of your situation.

AUTHENTICATION PROTOCOLS

Several different authentication protocols are available that can be used alone or in combination with one or more different authentication protocols that are designed to interoperate with those protocols. In this section, we examine some of the more popular versions of these protocols.

Password Authentication Protocol (PAP)

PAP is a subprotocol of the PPP protocol suite. PAP is a simple, standards-based password protocol. A user's ID and password are transmitted at the beginning of an incoming call, then validated by the receiving equipment using a central PAP database. The PAP password database is encrypted, but PAP does not encrypt the user ID or password on the transmission line. The password is transmitted in the clear, making it easy to snoop by tapping the line. The Password Authentication Protocol (PAP) can authenticate an identity and password for a peer, resulting in success or failure.

To establish communications over a point-to-point link, each end of the PPP link must first send LCP packets to configure the data link during the *Link Establishment* phase. Once the link has been established, PPP provides for an optional *Authentication* phase before proceeding to the *Network Layer Protocol* negotiation phase.

By default, authentication is not mandatory. If authentication of the link is desired, an implementation must specify the *Authentication Protocol Configuration* option during the *Link Establishment* phase.

PAP is generally used by devices that connect to a PPP network server over dial-up lines, but can also be used over dedicated links. The server can use the identification of the connecting host or router in the selection of options for Network layer negotiations.

PAP uses a rudimentary method for the peer to establish its identity using a two-way handshake. This is done only upon initial link establishment. After the *Link Establishment* phase is complete, a UserID/Password pair is repeatedly sent by the peer to the authenticating device until authentication is acknowledged or the connection is terminated.

PAP offers only minimal authentication, because the entire password is sent over the network medium "in the clear," and there is no protection from playback or repeated trial-and-error attacks. It is the peer that determines the frequency and timing of these attempts.

Any implementations consisting of more than one authentication protocol, like CHAP, must offer to negotiate with that protocol first before using PAP. This authentication method is generally reserved for applications that require a plaintext password or that do not have more powerful authentication protocols in common. PAP can be thought of as the "hey, something is better than nothing" authentication protocol.

PAP Configuration Options

00	01	02	03	04	05	06	07	08	09	10	11	12	13	14	15
Option								Length							
Data :::															

Option 8 bits.

Type	Length	Description
3	4	Authentication Protocol C023

Length   8 bits.
Data   Variable length.

Challenge Handshake Authentication Protocol (CHAP)

CHAP is another subprotocol of the PPP protocol suite. CHAP is a standards-based authentication service for periodically validating users with a sophisticated challenge-handshake protocol. (Different versions based on this same idea have been developed by Microsoft and would be represented using the MS-CHAP acronym.) The initial CHAP authentication is performed during the logon attempt. A network administrator can specify the rate of subsequent authentications. The use of repeated challenges is intended to limit the time of exposure to any single attack. CHAP transmissions are encrypted to afford greater protection. CHAP is secure against eavesdropping because it encrypts the password during transmission. However, one vulnerability is caused by the fact that a standard CHAP password database is kept in a plaintext form. Hence, it is not useful for large installations. CHAP is used to periodically verify the identity of the peer using a three-way handshake. This is done upon initial link establishment and may be repeated anytime after the link has been established.

1. After the *Link Establishment* phase is complete, the authenticating device sends a "challenge" message to the peer.

2. The peer responds with a value calculated using a one-way hash function.

3. The authenticating device checks the response against its own calculation of the expected hash value. If the values match, the connection will be authenticated. If they do not, the connection should be terminated.

4. At random intervals, the authenticating device sends a new challenge to the peer, after which it repeats steps 1 through 3.

To establish communications over a point-to-point link, each end of the PPP link must first send LCP packets to configure the data link during the *Link Establishment* phase. After the link has been established, PPP provides for an optional *Authentication* phase before proceeding to the *Network Layer Protocol* phase. CHAP is used by devices that connect to a PPP network server over dial-up lines, but can also be used over dedicated links. The server can use the identification of the connecting host or router in the selection of options for Network layer negotiations.

CHAP provides protection against a playback attack by the peer through the use of an incrementally changing identifier and a variable challenge value. The use of repeated challenges is intended to limit the time of exposure to any single attack. The authenticating device is in control of the frequency and timing of the challenges. This authentication method depends upon a "secret" known only to the authenticating device and that peer. The secret is not sent over the link. Although the authentication is only one-way, by negotiating CHAP in both directions, the same secret set may easily be used for mutual authentication. Since CHAP may be used to authenticate many different systems, name fields may be used as an index to locate the proper secret in a large table of secrets. This also makes it possible to support more than one UserID/Password pair per system and to change the secret in use at any time during the session.

CHAP requires that the secret be available in plaintext form. One-way hash encrypted password databases commonly available cannot be used. CHAP is not as useful for large installations, since every possible secret is maintained at both ends of the link.

The Challenge packet is used to begin the CHAP process. The authenticating device must transmit a CHAP packet with the Code field set to 1 (Challenge). Additional Challenge packets must be sent until a valid Code field set to 2 (Response) packet is received or an optional retry counter expires. A Challenge packet may also be transmitted at any time during the *Network Layer Protocol* phase to ensure that the connection has not been altered. The peer should expect Challenge packets during the *Authentication* phase and during the *Network Layer Protocol* phase. Whenever a Challenge packet is received, the peer must transmit a CHAP packet with the Code field set to 2 (Response). Whenever a Response packet is received, the authenticating device compares the Response Value with its own calculation of the expected value. Based on this comparison, the authenticating device must send a Success or Failure packet.

CHAP Packet Format

PPP Header	CHAP Header	Data

CHAP Header

00	01	02	03	04	05	06	07	08	09	10	11	12	13	14	15	16	17	18	19	20	21	22	23	24	25	26	27	28	29	30	31
Code								Identifier								Length															
Data																															
:::																															

▼ **Code** 8 bits. This field specifies the function to be performed.

Code	Description
1	Challenge
2	Response
3	Success
4	Failure

■ **Identifier** 8 bits. This field is used to match challenges, responses, and replies.

- ■ **Length** 16 bits. This is the size of the CHAP packet including the Code, Identifier, Length, and Data fields. Bytes outside of this range should be treated as Data Link layer padding and should be ignored on reception.
- ▲ **Data** Variable length. This can be zero or more bytes of data as indicated by the Length field.

The Authentication-Protocol for CHAP is C223 in hex.

The Extensible Authentication Protocol (EAP)

EAP is a general protocol for PPP authentication that supports multiple authentication mechanisms. EAP does not select a specific authentication mechanism at the *Link Control* phase; instead, it postpones this until the *Authentication* phase. This allows the authenticating device to request more information before determining the specific authentication mechanism. This also permits the use of a "back-end" server, which actually implements the various mechanisms, while the PPP authenticating device merely passes through the authentication exchange.

EAP Packet

PPP Header	LCP Header	EAP Header	Data

EAP Header

00 01 02 03 04 05 06 07	08 09 10 11 12 13 14 15	16 17 18 19 20 21 22 23 24 25 26 27 28 29 30 31
Code	Identifier	Length
Data :::		

- ▼ **Code** 8 bits. This field specifies the IPCP function to be performed.

Code	Description
1	Request
2	Response
3	Success
4	Failure

- ■ **Identifier** 8 bits. This is used to match EAP requests and replies.
- ■ **Length** 16 bits. This represents the size of the EAP packet including the Header and Data fields.
- ■ **Data** Variable length. This is zero or more bytes of data in the Length field.
- ▲ **Type** 8 bits.

The Authentication-Protocol for EAP is C227.

Type	Description
1	Identity
2	Notification
3	NAK (Response only)
4	MD5-Challenge
5	One-Time Password (OTP)
6	Generic Token Card
7	Reserved
8	Reserved
9	RSA Public Key Authentication
10	DSS Unilateral
11	KEA
12	KEA-VALIDATE
13	EAP-TLS
14	AXENT Defender Token
15	Windows 2000 EAP
16	Arcot Systems EAP
17	EAP-Cisco Wireless
18	Nokia IP Smartcard Authentication
19	SRP-SHA1 Part 1
20	SRP-SHA1 Part 2

The RADIUS Protocol

"RADIUS" stands for "Remote Authentication Dial-In User Service." Developed by Livingston Enterprises in collaboration with the Internet Engineering Task Force (IETF), RADIUS is an industry-standard protocol for authenticating remote users and protecting networks from unauthorized remote access.

Distributed Security

The impetus for the development of the RADIUS protocol was a need to provide a secure distributed security model for remote access. The traditional solution for authenticating a client is to require users to enter a username and password. If the username and password match a valid entry in the authenticating server's database, the login is completed, and the system provides access to the files and services that have been enabled for that user. Since the LAN is a closed system, it's easy for the network administrator to secure and manage the database of user accounts.

When remote connections are added to the network, however, the traditional login scheme becomes vulnerable to attack. This is because public telephone lines and third-party access devices are not under the control of corporate system administration, and this makes them potential points of unauthorized access. The RADIUS protocol enhances remote access security by overseeing the transactions that take place between the Network Access Server (NAS) of the Internet service provider (ISP) and an internal RADIUS authentication server located on the corporate network.

This distributed security model keeps all of the authentication technology separate from the underlying communications technology. All user authentication and network access information is kept safe within the corporate internal network. This way the corporate Network Access Server (NAS) becomes a client of the corporate RADIUS server whenever a remote user attempts to gain access to the network. This distributed approach to authentication ensures that the corporate network administrators maintain centralized control over all user authentication information. This requires any dial-in client wanting access to the corporate network to satisfy the prerequisites specified in the company's internal RADIUS server.

How RADIUS Works

Communication between a Network Access Server (NAS) and a RADIUS server is based on the User Datagram Protocol (UDP). Generally, the RADIUS protocol is considered a connectionless service. Issues related to server availability, retransmission, and timeouts are handled by RADIUS protocol instead of the TCP protocol. RADIUS is a client/server protocol. The RADIUS client is typically a NAS, and the RADIUS server is usually a router or a UNIX/Windows-based server.

The client sends user information to designated RADIUS servers and acts on the response it receives. The RADIUS servers receive user connection requests, authenticate the user, and then return the configuration information necessary for the client to gain access to the network. A RADIUS server can also act as a proxy client to other RADIUS servers or as a proxy client to another kind of authentication server.

Figure 15-1 shows the interaction between a dial-in user and the RADIUS client and server.

Here is a description of the steps that take place:

1. The user initiates a PPP authentication to the NAS.
2. The NAS prompts for the username and password when using PAP or in the case of CHAP, a challenge.
3. The user replies.
4. The RADIUS client sends the username and encrypted password to the RADIUS server.
5. The RADIUS server responds with *Accept, Reject,* or *Challenge.*
6. The RADIUS client acts upon services and service parameters bundled with the *Accept* or *Reject* response it receives.

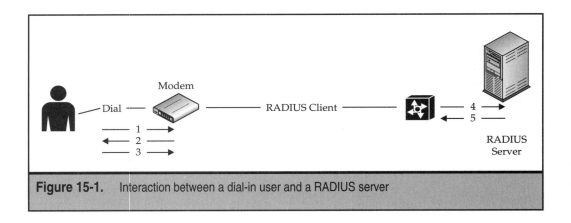

Figure 15-1.   Interaction between a dial-in user and a RADIUS server

Authentication and Authorization

The RADIUS server can support many different methods to authenticate a user. When it is provided with the username and original password given by the user, it can support PAP, CHAP, and other authentication mechanisms. Typically, a user login consists of a query (*Access Request*) from the NAS to the RADIUS server. The *Access Request* packet contains the username, an encrypted password, the IP address of the NAS, and the service port. The officially assigned port number for RADIUS is UDP 1812. (The early version used port 1645.) The format of the request also provides information about the type of session that the user wants to initiate. For example, if the query is presented in character mode, the inference is "Service Type = Exec User," but if the request is presented in PPP packet mode, the inference is "Service Type = Framed User" and "Framed Type = PPP."

When the RADIUS server receives the *Access Request* from the NAS, it searches a database for the username listed. If the username does not exist in the database, the RADIUS server immediately sends an *Access Reject* message. This *Access Reject* message can be accompanied by a text message indicating the reason for the refusal.

When using RADIUS, the authentication and authorization services are coupled together. If the username is found and the password is correct, the RADIUS server returns an *Access Response*, including a list of *attribute-value pairs* (AVP) (similar to the L2TP AVPs we talked about in Chapter 11) that describes the parameters to be used for this session. Typical parameters include service type (shell or framed), protocol type, IP address to assign the user (static or dynamic), Access Control List (ACL) to apply, or a static route to install in the NAS routing table. The configuration information in the RADIUS server defines what will be installed on the NAS. Figure 15-2 illustrates the RADIUS authentication and authorization sequence.

Accounting

RADIUS can also be used for accounting purposes. The accounting features of the RADIUS protocol can be used independently of RADIUS authentication or authorization. The RADIUS accounting functions allow data to be sent at the start and end of a session. They

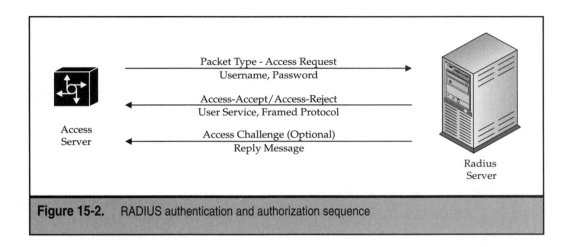

Figure 15-2.   RADIUS authentication and authorization sequence

can measure and record information related to the activity of that session. The information can include things like time, packet information, number of bytes, and so on, used
during the session. The *accounting port* for RADIUS is 1813.

Transactions between the client and RADIUS server are authenticated through the
use of a shared secret, which is never sent over the network. In addition, user passwords
are sent encrypted between the client and RADIUS server to guard against things like
packet sniffing.

RADIUS Packet

MAC Header	IP Header	UDP Header	RADIUS Header	Data

RADIUS Header

00	01	02	03	04	05	06	07	08	09	10	11	12	13	14	15	16	17	18	19	20	21	22	23	24	25	26	27	28	29	30	31
Code								Identifier								Length															
Authenticator - - -																															
Attributes :::																															

▼ **Code**   8 bits. This field identifies the type of RADIUS packet. If a packet is
received with an invalid Code field, it is silently discarded.

Code	Description
1	Access Request
2	Access Accept

Code	Description
3	Access Reject
4	Accounting Request
5	Accounting Response
11	Access Challenge
12	Status Server (Experimental)
13	Status Client (Experimental)
14–254	Unassigned
255	Reserved

- **Identifier**   8 bits. This field is used to match RADIUS request and reply packets.

- **Length**   16 bits. This field indicates the length of the packet including the Code, Identifier, Length, Authenticator, and Attribute fields. Bytes outside the range of the Length field should be treated as padding and should be ignored on reception. If the packet is shorter than the Length field indicates, it should be silently discarded. The minimum length is 20, and the maximum length is 4096.

- **Authenticator**   16 bytes. This field is the value that is used to authenticate the reply from the RADIUS server and is used in the password hiding algorithm.

- **Attributes**   Variable. RADIUS Attributes fields carry the specific authentication, authorization, and accounting details for the request and response. Some attributes may be included more than once. The effect of this is attribute specific and is specified in each attribute description. The end of the list of attributes is indicated by the Length of the RADIUS packet.

Type	Length	Value...

- **Type**   8 bits.

- **Length**   8 bits. This field indicates the length of this attribute including the Type, Length, and Value fields. If an attribute is received in an Accounting Request packet with an invalid Length, the entire request should be silently discarded.

- **Value**   Variable. This field contains information specific to the attribute. The format and length of this field are determined by the Type and Length fields. The format of the field can be one of the following data types:

  - **String**   0 to 253 bytes.
  - **Address**   32 bits.
  - **Integer**   32 bits.

- **Time**   32 bits. This is how many seconds since 00:00:00 GMT, January 1, 1970.

The valid attributes are

Type	Length	Description
1	>= 3	Username
2	18 to 130	User Password
3	19	CHAP Password
4	6	NAS IP Address
5	6	NAS Port
6	6	Service Type
7	6	Framed Protocol
8	6	Framed IP Address
9	6	Framed IP Netmask
10	6	Framed Routing
11	>= 3	Filter Id
12	6	Framed MTU
13	6	Framed Compression
14	6	Login IP Host
15	6	Login Service
16	6	Login TCP Port
17		Unassigned
18	>= 3	Reply Message
19	>= 3	Callback Number
20	>= 3	Callback Id
21		Unassigned
22	>= 3	Framed Route
23	6	Framed IPX Network
24	>= 3	State
25	>= 3	Class
26	>= 7	Vendor Specific
27	6	Session Timeout
28	6	Idle Timeout
29	6	Termination Action

Type	Length	Description
30	>= 3	Called Station Id
31	>= 3	Calling Station Id
32	>= 3	NAS Identifier
33	>= 3	Proxy State
34	>= 3	Login LAT Service
35	>= 3	Login LAT Node
36	34	Login LAT Group
37	6	Framed AppleTalk Link
38	6	Framed AppleTalk Network
39	>= 3	Framed AppleTalk Zone
40	6	Acct Status Type
41	6	Acct Delay Time
42	6	Acct Input Octets
43	6	Acct Output Octets
44	>= 3	Acct Session Id
45	6	Acct Authentic
46	6	Acct Session Time
47	6	Acct Input Packets
48	6	Acct Output Packets
49	6	Acct Terminate Cause
50	>= 3	Acct Multi Session Id
51	6	Acct Link Count
52	6	Acct Input Gigawords
53	6	Acct Output Gigawords
54		Unassigned
55	6	Event Timestamp
56 – 59		Unassigned
60	>= 7	CHAP Challenge
61	6	NAS Port Type
62	6	Port Limit

Type	Length	Description
63	>= 3	Login LAT Port
64	6	Tunnel Type
65	6	Tunnel Medium Type
66	>= 3	Tunnel Client Endpoint
67	>= 3	Tunnel Server Endpoint
68		Acct Tunnel Connection
69	>= 5	Tunnel Password
70	18	ARAP Password
71	16	ARAP Features
72	6	ARAP Zone Access
73	6	ARAP Security
74	>= 3	ARAP Security Data
75	6	Password Retry
76	6	Prompt
77	>= 3	Connect Info
78	>= 3	Configuration Token
79	>= 3	EAP Message
80	18	Message Authenticator
81	>= 3	Tunnel Private Group ID
82	>= 3	Tunnel Assignment ID
83	6	Tunnel Preference
84	10	ARAP Challenge Response
85	6	Acct Interim Interval
86		Acct Tunnel Packets Lost
87	>= 3	NAS Port Id
88	>= 3	Framed Pool
89		Unassigned
90	>= 3	Tunnel Client Auth ID
91	>= 3	Tunnel Server Auth ID
92		Unassigned
93		
94		Originating Line Info
95–191		Unassigned

Type	Length	Description
192–223		Experimental
224–240		Implementation Specific
241–255		Reserved

TERMINAL ACCESS CONTROLLER ACCESS CONTROL SYSTEM (TACACS), XTACACS, AND TACACS+

Unlike the peer relations designed into PAP and CHAP, TACACS, like RADIUS, is designed to function as a client/server system, which affords it more flexibility, especially in security management.

A client/server (protocol and server) architecture places all security information on a single, central database, instead of being dispersed around a network in different devices. This is especially useful if there are thousands of users who are using thousands of access servers distributed around the network. This makes it much more scalable for large enterprises. TACACS and other remote access security protocols are designed to support thousands of remote connections. In a large network, the user database is usually large and is best kept on a centralized server. This saves memory in all the access devices and eliminates the need to update every access server when new users are added or when passwords are modified or changed.

Central to the operation of TACACS and RADIUS is an authentication server. Typically, a TACACS authentication server handles requests from authentication client software that's installed at a gateway or network entry point. The authentication server maintains a database of user IDs, passwords, PINs, and secret keys, which it uses to grant or deny network access requests. All authentication, authorization, and accounting data is handled by the centralized server when a user tries to log in.

Three versions of TACACS are available today: TACACS, Extended TACACS, and the most common of the three, TACACS+. These protocols are closely associated with Cisco Systems because they were the company that took the ball and ran with it in terms of development and addition to the standard. Although TACACS has been described in an IETF RFC, and the source code is freely available for anyone to implement, most vendors view TACACS as proprietary.

TACACS and XTACACS transmit all data in the clear between the user and the server, whereas TACACS+ adds a message-digest function to eliminate the plaintext transmission of passwords. TACACS+ also supports multiprotocol logins, meaning that a single username and password pair can authenticate a user for multiple devices and networks.

TACACS has proxy capabilities that make it easier for a corporate client to share VPN security data with an ISP, which is necessary when a VPN is outsourced. This way the ISP runs a proxy server to control dial-in access based on access rights managed by the corporate customer on its own secure server. But transmitting authentication packets between the parent server and the proxy server across a public network poses a security risk. TACACS and XTACACS use UDP for the most part, but can use TCP. TACACS+ uses TCP exclusively, and RADIUS uses UDP only.

TACACS

TACACS forwards username and password information to a centralized server. The centralized server can either be a TACACS database or a database like the UNIX password file with TACACS protocol support. For example, the UNIX server with TACACS passes requests to the UNIX database and sends the accept or reject message back to the access server.

XTACACS (Extended TACACS)

XTACACS defines the extensions that support these added advanced features:

▼ Multiple TACACS servers

■ syslog—Sends accounting information to a UNIX host

▲ connect—Where the user is authenticated into the access server "shell" and can Telnet or initiate SLIP or PPP or ARA after initial connection

XTACACS is multiprotocol and can authorize connections with:

▼ SLIP

■ enable

■ PPP (IP or IPX)

■ ARA

■ EXEC

▲ Telnet

TACACS+

TACACS+ allows a separate access server (the TACACS+ server) to provide the services of authentication, authorization, and accounting independently. Each service can be tied into its own database or can use the other services available on that server or on the network.

Authentication

The TACACS+ protocol forwards many types of username password information. This information is encrypted over the network with MD5. TACACS+ can forward the password types for ARA, SLIP, PAP, CHAP, and standard Telnet. This allows clients to use the same username password for different protocols.

Authorization

TACACS+ provides a mechanism to tell an access server which access list that a user connected to port 1 uses. The TACACS+ server and location of the username/password information identify the access list through which the user is filtered. The access list(s) reside on the access server. The TACACS server responds to a username with an accept and an Access List number, which causes that list to be applied.

Accounting

TACACS+ provides accounting information to a database through TCP to ensure a more secure and complete accounting log. The accounting portion of the TACACS+ protocol contains the network address of the user, the username, the service attempted, protocol used, time and date, and the packet-filter module originating the log. For Telnet connections, it also contains source and destination port, log, alert type, and action carried, such as communication accepted or rejected.

Let's take a look at the different header formats for each packet format. TACACS and XTACACS share similar features, so we will show you those two first, followed by the individual field definitions of both.

The TACACS+ header is a little different, so after the field definitions of TACACS and XTACACS will come the header format for TACACS+ followed by the individual field definitions.

TACACS Packet Format

MAC Header	IP Header	UDP Header	TACACS Packet

TACACS Header

00	01	02	03	04	05	06	07	08	09	10	11	12	13	14	15	16	17	18	19	20	21	22	23	24	25	26	27	28	29	30	31
Version								Type								Nonce															
User Length/Response								Pass Length/Reason								Data :::															

XTACACS Header

00 01 02 03 04 05 06 07	08 09 10 11 12 13 14 15	16 17 18 19 20 21 22 23	24 25 26 27 28 29 30 31
Version	Type	Nonce	
Username Length	Password Length	Response	Reason
Result 1			
Destination Address			
Destination Port		Line	
Result 2			
Result 3		Data :::	

▼ **Version** 8 bits. This field must be set to 0 for simple form, 128 for extended form.

■ **Type** 8 bits.

Version	Description
1	LOGIN
2	RESPONSE (Server to Client Only)
3	CHANGE
4	FOLLOW
5	CONNECT
6	SUPERUSER
7	LOGOUT
8	RELOAD
9	SLIPON
10	SLIPOFF
11	SLIPADDR
12–128	Unassigned
129–255	Local Use

■ **Nonce** 16 bits. This field is set by the client to an arbitrary value. It allows clients that may have multiple outstanding requests to identify which request a response is for. The server must copy this value to the reply unaltered.

- **Username Length**   8 bits, 0 to 255. This field is set by the client to the length of the username in characters. The server must copy this value to the reply unaltered.

- **Response**   8 bits. In this field the server sets the value to one of the following:

Response	Description
0	Accepted
1	Rejected

- **Password Length**   8 bits, 0 to 255. This field is set by the client to the length of the password in characters. The server must copy this value to the reply unaltered.

- **Reason**   8 bits.

Reason	Description
1	Expiring
2	Password
3	Denied
4	Quit
5	Idle
6	Drop
7	Bad

- **Result 1**   32 bits. This field is set by the client to zero. For LOGIN or CONNECT requests, it is set by the server as specified in the request description. For all other requests, it should be set by the server to zero.

- **Destination Address**   32 bits. This field is set by the client. On CONNECT, SLIPON, and SLIPOFF requests, it specifies an IP address. It should be set to zero on all other requests. For SLIPON and SLIPOFF requests, this value should be the IP address assigned to the line. For CONNECT requests, this value is the IP address of the host that the user is attempting to connect to. The server copies this value to the reply.

- **Destination Port**   16 bits. This field is set by the client. On CONNECT requests, it specifies the port number that the user is attempting to connect to. It should be set to zero on all other requests. The server copies this value to the reply.

- **Line**   16 bits. This field is set by the client to the line number that the request is for. The server copies this value to the reply.

- **Result 2**   32 bits. This field is set by the client to zero. For LOGIN or CONNECT requests, it is set by the server as specified in the request description. For all other requests, it should be set by the server to zero.

■ **Result 3**   16 bits. This field is set by the client to zero. For LOGIN or CONNECT requests, it is set by the server as specified in the request description. For all other requests, it should be set by the server to zero.

▲ **Data**   Variable length. This field contains just the text of the username and password, with no separator characters. (You use Username Length and Password Length to sort them out.) The server does not copy the values to the reply. (However, the server does copy the Username Length and Password Length fields to the reply.) The username data may be in uppercase. Comparisons should be case insensitive.

TACACS+ Header

The format of the header is shown in the following illustration:

00	01	02	03	04	05	06	07	08	09	10	11	12	13	14	15	16	17	18	19	20	21	22	23	24	25	26	27	28	29	30	31
Maj. Ver.				Min. Ver.				Packet Type								Sequence Number								Flags							
Session ID (4 bytes)																															
Length (4 Bytes)																															

▼ **Major Version**   This is the major TACACS+ version number.

■ **Minor Version**   This is the minor TACACS+ version number. This is intended to allow revisions to the TACACS+ protocol while maintaining backward compatibility.

■ **Packet Type**   Possible values are:
   ■ TAC_PLUS_AUTHEN = 0x01 (Authentication)
   ■ TAC_PLUS_AUTHOR = 0x02 (Authorization)
   ■ TAC_PLUS_ACCT = 0x03 (Accounting)

■ **Sequence Number**   This is the sequence number of the current packet for the current session. The first TACACS+ packet in a session must have the sequence number 1, and each subsequent packet will increment the sequence number by one. Thus, clients only send packets containing odd sequence numbers, and TACACS+ daemons only send packets containing even sequence numbers.

■ **Flags**   This field contains various flags in the form of bitmaps. The flag values signify whether the packet is encrypted.

■ **Session ID**   This is the ID for this TACACS+ session.

▲ **Length**   This is the total length of the TACACS+ packet body (not including the header).

Kerberos

The Kerberos authentication model is based on the Needham and Schroeder key distribution protocol. Kerberos was created by MIT as a solution to these network security problems.

The Kerberos protocol uses strong (secret-key) cryptography so that a client can prove its identity to a server (and vice versa) across an insecure network connection. After a client and server have used Kerberos to prove their identity, they can also encrypt all of their communications to assure privacy and data integrity as they go about their business.

When a user requests a service, her/his identity must be established. To do this, a ticket is presented to the server, along with proof that the ticket was originally issued to the user, not stolen. There are three phases to authentication through Kerberos:

▼ **First phase**   The user obtains credentials to be used to request access to other services.

■ **Second phase**   The user requests authentication for a specific service.

▲ **Third phase**   The user presents those credentials to the end server.

Kerberos Credentials

The Kerberos authentication model uses two types of credentials:

▼ Tickets

▲ Authenticators

Both are based on private-key encryption using the Digital Encryption Standard (DES), but they are encrypted using separate keys.

Kerberos Ticket

A *ticket* is used to securely pass the identity of the person to whom the ticket was issued between the authentication server and the end server. A ticket also passes information that can be used to make sure that the person using the ticket is the same person to whom it was issued.

A ticket is good for a single server and a single client. It contains the name of the server, the name of the client, the Internet address of the client, a timestamp, a lifetime, and a random session key. This information is encrypted using the key of the server for which the ticket will be used. Once the ticket has been issued, it may be used multiple times by the named client to gain access to the named server until the ticket expires. Note that because the ticket is encrypted in the key of the server, it is safe to allow the user to pass the ticket on to the server without having to worry about the user modifying the ticket.

Kerberos Authenticator

The *authenticator* contains the additional information which, when compared against what is found in the ticket, proves that the client presenting the ticket is the same client to which the ticket was issued.

Unlike the ticket, the authenticator can only be used once. A new one must be generated each time a client wants to use a service. This does not present a problem, because the client is able to build the authenticator itself. An authenticator contains the name of the client, the workstation's IP address, and the current workstation time. The authenticator is encrypted in the session key that is part of the ticket.

Getting the Initial Kerberos Ticket

When the user wants access to a workstation, only one piece of information can be used to prove identity, and that is the user's password. The initial exchange with the authentication server is designed to minimize the chance that the password will be compromised, while at the same time not allowing users to properly authenticate themselves without knowledge of that password.

The first thing that happens is that the workstation will prompt for the username. Once it has been entered, a request is sent to the authentication server containing the user's name and the name of a special service known as the *ticket-granting service*.

The authentication server checks that it knows about the client. If so, it generates a random session key that will later be used between the client and the ticket-granting server. It then creates a ticket for the ticket-granting server that contains the client's name, the name of the ticket-granting server, the current time, a lifetime for the ticket, the client's IP address, and the random session key just created. This is all encrypted in a key known only to the ticket-granting server and the authentication server.

The authentication server then sends the ticket, along with a copy of the random session key and some additional information, back to the client. This response is encrypted in the client's private key, known only to Kerberos and the client, which is derived from the user's password.

Once the response has been received by the client, a new prompt asks for the user's password. The password is then converted to another DES key. It is this key that is used to decrypt the response from the authentication server. The ticket and the session key, along with some of the other information, are stored for future use, and the user's password and DES key are erased from memory.

Once the exchange has been completed, the workstation will use the information to prove the identity of its user for the lifetime of the ticket-granting ticket. As long as the software on the workstation had not been previously tampered with, no information exists that will allow someone else to impersonate the user beyond the life of the ticket.

Requesting a Kerberos Service

Now, let's assume that the user already has a ticket for the desired server. To gain access to the server, the application builds an authenticator containing the client's name and IP address and the current time. The authenticator is then encrypted using the session key that was received with the ticket for the server. The client then sends the authenticator along with the ticket to the server.

Once the authenticator and ticket have been received by the server, the server decrypts the ticket, uses the session key included in the ticket to decrypt the authenticator, and compares the information in the ticket with that in the authenticator, the IP address from which the request was received, and the present time. If everything matches, it allows the request to proceed.

It is assumed that clocks are synchronized to within several minutes. If the time in the request is too far in the future or the past, the server treats the request as an attempt to replay a previous request. The server will keep track of all past requests with timestamps that are still valid. By doing this, it can avoid replay attacks. Any request received with the same ticket and timestamp as one already received will be discarded.

Finally, if the client specifies that it wants the server to prove its identity, too, the server adds one to the timestamp the client sent in the authenticator, encrypts the result in the session key, and sends the result back to the client. At the end of this exchange, the server is certain that the client is who it says it is. If mutual authentication occurs, the client is also convinced that the server is authentic. Also, the client and server share a key that no one else knows. This way both the server and the client can safely assume that any reasonably recent message encrypted in that key originated with the other party.

SUMMARY

In this chapter, we learned about authentication, access control, and the many different methods used to authenticate humans and machines.

We talked about the different kinds of authentication systems available and then discussed in detail the most common protocols used by these systems, such as PAP, CHAP, EAP, RADIUS, TACACS, and Kerberos.

This chapter completes our section on secure communications. We will now move on to our next section to learn about the security standard designed to work with current and future Internet communications, IPSec.

PART V

IPSec

CHAPTER 16

IPSec Components

This is our first chapter in a three-chapter section covering IPSec. You have seen IPSec mentioned throughout the book so far, and as promised we will be covering IPSec in some detail. This is where everything you learned about in the previous section will be applied.

There are three main components of the IPSec architecture. They are

▼ The Authentication Header (AH) protocol

■ The Encapsulating Security Payload (ESP) protocol

▲ The Internet Key Exchange (IKE) protocol

IPSec is short for IP Security and is a suite of protocols designed to provide security at the Network layer. Within the layered communications stack model, the Network layer is the lowest layer that can provide end-to-end security. This means that it can provide protection for any upper-layer data carried in the payload of an IP datagram, without requiring a user to modify the applications. The importance of applying security at this layer will start to become very evident.

In this chapter, we will cover the IPSec framework in detail along with its two main protocols, namely, the Authentication Header (AH) protocol and the Encapsulating Security Payload (ESP) protocol. In addition, we will examine their respective header formats, cryptographic features, and the two different modes of communication. Along with a little history and background, we will also be discussing some of the other important concepts and components that complete the architecture for IPSec.

In the following chapter, we will cover the third IPSec component, the Internet Key Exchange (IKE) protocol, formerly referred to as ISAKMP/Oakley. This details how the keys are exchanged securely over a public medium like the Internet. IPSec was designed for interoperability. IPSec is independent of the current cryptographic algorithms and can accommodate new ones as they become available. It was designed to work with the current IPv4 protocols and as a mandatory component of the future IPv6 protocols.

Finally, in the last chapter in this section, armed with knowledge of the protocols and the concepts behind their operation, we will talk about the different implementations and how they are combined to meet virtually any VPN requirement an organization might need.

IPSEC BACKGROUND

When we look back at some of the things that have affected the progress of human interaction, whether it be the homing pigeon, the printing press, the automobile, telephone, or anything else, nothing has had a more profound affect than the advent of the Internet. We have also never seen a time when technology actually outpaced our ability to use it.

We discussed in earlier chapters how those who developed the protocols that make up the Internet's infrastructure never imagined that it would become as big as fast as it has. That is the reason that first implementation (known as IP version 4 or, more commonly, IPv4) of these protocols wasn't developed with security as a major factor. The demand and necessity for security on the Internet has been increasing in the same proportions as its growth over the last few years.

The lack of secure communication mechanisms on the Internet has been an obstacle to companies and users alike taking advantage of its ubiquity to complete financial transactions and many other activities that required such security. During this time several approaches have been developed to deal with these problems.

Encryption and authentication controls can be implemented at several layers in a network infrastructure. This can be seen in Figure 16-1.

Many of these approaches provided only partial solutions to these problems. Some of these approaches were implemented in the Application and Transport layers. We have

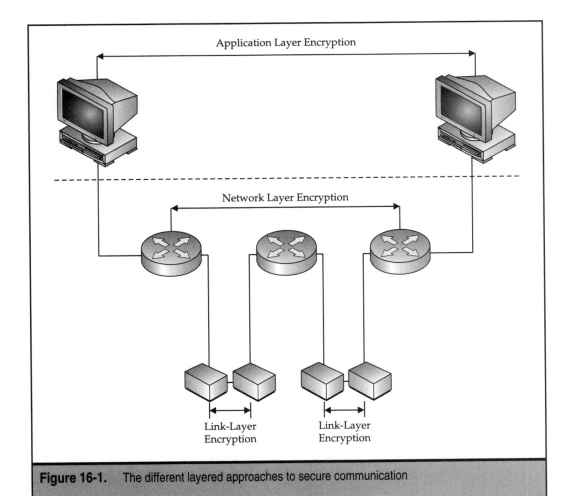

Figure 16-1.   The different layered approaches to secure communication

talked about many of these protocols in previous chapters. Examples of these protocols are SSL, Secure Shell (SSH), Secure Hyper-Text Transfer Protocol (SHTTP), Pretty Good Privacy (PGP), Kerberos, and so on. SSL is a Transport layer protocol, while the others are Application layer protocols. All of these methods are capable of key negotiation and other security services. Applications are enhanced to call into this system to use their security mechanism.

For example, SSL provides application encryption for web browsers and other applications. SSL protects the confidentiality of data sent from each application that uses it, but it does not protect data sent from other applications. Every system and application must be protected with SSL in order for it to work efficiently.

Other approaches involved link layer schemes. Institutions such as the military have been using link layer encryption for years. With this scheme, every communications link is protected with a pair of encrypting devices. Those devices would have to share the same scheme on each end of the link. While this system provides excellent data protection, it is harder to manage. It also requires that each end of every link in the network is secure. That idea is out of the question when it involves the Internet, where few (if any) of the intermediate links are accessible or trusted by the user.

IPSec uses Network layer encryption and authentication techniques providing an end-to-end security solution in the network architecture itself. This way the end systems and applications do not need to implement any changes to benefit from the security IPSec provides. Because the encrypted packets look like any other ordinary IP packets, they can be easily routed through any IP network, such as the Internet, without any changes to the intermediate networking equipment either. The only devices that require an IPSec implementation are the communicating endpoints. This feature greatly reduces both implementation and management costs.

The need for a new version of IP and a better solution for a comprehensive and unobtrusive approach to security soon became very apparent. This caused the Internet Engineering Task Force (IETF), the international community of network designers, operators, vendors, and researchers concerned with the evolution of the Internet architecture, to start work on a new version right away.

The development of IPSec started out almost at the same time as the development of the Internet Protocol version 6 (IPv6). IPv6 was specified from day one with the IPSec protocol as an integral part of the protocol. (Although there have been some beta releases not implementing IPSec.) But due to the broad use of the current IP protocol, IPv4, and the inherent slow development of IPv6, it was clear that the IPSec protocol had to include a specification that could be used with IPv4 as well.

Aside from the incredible logistical problems in converting the entire existing IPv4 infrastructure to IPv6, another reason for the slow development is that many of the cryptographic algorithms involved have been plagued with copyright and export restrictions. Many countries, the United States in particular, categorize these algorithms as military armaments and have very strict export regulations on any programs using these algorithms. Cryptographic functions we now see being deployed on the Internet have mostly been restricted to military and governmental use. This creates very tricky situations for multinational companies.

This is another example of technology outpacing legislation. Existing laws are very ambiguous and difficult to enforce and will almost certainly have to undergo some revisions, because in most cases it is legal to export and publish the cryptographic theory of these algorithms but not their implementations. Obviously, they can instead be implemented in a country without export restrictions and then exported to the rest of the world.

This also creates a lot of problems for any foreigners receiving updates or modifications from U.S. citizens, because this might put them under the export restrictions. Worse, this could cause a division of the Internet and isolate many countries. This could erode the ubiquity and unfettered access that has made the Internet the global phenomenon it is today.

Since 1992, many drafts for IPSec had been proposed and rejected. Some ideas were adopted and other ideas were merged. This has resulted in an entire architecture consisting of a number of different protocols, algorithms, and their hybrids. One thing that makes IPSec such a viable solution is that it is a very flexible and open architecture. It can be used with the myriad of cryptographic algorithms, open and proprietary, that exist today and with any others yet to be developed.

A Quick Note on IPv4 Versus IPv6

IPv4 is the current standard of the TCP/IP protocol architecture and the one we have come to know and love. However, due to the phenomenal growth of the Internet over the past few years, problems were starting to develop, such as poor security and depletion of address space.

A new version would have to be created to address these problems. This new version would first be called IPng (IP next generation) and eventually called IPv6. (I have no idea where version 5 ended up. Hey! If you hurry, this would be a perfect opportunity to create your own version called IPv5.)

In the future, a lot of new addresses will be required, and the need won't be limited to just computers. Almost every kind of electronic device you use might need its own IP address. This will include things like telephones, microwave ovens, refrigerators, freezers, and any other kind of appliance you can imagine. This will require an enormous number of new addresses.

IPv4 is based on a 32-bit addressing scheme. IPv6 will be based on a 128-bit addressing scheme. The maximum theoretical number of possible addresses for IPv4 is 2^{32} or about 4 billion. (In reality, the number is much smaller due to the waste caused by A.B.C.D.E. classifications.) The maximum theoretical number of possible addresses for IPv6 is 2^{128} or a 39-digit number allowing for...wait, give me a second...let's see, you take your 2...carry the 5, multiply by the hypotenuse...boat traveling up a river at 6 knots against a river traveling...GOT IT! It's 6.7 zillion!...well, it's something like that...it's very many.

In addition to more address space, IPv6 solves three essential security problems:

▼ Since IPSec is mandatory to all IPv6 implementations, every device using IPv6 will have security features available from the beginning.

■ No need exists for Address Resolution (ARP) of the MAC address. This has been replaced with IPv6 Neighbor Discovery, which runs over IPv6 instead of directly over the MAC layer and can now be secured with IPSec.

▲ IPv6 headers have been constructed to make it possible to authenticate many of the headers that could not be authenticated in IPv4.

IPSec promises to solve the security problems with the current version of the Internet Protocol (IPv4). IPSec support for IPv4 will alleviate a lot of the security issues and, along with some of the "Band-Aid" solutions for the addressing problem, may delay the deployment of IPv6 to the Internet. However, the expansion of the Internet and the inevitable exponential need for more IP addresses will make migration to IPv6 unavoidable.

That being said, discussion of IPSec as it relates to IPv6 is out of the scope of this book. For our purposes, the remainder of our discussion regarding IPSec will be as it applies to the IPv4 standard used today.

IPSEC COMPONENTS AND CONCEPT OVERVIEW

IPSec is a very complex subject, and it is important that you be able to recognize the components and acronyms you will be seeing as we move forward. Before we get into any detail on the subtleties of IPSec, we are going to quickly define those components and acronyms and provide an overview of some concepts that are unique to IPSec and some other concepts that share similarities with other technologies.

Once we have established an overall picture of all the items and concepts used in the IPSec architecture, we will devote the rest of this chapter, and the two chapters that follow, to learning about each one of them in great detail, giving you a very solid understanding of IPSec as a whole. This should help clear up any uncertainties you may have had with respect to all of the IPSec references in our previous chapters.

IPSec is like a suite of protocol suites and a structured process for machines to communicate with each other.

IPSec Device Nomenclature

IPSec devices exist and generally are referred to in two ways. Descriptions of these devices will usually take the form of a host or gateway (or more formally a security gateway). (We will use "gateway" because we are so easy-going.)

Gateway   A *gateway* will most often be an IPSec-compliant router, IPSec-capable firewall, or a VPN concentrator. A gateway can act or play the role of host when it is the final endpoint. It all depends on the security association combinations being employed.

Host   This represents an IPSec endpoint only or the device of the final destination in an IPSec tunnel.

The Three Main Pieces of IPSec

The IPSec protocol suite provides three main pieces overall.

The Authentication Header (AH) Protocol   This protocol is used for authentication purposes only. It ties data in each packet to a verifiable signature that allows you to verify both the identity of the person sending the data and that the data has not been altered.

The Encapsulating Security Payload (ESP) Protocol   This protocol is used for encryption and authentication purposes. The authentication portion of this protocol is not as complete as the AH protocol, and this is why they are often used together. ESP does, however, protect the data from being monitored during transit.

Internet Key Exchange (IKE) Protocol   The earlier mentioned protocols use symmetric-key algorithms to accomplish their goals. The reason for this is that it requires far less bandwidth during transmission. To exchange these symmetric keys securely, a powerful negotiation protocol is used that allows users to agree on authentication methods, encryption methods, and the keys to use. It also specifies how long the keys can be used before changing them, and that allows smart, secure key exchange.

Transforms   IPSec uses cryptographic algorithms. The specific implementation of an algorithm for use by an IPSec protocol is often called a *transform*. For example, the DES algorithm used in ESP is called the ESP DES-CBC transform. It is called a transform for obvious reasons in that these algorithms *transform* the original data into ciphered data.

Security Associations

The concept of a *security association* (SA) is fundamental to IPSec. An IPSec connection can be created using a choice of many options for performing a secure network connection. An SA must keep track of and manage quite a bit of information. The security association is a locally significant method that IPSec uses to track all the particulars concerning a given IPSec communication session.

So, an IPSec security association (SA) is a *unidirectional* relationship (IKE SAs are bidirectional) between two or more entities that describes how the entities will use the security services to communicate securely. These definitions can get a little confusing at times, because SAs are used for more than just IPSec; they are also used with the IKE protocol. Therefore, IKE SAs describe the security parameters between two IKE devices. (This is one of the crossover concepts we spoke of at the beginning of the section.)

The IPSEC SA includes a *Security Parameter Index* value followed by a *Destination Address*. Finally, the SA describes the *Security Protocol* and mode used by the traffic it carries. This will indicate exactly which algorithms to use, like DES or IDEA for encryption and MD5 or SHA for integrity/authentication. After deciding on the algorithms, the two devices must then share session keys. Each IPSec connection can provide for two security services:

▼ Either *encryption only* or *integrity/authentication only*

▲ A combination of both *encryption* and *integrity/authentication*

These algorithms would determine the nature of the *Security Protocol* value and are represented in *one* of the following modes (but *not both*):

▼ AH in Transport or Tunnel mode

▲ ESP in Transport or Tunnel mode

In other words, should a connection require the protection of both AH *and* ESP, then *two* SAs must be defined for each direction. In this case, the set of SAs that define the connection would be referred to as an *SA bundle*. The SAs in the bundle do not have to terminate at the same endpoint. For example, a mobile host could use an SA bundle consisting of an AH SA between itself and a gateway and a nested ESP SA that extends to a host behind the gateway. To sum up, an SA is a locally significant, unidirectional, logical connection between two IPSec systems that is uniquely identified using these three items represented in a format called a *triple*:

▼ Security Parameter Index (SPI)

■ IP Destination Address

▲ Security Protocol

SA Triple   A security association (SA) is uniquely identified by the following format, called a *triple,* and represented as:

<Security Parameter Index, IP Destination Address, Security Protocol>

Security Parameter Index (SPI)   This is a 32-bit value used to identify different SAs with the same destination address and security protocol. The SPI is carried in the header of the security protocol (AH or ESP). The SPI has only local significance, as defined by the creator of the SA. All SPI values in the range of 1 to 255 are reserved by the Internet Assigned Numbers Authority (IANA). The SPI value of 0 must be used for local, implementation-specific purposes only. Generally, the SPI is selected by the destination system during the SA establishment.

IP Destination Address   This address may be a unicast, broadcast, or multicast address. However, most SA management mechanisms are defined only for unicast addresses.

Security Protocol   This is dependent on the security protocol and communication mode being used in the SA and can include one of the following representations:

▼ The Authentication (AH) protocol in Transport mode

■ The Authentication (AH) protocol in Tunnel mode

■ The Encapsulation Security Payload (ESP) protocol in Transport mode

▲ The Encapsulation Security Payload (ESP) protocol in Transport mode

   This is dependent on the mode of the protocol in that SA. You can find the explanation of these protocol modes later in this chapter. SAs are used to describe traffic flowing in one direction from source to destination (unidirectional communication). For bidirectional communication between two IPSec systems, there must be two SAs defined, one for each direction.

IPSec Communication Mode

The type of operation for IPSec connectivity is directly related to the role the system plays in the SA status. There are two modes of operation for both AH and ESP.

Transport Mode   Transport mode is used to protect upper-layer protocols and only affects the IP packet payload. Transport mode is established when the endpoint is a host or when communications are terminated at the endpoints. If the gateway in a gateway-to-host communication were to use Transport mode, it would act as a host system, which can be acceptable for direct protocols to that gateway.

Tunnel Mode   Tunnel mode encapsulates the entire IP packet to tunnel the communications in a secured communication. Tunnel mode is established for gateway services and is fundamentally an IP tunnel with authentication and encryption. This is the most common mode of operation. Tunnel mode is required for gateway-to-gateway and host-to-gateway communications.

Combining Security Associations

Through the use of a single SA, IP packets may be provided security by either the AH or ESP protocols, but not both. However, there is no restriction on the use of multiple SAs to implement a security policy.

SA Bundles   An *SA bundle* is a pair of IPSec SAs with different security protocols and transforms combined together to provide the desired security. For example, if one SA protects the integrity of packets and the other provides secrecy, then the two SAs can be combined to protect both integrity/authentication and encryption. Note that the two SAs in a bundle are shared between the same two systems and that the destination addresses of the two SAs must be the same. Like an IPSec SA, an IPSec SA bundle is unidirectional, so a pair of SA bundles is needed for bidirectional security. The actual order in which the SAs are bundled is defined by the security policy. However, the IPSec protocol does define two different ways in which SAs may be combined, which are known as *transport adjacency* and *iterated tunneling.*

Transport Adjacency   *Transport adjacency* refers to the process of applying multiple transport SAs to the same IP packet, without making use of tunneling SAs. This level of combination allows both AH and ESP to be applied to IP packets, and no further nesting. The idea is that strong algorithms are used in both AH and ESP; hence, further nesting would yield no more benefits. The IP packet is only processed once, and that is at its final destination.

Iterated Tunneling   *Iterated tunneling* refers to the process of applying multiple, or layered, security protocols through IP tunneling. This approach allows for multiple levels of nesting. Each tunnel may originate or terminate at a different IPSec site along the path. The IPSec protocol illustrates three basic cases of iterated tunneling.

The earlier mentioned security policies (transport adjacency and iterated tunneling) may also be combined. This means that an SA bundle could be constructed from one tunnel SA and one or two transport SAs, applied in sequence.

Security Databases

Within the IPSec protocol two nominal databases always exist. These databases are the *Security Policy Database* (SPD) and the *Security Association Database* (SAD).

Security Policy Database (SPD)   We have established that a security association is simply a management construct that is used to enforce a security policy. A *Security Policy Database* (SPD) is responsible for all IP traffic; it must be consulted during the processing of all traffic. This includes all inbound and outbound traffic, whether it be IPSec or non-IPSec traffic. To support this, the SPD requires distinct entries for inbound and outbound traffic, which are defined by a set of IP and upper-layer protocol field values, called *selectors*. These entries can specify that some traffic not go through IPSec processing, some be discarded, and others go through IPSec processing. Entries in this database are very similar to the firewall rules, ACLs, or packet filters you've read about.

SA Selectors   *SA selectors,* or selector values, constitute the field portion of the SPD entry that specifies or selects which traffic will be affected by the database policy, which can specify one or many SAs. Acceptable selectors might include entries like any host, a specific source address, a destination address, a specific user ID, a network IP range, the specific protocol used, the high-level port number specified in the connection, and so on. Almost any criteria that can be used to identify a packet can be used as an SA selector for the SPD.

Security Association Database (SAD)   The *Security Association Database* (SAD) contains parameter information about each SA, such as AH or ESP algorithms and keys, sequence numbers, protocol mode, and SA lifetime. For outbound processing, an SPD entry points to an entry in the SAD. That is, the SPD determines which SA is to be used for a given packet. For inbound processing, the SAD is consulted to determine how the packet must be processed.

Key Management and Key Exchange

As with any security protocol, IPSec requires a means of negotiating which protocols, encryption algorithms, and keys to use. In addition, IPSec also requires a means by which it can keep track of all these agreements between various entities. IPSec provides specification for both *manual* and *automated* key SAs and cryptographic key management.

Manual Management   Manual key and SA management is the simplest form of management. This is where the system administrator manually configures each system with keys and security association management data necessary to communicate securely with peer systems. Manual management can work just fine in small, static environments. However, for larger size networks, it is not very practical.

Automated Management   Automated key-management protocols can create keys as they're needed for SAs. Automated management also provides a great deal of scalability for the larger, distributed systems used in medium- to large-sized organizations. Several protocols exist that can provide automated management; however, IKE is the protocol used by IPSec.

THE AUTHENTICATION HEADER (AH) PROTOCOL

This section will explain the services provided by the AH protocol and how they work. The Authentication header is designed to provide data integrity, authentication, and replay protection for the IP datagram.

Actual data integrity is assured by the hash value generated by a *Message Authentication Code* (MAC) using something like the MD5 algorithm we learned about in the previous section of the book (Section IV).

Now, if we want to be able to authenticate the origin of the data, we simply include a secret shared key along with the data we just hashed. The algorithms used by the AH protocol are known as *hashed message authentication codes* (HMAC). HMAC applies a conventional keyed-message authentication code two times in succession, as shown in schematic form in Figure 16-2.

The naming convention you would see for this is derived very easily. You will notice that the underlying Message Authentication Code in Figure 16-2 is MD5 or SHA-1. You

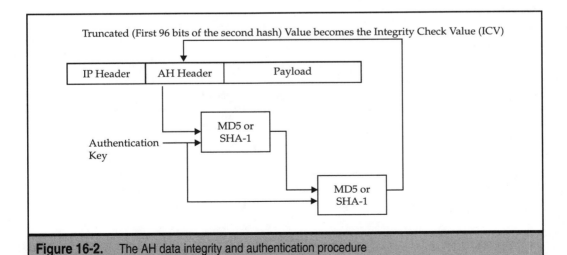

Figure 16-2.   The AH data integrity and authentication procedure

will also notice that we mention in the figure that we take the first 96 bits of the second hash function. MD5 and SHA-1 are 128-bit and 160-bit algorithms, respectively. The reason we take only the first 96 bits is for backward compatibility with the original IPSec specification. These algorithms would be referred to as HMAC-MD5-96 and HMAC-SHA-1-96, respectively. "HMAC" means it is a Hashed MAC; the next thing listed is the algorithm being used, namely MD5 and SHA-1 in this case; and the last amount represents the number of bits being used. This is because IPSec will allow for all future versions and any other versions that don't choose to be backward compatible.

Replay protection is provided by using the Sequence Number field within the AH header whose value is covered by the authentication procedure. The sender must always use the replay service, but it is up to the receiver whether the replay information is validated. These services are connectionless, which means that they work on a per-packet basis.

In IPSec, all three of these procedures are lumped together and simply referred to by the name "authentication."

AH authenticates as much of the IP datagram as possible. However, some fields in the IP header change on the way to their destination. These fields that are subject to such change are called *mutable fields*. An example of that would be the Time to Live (TTL) field, which must be decremented by each router along the way. Obviously, that can't be included in the AH *Integrity Check Value* (ICV). So, when calculating an ICV, the mutable fields are treated as if they contained all zeros and are carried in the AH Header field, as shown in Figure 16-2. The mutable IPv4 fields that can't be protected by AH are

▼ Type of Service (TOS)

■ Flags

■ Fragment Offset

■ Time to Live (TTL)

▲ Header Checksum

When protection of these fields is required, tunneling should be used. The payload of the IP packet is considered immutable and is always protected by AH. AH is identified by protocol number 51, assigned by the IANA in RFC 1700. The protocol header immediately preceding the AH header contains this value in its Protocol field.

AH processing is applied only to nonfragmented IP packets. However, an IP packet with AH applied can be fragmented by intermediate routers. In this case, the destination first reassembles the packet and then applies AH processing to it. If an IP packet that appears to be a fragment (when the Offset field is something other than zero, or the More Fragments bit is set) is input to AH processing, it will be discarded. This is to provide protection from overlapping fragment attacks. These kinds of attacks exploit weaknesses in the fragment reassembly algorithm to create forged packets and force them through a firewall.

Packets that failed authentication are discarded and never delivered to upper layers. This mode of operation greatly reduces the chances of successful denial of service attacks that try to block the communication of a host or gateway by flooding it with bogus packets.

AH Transport and Tunnel Modes

AH can be applied in either of two modes:

▼   AH Transport mode

▲   AH Tunnel mode

Figure 16-3 shows how the AH protocol operates on an original IP datagram in each of these two modes.

▼   In Transport mode, the original datagram's IP header is the outermost IP header, followed by the AH header, and then the payload of the original IP datagram. The entire original datagram along with the AH Header itself is authenticated. Should changes or tampering with any of the fields occur, with the exception of the mutable fields, it will be detected. Always remember that all information in the datagram is in cleartext form and therefore is subject to eavesdropping while it is in transit.

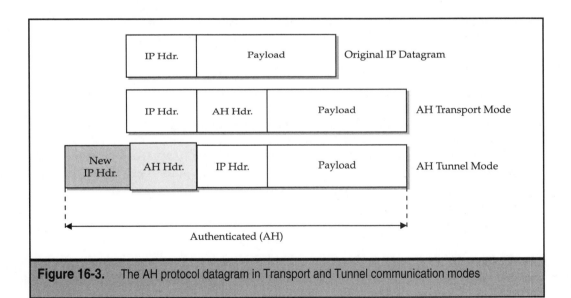

Figure 16-3.   The AH protocol datagram in Transport and Tunnel communication modes

■ In Tunnel mode, a new IP header is generated for use as the outer IP header of the resultant datagram. The source and destination address of the new header will generally differ from those used in the original header. Since routing protocols operate only on the outside IP header, a common case is to have the destination address of the new outside header be the corporate edge router or firewall and remove the outside header before sending it to its intended internal address specified in the new header. The new header is then followed by the AH header, and then by the original datagrams in its entirety, both its IP header and the original payload. The entire datagram (new IP header, AH header, IP header, and IP payload) is protected by the AH protocol. Any change to any field (except the mutable fields) in the Tunnel mode datagram can be detected. Again, always remember that this is only an authentication protocol and that all information in the datagram is in cleartext form and subject to eavesdropping while it is in transit.

▲ AH may be applied alone, in combination with ESP, or even nested within another instance of itself. With these combinations, authentication can be provided between a pair of communicating hosts, between a pair of communicating firewalls, or between a host and a firewall.

The Authentication Header Format

The Authentication header is placed between the main IP header and the upper-layer protocol header. It consists of five fixed-length fields and one variable-length field, making the shortest possible Authentication header 128 bits. You can see the AH protocol header format in Figure 16-4.

The parts of the Authentication header are:

▼ **Next Header** The 8-bit Next Header field in the AH is a typical solution inherited from the development of the IPv6 protocol. In the IPv6 protocol all IP (extension) headers contain this field. This field simply contains a numeric value identifying the next following header. The values for this field are defined by the Internet Assigned Number Authority (IANA).

■ **Payload Length** This 8-bit field specifies the length of the AH expressed in 32-bit words, minus 2. In the standard case, we have 96 bits (3×32 bits) of authentication data, and a 3×32-bit fixed portion, minus 2; this results in a Payload Length field of 4. The value "null" is only allowed for debugging purposes.

■ **Reserved** This 16-bit field is reserved for future use and set to zero.

■ **Security Parameter Index (SPI)** This 32-bit field is used together with the IP destination address and security protocol to identify a security association. The SPI is an arbitrary value chosen during establishment of the SA with the understanding that the values of 1 through 255 have been reserved for future use by the IANA. The value 0 has been set aside for internal use in the IP stack and may be used to indicate that an SA has not yet been established for this connection. The SPI has only local significance, as defined by the creator of the SA.

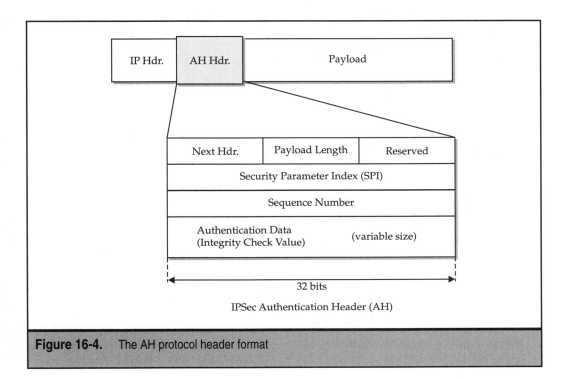

Figure 16-4.   The AH protocol header format

■ **Sequence Number**   This 32-bit field is a progressively increasing (one at a time) counter that is used for replay protection. The value is increased by one for each sent packet, and it is mandatory that the sender use this field. Replay protection is optional; however, it is at the discretion of the receiver whether to process it. At the establishment of an SA, the sequence number is set to zero. The first packet transmitted using the SA has a sequence number of 1. Sequence numbers are not allowed to repeat. This means that for every SA being transmitted, the maximum number of packets is 2^{32} minus 1. So, once the highest possible sequence number has been used, a new SA and consequently a new key must be established. Antireplay is enabled at the sender by default. If upon SA establishment the receiver chooses not to use it, the sender is not concerned with the value in this field anymore. Some points to note:

1. Typically, the antireplay mechanism is not used with manual key management.

2. The original AH specification in RFC 1826 did not discuss the concept of sequence numbers. Older IPSec implementations that are based on that RFC can therefore not provide replay protection.

▲ **Authentication Data**   This is a variable-length field, also called Integrity Check Value (ICV). The ICV for the packet is calculated with the algorithm selected at the SA initialization. For flexibility purposes, the length of this field is not fixed, but still must be a multiple of 32 bits. If the output of the algorithm is not a multiple of 32 bits, padding must be used to fill in the remaining bits. The length of the ICV is a value *not present* in the Authentication header, but is specified by the algorithm specification. As the name implies, the ICV is used by the receiver to verify the integrity of the incoming packet. In theory, any MAC algorithm can be used to calculate the ICV. The specification requires that HMAC-MD5-96 and HMAC-SHA-1-96 must be supported (for backward compatibility). The original RFC 1826 required Keyed MD5. In practice, Keyed SHA-1 is also used. Implementations usually support two to four algorithms, but the choice is up to the implementer as to how many.

THE ENCAPSULATING SECURITY PAYLOAD (ESP) PROTOCOL

The *Encapsulating Security Payload* (ESP) header is designed to provide security services and confidentiality to the protocols above the IP layer. In addition to confidentiality, ESP can also provide some of the services provided by the AH protocol, such as data origin authentication, connectionless integrity, and antireplay service. Due to the encryption provided by ESP, it can also provide some protection network analyzers (sniffers). What kind of services are to be used is decided at the time of the security association (SA) establishment. Confidentiality may be selected by itself, but it is best when authentication is combined together with encryption.

ESP encrypts the payload of an IP packet using symmetric-key algorithms. The Next Header field actually identifies the protocol carried in the payload. ESP provides data origin authentication, data integrity, and replay protection in ways that are similar to the methods used with the AH protocol. However, the protection of ESP does not extend over the whole IP datagram as it does in the AH protocol.

ESP adds approximately 24 bytes per packet that can be a factor when considering throughput calculation, fragmentation, and path MTU discovery. ESP is used to provide integrity check, authentication, and encryption to IP datagrams. Optional replay protection is also possible. These services are connectionless and operate on a per-packet basis. The desired services are selectable upon SA establishment. However, some restrictions apply:

▼ Integrity check and authentication must go together.

■ Replay protection can only be selected with integrity check and authentication.

▲ Replay protection is at the discretion of the receiver.

Encryption can be selected by itself, but it is highly recommended that if encryption is enabled, that integrity check and authentication be turned on. If only encryption is used, intruders could forge packets in order to mount cryptanalytic attacks. This is infeasible when integrity check and authentication are in place.

Although both authentication with integrity check and encryption are optional, at least one of them is always selected. Otherwise, it wouldn't make much sense to use ESP. ESP is identified by protocol number 50 as defined in RFC 1700 by the IANA. The protocol header immediately preceding the AH header will contain this value in the IPv4 Protocol field.

ESP processing will only be applied to nonfragmented IP packets. An IP packet with ESP applied can be fragmented by intermediate routers, however. In this case, the destination will first reassemble the packet and then apply the ESP processing to it. If an IP packet that appears to be a fragment (when the Offset field is something other than zero, or the More Fragments bit is set) is processed by ESP, it will be discarded. This prevents the overlapping fragment attack mentioned earlier in the "Authentication Header (AH) Protocol" section.

If both encryption and authentication with integrity check are selected, then the receiver first authenticates the packet, and only if this step was successful, proceeds with decryption. This mode of operation saves computing resources and reduces the vulnerability to denial of service attacks.

ESP Transport and Tunnel Modes

Like AH, ESP can be used in two ways:

▼ Transport mode

▲ Tunnel mode

Take a look at Figure 16-5 to see the ESP Transport and Tunnel datagrams.

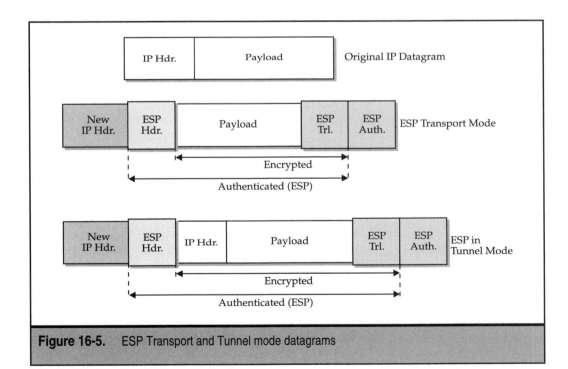

Figure 16-5.   ESP Transport and Tunnel mode datagrams

ESP in Transport Mode

In this mode, the original IP datagram is taken, and the ESP header is inserted right after the IP header, as shown in Figure 16-6. If the datagram already has IPSec headers, then the ESP header is inserted before any of those. The ESP trailer and the optional authentication data are appended to the payload.

ESP in Transport mode provides neither authentication nor encryption for the IP header. This is a disadvantage, since false packets might be delivered for ESP processing. The advantage of Transport mode is the lower processing overhead. As in the case of AH, ESP in Transport mode is used by hosts and does not require the use of security gateways.

ESP in Tunnel Mode

As expected, this mode applies the tunneling principle. A new IP packet is constructed with a new IP header, and then ESP in Transport mode is applied, as illustrated in Figure 16-5. Since the original datagram becomes the payload data for the new ESP packet, its protection is total if both encryption and authentication are selected. However, the new IP header is still not protected.

Tunnel mode is used whenever either end of a security association is a gateway. Thus, between two firewalls, the Tunnel mode is always used. Although gateways are

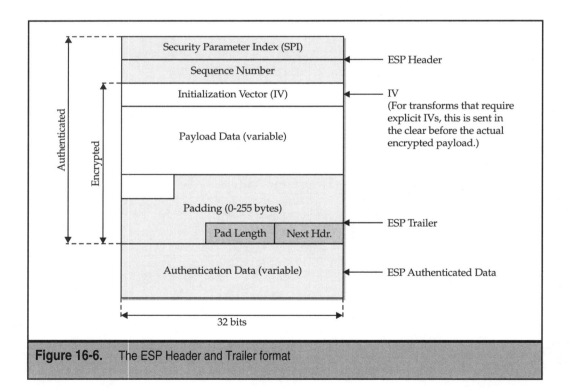

Figure 16-6. The ESP Header and Trailer format

supposed to support Tunnel mode only, often they can also work in Transport mode. This mode is allowed when the gateway acts as a host, that is, in cases when traffic is destined to itself. Examples are SNMP commands or ICMP echo requests.

In Tunnel mode, the outer header's IP address does not need to be the same as the inner header's address. For example, two security gateways may operate an ESP tunnel that is used to secure all traffic between the networks they connect together. Hosts are not required to support Tunnel mode, but often they do.

The advantages of the Tunnel mode are total protection of the encapsulated IP datagram and ability to use private addresses. Extra processing is associated with this mode, however.

ESP Packet Format

The format of the ESP packet is more complicated than that of the AH packet. Actually, there is not only an ESP header, but also an ESP trailer and ESP authentication data. The payload is located (encapsulated) between the header and the trailer, hence the name of the protocol. You can see the ESP Header and Trailer format in Figure 16-6.

The following fields are part of an ESP packet:

▼   **Security Parameter Index (SPI)**   This 32-bit field is used together with the IP Destination Address and Security Protocol to identify a security association. The SPI is an arbitrary value chosen during establishment of the SA with the understanding that the values of 1 through 255 have been reserved for future use by the IANA. The value 0 has been set aside for internal use in the IP stack and may be used to indicate that an SA has not yet been established for this connection. The SPI has only local significance, as defined by the creator of the SA.

■   **Sequence Number**   This 32-bit field is a progressively increasing (one at a time) counter that is used for replay protection. The value is increased by one for each sent packet, and it is mandatory to use this field by the sender. Replay protection is optional; however, it is at the discretion of the receiver whether to process it. At the establishment of an SA, the sequence number is set to zero. The first packet transmitted using the SA has a sequence number of 1. Sequence numbers are not allowed to repeat. This means that for every SA being transmitted, the maximum number of packets is 2^{32} minus 1. So, once the highest possible sequence number has been used, a new SA and consequently a new key must be established. Antireplay is enabled at the sender by default. If upon SA establishment the receiver chooses not to use it, the sender is not concerned with the value in this field anymore. Some points to note:

  ■   Typically, the antireplay mechanism is not used with manual key management.

  ■   The original ESP specification in RFC 1827 did not discuss the concept of sequence numbers. Older IPSec implementations that are based on that RFC can therefore not provide replay protection.

■ **Payload Data**   The Payload Data field is mandatory. It consists of a variable number of bytes of data described by the Next Header field. This field is encrypted with the cryptographic algorithm selected during SA establishment. If the algorithm requires initialization vectors, these are also included here. The ESP specification requires support for the DES algorithm in CBC mode (DES-CBC transform). Often, other encryption algorithms are also supported, such as Triple DES.

■ **Padding**   Most encryption algorithms require that the input data must be an integral number of blocks. Also, the resulting ciphertext (including the Padding, Pad Length, and Next Header fields) must terminate on a 4-byte boundary, so that Next Header field is right-aligned. That is why this variable-length field is included. It can be used to hide the length of the original messages, too. However, this could adversely affect the effective bandwidth. Padding is an optional field. The encryption covers the Payload Data, Padding, Pad Length, and Next Header fields.

■ **Pad Length**   This 8-bit field contains the number of the preceding padding bytes. It is always present, and the value of 0 indicates no padding.

■ **Next Header**   The Next Header is an 8-bit mandatory field that shows the data type carried in the payload, for example, an upper-level protocol identifier such as TCP. The values are chosen from the set of IP Protocol Numbers defined by the IANA.

▲ **Authentication Data**   This field is variable in length and contains the ICV calculated for the ESP packet from the SPI to the Next Header field inclusive. The Authentication Data field is optional. It is included only when integrity check and authentication have been selected at SA initialization time. The ESP specifications require two authentication algorithms to be supported: HMAC with MD5 and HMAC with SHA-1. Often, the simpler keyed versions are also supported by the IPSec implementations. Some points to note:

1. The IP header is not covered by the ICV.

2. The original ESP specification in RFC 1827 discusses the concept of authentication within ESP in conjunction with the encryption transform. That is, there is no Authentication Data field, and it is left to the encryption transforms to eventually provide authentication. The current RFC is 2406.

WHY TWO AUTHENTICATION PROTOCOLS?

Knowing about the security services of ESP, you might ask if there is really a requirement for AH. Why does ESP authentication not cover the IP header as well? There is no official answer to these questions, but here are some points that justify the existence of two different IPSec authentication protocols:

▼ ESP requires strong cryptographic algorithms to be implemented, whether it will actually be used or not.

■ Often, only authentication is needed. While ESP could have been specified to cover the IP header as well, AH performs better when compared to ESP with authentication only, because of the simpler format and lower processing overhead. It makes sense to use AH in these cases.

■ Having two different protocols means finer-grade control over an IPSec network and more flexible security options. By nesting AH and ESP, for example, you can implement IPSec tunnels that combine the strengths of both protocols.

▲ The ESP protocol allows the payload to be encrypted without being authenticated. In virtually all cases, encryption without authentication is not useful.

SECURITY ASSOCIATIONS AND POLICIES

It is important to keep in mind that IPSec only covers security at the IP layer. That means it involves the communication *between* computers and other network devices. Recall in the Chapter 15 when we spoke about authentication of people and machines. IPSec does not address security issues related on the user level, for example, direct physical access to a host.

These issues must also count as a part of the whole security policy if the network is supposed to be secure. As an example, a company has deployed a VPN network between its different regional departments. All traffic between the departments is secured by IPSec-compliant security gateways, using manually distributed keys. The keys are stored on the security gateway computer without thorough protection of the key directory. Even though the communication itself is secure, the keys are easy to obtain, thus making the system very insecure.

The services IPSec *can* provide are

▼ Access control

■ Packet replay protection

■ Connectionless integrity

■ Data origin authentication

■ Confidentiality

▲ Partly (non-mutable fields) traffic analysis protection

As the services are implemented at the IP layer, they can also be used by upper-layer protocols such as TCP, UDP, ICMP, and so on. In addition, some of the encryption algorithms can provide data compression, both saving bandwidth and adding diffusion to the encrypted data.

Security Associations

The use of a *security association* (SA) is a fundamental part of the IPSec implementation. The SA contains information on what security service is used for a given connection. The term *security association* is a concept and can be implemented in several ways. The security association concept is used by both the AH and ESP, and by one of the major functions of the Internet Key Exchange (IKE) as well.

A security association (SA) is a unidirectional (the unidirectionality is only for the IPSEC SAs and not for IKE SAs) "channel" offering either ESP or AH security to it. The simplest scenario is two security gateways or hosts with a bidirectional communication using two SAs, one in each direction. Note that a single SA cannot offer both ESP and AH security; this must be done by using multiple security associations. This means that two hosts communicating using both AH and ESP must use four SAs, one for each security service, two in each direction. Take a look at Figure 16-7 to see the SAs for communicating hosts using both AH and ESP.

 This icon represents a *logical connection* using one or more security associations (AH or ESP, Transport or Tunnel).

 This icon represents a device supporting IPSec.

 This icon represents a physical connection.

The SA may be applied in either Transport or Tunnel mode. In IPv4, the security protocol header is placed between the main IP header and the Transport layer protocol. If using ESP Transport mode, security is only provided for the higher-layer protocols, not the main IP header. Considering that several of the fields in the IP header may change in transit, this is rather obvious. It would make little sense having to decrypt the packet before performing the alterations needed in the IP header. If the desire is to secure correct routing, AH is used. AH is extended to selected portions of the main IP header. The reason that AH doesn't cover the entire datagram is that some of the fields may be altered under transit and would thus result in the authentication check failing. But the most important fields such as the source and destination are covered.

Tunnel and Transport Mode

In essence, the difference between Transport mode and Tunnel mode is that Tunnel mode keeps the original IP datagram and packs it into a new packet, while Transport only does some modifications to the datagram. It can be compared to the situation where you want to send a fragile object with the public mail service. Two solutions are available:

▼ **Transport mode** Modify the object itself so it can withstand the rough handling.

▲ **Tunnel mode** Keep the object as it is, but pack it into a solid box with soft material inside, so the object is protected from rough handling.

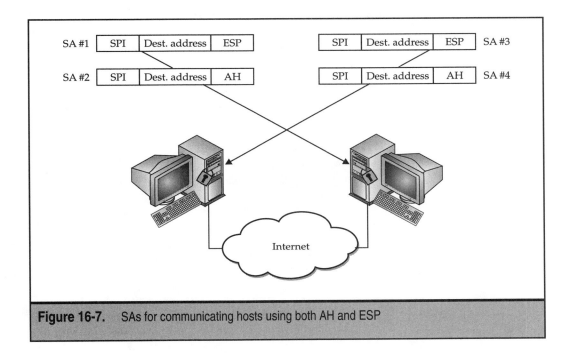

Figure 16-7.   SAs for communicating hosts using both AH and ESP

As a *security gateway* (SG) doesn't send any data itself, but acts on behalf of a trusted subnet, an SA between two SGs is always in Tunnel mode. As a rule, Transport mode is only used between two hosts. Two hosts *can* use Tunnel mode, but this is rarely done, because it adds additional overhead and doesn't provide any additional security. Further, it is worth noting that if traffic is destined for the security gateway, the gateway is actually acting as a host, and Transport mode is allowed.

In general, Tunnel mode means adding a new outer header, encapsulating the original datagram and header. When an SG receives an outbound datagram, it encapsulates the original datagram and adds an outer IP header. The outer header is then addressed to the unit at the end of the secure tunnel that performs the IP processing, while the inner header specifies the ultimate destination. This means that when using ESP tunneling, the whole original (inside) datagram is protected, and the ultimate address is not available in cleartext. This method can, to some extent, provide traffic analysis (sniffer) protection as it hides the ultimate destination address.

Even if the data itself is encrypted, a middleman can glean a lot of information just from studying traffic patterns. It is not just the content of a message that is important; sometimes the mere existence of the message is sensitive. This means that many times you will want to hide both the sender's and receiver's address, size of packet, and what time the packet was sent.

The set of security services offered by a security association depends on the following factors:

▼ The selected security protocol, AH or ESP

■ The selected mode, Transport or Tunnel

▲ The election of optional services selected for use with the protocol, for example, compression

ESP, like AH, can also provide authentication services. The difference is that when using ESP for authentication, not as much of the packet is authenticated. This means that fields like the outer IP header are not authenticated. Another drawback of using ESP only for authentication is that ESP produces more overhead than just using AH. If only the upper-layer data needs authentication, ESP might be a solution. Antireplay service, similar to the AH antireplay, can also be incorporated when using the ESP.

In addition to keeping the ultimate destination address confidential, ESP (in Tunnel mode) can also be used to hide the actual size of the packet. This is done by using padding in the Payload field. This is an approach that can be used to protect against traffic analysis attacks. The drawback is, of course, that it is not very efficient since it consumes both computing resources and bandwidth.

As mentioned earlier, a single SA can only support one security protocol (either AH or ESP). In some cases, it might be required by the security policy to use more than one of the security protocols between two points in a network. It is then necessary to apply two or more SAs between these two points. This is referred to as an *SA bundle,* which is a series of SAs the traffic must use in order to satisfy the security policy. All the SAs making up an SA bundle do not have to terminate at the same destination. When you have a mobile host, it can have one SA between the mobile host and gateway, and another nested (tunneled) SA to a host within the corporate network.

Combining IPSec Protocols

The AH and ESP protocols can be applied alone or in combination. Given the two modes of each protocol, quite a number of combinations are possible. To make things even worse, the AH and ESP SAs do not need to have identical endpoints, so the picture becomes rather complicated. Luckily, out of the many possibilities, only a few make sense in real-world scenarios.

There are two approaches for an SA bundle creation:

▼ **Transport adjacency**   Both security protocols are applied in Transport mode to the same IP datagram. This method is practical for only one level of combination.

▲ **Iterated "nested" tunneling**   The security protocols are applied in Tunnel mode in sequence. After each application, a new IP datagram is created, and the next protocol is applied to it. This method has no theoretical limit in the nesting levels. However, more than three levels are impractical, to say the least.

These approaches can be combined; for example, an IP packet with *transport adjacency* IPSec headers can be sent through *iterated tunnels.* When designing a VPN, limit the IPSec processing stages applied to a certain packet to a reasonable level. Two stages are sufficient for almost all the cases, and you shouldn't exceed three cases.

To be able to create an SA bundle in which the SAs have different endpoints, at least one level of tunneling must be applied. Transport adjacency does not allow for multiple source or destination addresses, because only one IP header is present.

The practical principle of the combined usage is that upon the receipt of a packet with both protocol headers, the IPSec processing sequence should be authentication followed by decryption. Why bother with the decryption of packets of uncertain origin?

Following the preceding principle, the sender first applies ESP and then AH to the outbound traffic. In fact, this sequence is an explicit requirement for Transport mode IPSec processing. Now, you might ask, "When using both AH and ESP, should ESP authentication be turned on? I mean, AH authenticates the packet anyway."

The answer is simple. Turning on ESP authentication makes sense only when the ESP SA extends beyond the AH SA. In this case, not only does it make sense to use ESP authentication, it is highly recommended to do so to avoid spoofing attacks in the intranet. As far as the modes are concerned, the most common way is to use Transport mode between the endpoints of a connection and to use Tunnel mode between two machines when at least one of them is a gateway.

SECURITY DATABASES

The *Security Policy Database* (SPD) and the *Security Association Database* (SAD) have two quite different functions. The former is used for *establishing* a connection, while the latter is used for managing *active* connections. In other words, the SAD is empty or does not exist until some connections are set up. The SAD contains all parameters relevant for a given SA connection.

The Security Policy Database (SPD)

The IPSec implementation runs on a host or security gateway environment affording protection to IP traffic. The security features and security requirements for the different connections are described in a Security Policy Database (SPD). The SPD is set up and maintained by the system administrator. It can have the same function as filtering rules for firewalls. A firewall would normally have two outcomes of the filtering rules: either forward or discard. The SPD has a third possibility: apply IPSec services. The decision of which action to take is based on IP and Transport layer information, which is matched against entries in the SPD similar to the following table:

Address	Protocol	Action
All	HTTP	Forward
172.130.x.x	TCP	Use AH
192.168.10.x	TCP	Use ESP
All	All	Discard

When searching such a database, the database is always searched in the same order to maintain consistency upon choosing an action. The SPD is similar to the packet filtering criteria used by routers to select different outcomes based on a packet's IP address.

IPSec provides security services and enables a system to select the appropriate security protocols, determine the algorithms, and put in place any required cryptographic keys required to provide the requested services.

The SPD specifies what services should be offered to inbound and outbound traffic, and in what fashion. This section only describes a minimum functionality that it has to provide. An SPD lookup for *outbound* traffic has in general three outcomes:

▼ Discard the data

■ Bypass IPSec processing

▲ Be processed by IPSec

The first outcome is traffic that is not allowed to traverse the security gateway. For example, this can be a request from a host in the Internet that is not allowed access to the corporate LAN.

The second outcome is traffic that does not need any authentication or confidentiality services and is therefore processed as normal IP traffic.

The third outcome is the traffic that is offered to IPSec services, and the SPD entry specifies the protocols to be employed, the algorithms, and in what mode.

IPSec requires the SPD to make lookups on a packet-by-packet basis. Every packet is subject to IP processing, and the SPD must specify what action to take in each case. This may not be necessary on a host utilizing a socket or port interface, as it can use the port number to specify what action to take. On some host systems, it is possible to let the user or application override the default settings for the traffic. The application may, for example, perform its own encryption, and it is therefore not desirable or necessary to perform lower-layer encryption.

The SPD is used to control all traffic through a system, including security and key management. Key-management traffic requires special treatment so it won't get discarded. The granularity of an SA is defined in the SPD and should be configurable according to the following parameters:

▼ **Destination address** The address can be a single address or a range of addresses using wildcards. If the packet is a tunneled packet, the final destination addressed must be revealed before making the SPD lookup. When an IP packet arrives at its destination endpoint, the combination of destination address, SPI, and protocol is used to make a lookup in the SAD to find the corresponding parameters and keys. The packet is then processed accordingly. If it is a tunneled packet, the ultimate destination address is revealed after the IPSec processing, and a lookup in SPD can be made to determine further processing.

■ **Source address** The address can be a single address or a range of addresses using wildcards.

- **Name**   There are two instances in this case: Either fully qualified DNS (FQDN) user ID, like *robert@mirando.com,* or a DNS system name, for example, *mirando.com.* OPAQE should be supported for this entry.

- **Transport layer protocol**   This is information in the IPv4 Protocol field. If several headers are used, the header chain must be parsed until the proper header is found. OPAQE should be supported for this entry.

- ▲ **Source and Destination ports**   These would be the TCP/UDP ports. This selector is not required by IPSec and is therefore optional to support.

NOTE: If using the ESP, some of these parameters will not be in cleartext. It is therefore required that an entry can hold the value "OPAQE," signifying that this field is not readable or inaccessible.

Security Association Database

An entry in the SAD defines the parameters connected to one single SA. All SAs have an entry in the SAD. Just like the IPSec standard that specifies a database in order to keep track of the services available and how policies are applied, the standard also specifies a different database that tracks every active SA. This database entry contains a record of all the values negotiated at the time each SA was initially created. This database is used by the SPD to quickly link packets to an existing security association in order to process them according to existing policies and conforming to the information contained in the SPI. This database is called the *Security Association Database* (SAD). The SAD is responsible for each SA in the communications defined by the SPD. Each SA has an entry in the SAD.

For outbound traffic, the SA is pointed to by an entry in the SPD. If the outbound traffic matches an entry in the SPD that does not point to an existing entry in the SAD, a new entry is created according to the parameters specified by the SPD.

For an inbound datagram, the entry in the SAD is specified by the following three fields:

- ▼ **Outer Header's Destination Address**   This address can either be IPv4 or IPv6. On a host, this address will always be the same.

- **IPSec Protocol**   This would be either AH or ESP.

- ▲ **Security Parameter Index (SPI)**   This is a 32-bit integer value used to distinguish between SAs having the same destination address and IPSec protocol.

The SAD database also contains these parameters for processing IPSec protocols and the associated SA:

- ▼ Sequence number counter for outbound communications.

- Sequence number overflow counter that sets an option flag to prevent further communications utilizing the specific SA.

■ A 32-bit antireplay window that is used to identify the packet for that point in time traversing the SA, and which provides the means to identify that packet for future reference.

■ Lifetime of the SA that is determined by a byte count or time frame, or a combination of the two.

■ The algorithm used in the AH.

■ The algorithm used in the authenticating ESP.

■ The algorithm used in the encryption of the ESP.

■ IPSec mode of operation, Transport or Tunnel mode.

▲ Path MTU (PMTU). This is data that is required for ICMP data over an SA.

Each of these parameters is referenced in the SPD for assignment to policies and applications. The SAD maintains a record of all existing security associations for as long as they live, and it is always used in conjunction with the SPD to correctly sort and process all IPSec traffic. Any of the core components of an SA can become a selector parameter for the SAD, including destination IP, IPSec protocol (AH or ESP), and SPI.

SUMMARY

We have talked about an overview of IPSec as a whole, and in this chapter we have discussed AH, ESP, security associations, and the databases used in processing the data. In the next chapter, we will talk about key management and the IKE protocol used by IPSec.

CHAPTER 17

Key Management

When you transmit data over a public network such as the Internet, be aware that anybody can, and will, examine everything you send. The only effective tool available to ensure the security of communication is cryptography. Modern cryptography relies on secrecy existing entirely in the encryption keys, not the algorithms. In fact, the best algorithms are put under public scrutiny and have already been disclosed. This practice makes the keys, their method of exchange, and key management the key (pardon the pun) to secure communication in public networks.

During key exchange and key management, several layers of activity and procedures must take place prior to the establishment of a security association (SA), and several mechanisms are used to accommodate these procedures. With this in mind, a couple of things become evident:

▼ Key exchange is fundamentally a complicated business.

▲ Key exchange gets more complicated as the group of communicating players expands.

Just because a system says it does encryption, that alone doesn't mean it's appropriate for your needs, even if the system does encryption well. Any proposed secure VPN solution is only as good as its method of key exchange, especially in the larger enterprise implementations. IPSec supports those large enterprise needs, and its industrial-strength key exchange and protocol negotiation scheme set it apart from all other security systems.

In this chapter, we discuss each of these mechanisms as they relate to key exchange and key management. We provide a basic overview of all the elements involved in key exchange and key management and of how they work together. Then, in Chapter 18, we discuss key exchange and key management protocols in more detail.

KEY MANAGEMENT CONCEPTS AND OVERVIEW

Key management terms and definitions are anything but obvious and can be very confusing. The confusion is understandable when you consider that IKE is a product of several different protocols. Many of the terms are often used interchangeably, and the differences between them can be subtle. This can only add to the misunderstandings. In this section, we cover the different protocols that are used to get keys and data from one system to another. We discuss the protocols in more detail in Chapter 18.

The Internet Key Exchange (IKE) protocol is a hybrid of four protocols that are combined to provide an IPSec-specific key-management platform. The four protocols are

▼ **Internet Security Association and Key Management Protocol (ISAKMP)**   ISAKMP provides the general framework for authentication and key exchange, but does not define them. ISAKMP is designed to support any existing (and future) key-exchange protocols using a series of what ISAKMP calls *exchanges*.

- **Oakley**   Oakley describes a series of key exchanges similar to ISAKMP that it refers to as *modes* and details the services provided by each. Oakley also includes things like Perfect Forward Secrecy for key generation, identity protection, and authentication.

- **Secure Key Exchange Mechanism (SKEME)**   SKEME describes a versatile key-exchange technique that provides anonymity, nonrepudiation, and quick key refreshment.

- **Photuris**   Photuris was designed for use with Perfect Forward Secrecy privacy protection. This introduced the concept of the *anti-clogging token* (ACT) more commonly referred to as a *cookie*. The concept of cookies, combined with elements from Diffie's *Station-to-Station* (STS) protocol, was brought to us by the creators of Photuris (Phil Karn and William Simpson) and was designed to protect against resource clogging, better known as *denial of service* (DoS) attacks.

The Internet Security Association and Key Management Protocol (ISAKMP) protocol defines the procedures for authenticating a communicating peer and for key generation techniques. All of these are necessary to establish and maintain an SA in an Internet environment. ISAKMP defines payloads for exchanging key and authentication data. These formats provide a consistent framework, which is independent of the encryption algorithm, authentication mechanism being implemented, and security protocol, such as IPSec.

Oakley and SKEME each define a method to authenticate and establish a key exchange, and IKE implements both methods combined with ISAKMP to negotiate and set up keys for the IPSec security association. Parts of the SKEME protocol are used for fast rekeying using an exchange of random numbers referred to as *nonces*.

Before two network nodes can use IPSec to communicate, they must first perform authentication to prove their identity, and then establish two different kinds of security associations (SAs):

- ▼ An ISAKMP SA

- ▲ An IPSec SA

The negotiation of these two kinds of security associations is performed in two phases:

- ▼ **Phase 1**   This involves authentication and establishment of the ISAKMP SA. This can be considered as the "secure channel" setup phase, where the two parties agree on how they intend to protect further negotiations traffic between them.

- ▲ **Phase 2**   This involves negotiation of one or many security associations on behalf of the services in security protocols such as AH and/or ESP in the case of IPSec or any other security protocols that need keying material and/or parameter negotiation.

Different portions of each of these protocols work in conjunction to securely provide keying information specifically for the IETF *Domain of Interpretation* (DOI) specification for IPSec.

Public Key Infrastructure (PKI) is a suite of protocols that provides several areas of secure communication based on trust and digital certificates. PKI integrates digital certificates, public-key cryptography, and certificate authorities into a total, enterprisewide network security architecture that may be utilized by IPSec.

The terms *IKE* and *ISAKMP/Oakley* are used interchangeably by various vendors. Many also use *ISAKMP* terms (for example, *exchanges*) interchangeably with IKE terms (for example, *modes*) to describe the individual key management and exchange functions.

While this is not improper, you should always keep this fact firmly in mind: *ISAKMP*, in and of itself, is designed to addresses the *procedures* or *framework* for key management/exchange operations and *not* the *details* of the operations as they pertain to IPSec. *IKE* is the proper term that best represents the IPSec implementation of key management/exchange operations.

The *exchange* term is a part of the ISAKMP protocol. The *mode* term is taken from the Oakley protocol and used in IKE to describe the function provided in some ISAKMP *exchanges*. The *phase* term is a part of the ISAKMP protocol.

If you're finding these terms frustrating, fear not! We aren't trying to confuse you. We are interchanging these terms to make you aware that the same thing is being done in the marketplace and the real world every day. We are doing this so you will recognize these terms as you encounter them later in this chapter, in Chapter 18, and in the future. This way, when you hear these terms used or misused in the marketplace, you will understand the underlying intent. Rest assured, all of this will be sorted out shortly.

PERFECT FORWARD SECRECY (PFS)

One of the dangers of sending data over a public network infrastructure is the constant and unfettered opportunities an attacker has to get hold of that data, encrypted or not. You can minimize these risks by using larger and larger keys to encrypt the data. Large keys can seriously impair network performance, because using large keys makes the encryption process much more complex.

You can avoid this problem by using reasonably sized keys and constantly changing them during the entire communication session. But this solution comes with its own difficulties.

You need a way to generate those new keys often enough to be effective and to have the other peer accept those new keys as well. Unfortunately, you can use neither the existing key you're changing from, nor the material that was used to generate that key. The danger is that, if you did, anyone who manages to get hold of that current key could easily use that information to determine what your new key would be.

So, how do you generate these new keys and do it in a way that is not dependent on the value of the current key? Your method should ensure that even if someone obtains a current key and breaks it, it would only give them a small part of the overall picture. This would force them to get and break yet another entirely unrelated key to get the next part. If this were to happen constantly and often enough, there would be no time for them to do anything with that information.

The solution to all of these difficulties is what cryptographers call *Perfect Forward Secrecy*. IKE uses a scheme called Diffie-Hellman to do this.

DIFFIE-HELLMAN

We discussed this algorithm in Chapter 13, but we will quickly recap it here. In a nutshell, a Diffie-Hellman exchange works like this:

▼   Two entities independently and randomly generate values similar to a public-key private-key pair.

■   Each entity sends its public-key value to the other. (To protect against any man-in-the-middle attacks, they would also need to use an authentication method. This added step is not addressed using Diffie-Hellman by itself.)

■   Each entity uses the public key received from the other to generate a private key.

▲   Each entity will combine the public key received from the other with the private key they just generated using the Diffie-Hellman combination algorithm.

The resulting value is the same on both sides and then can be used for fast symmetric-key encryption algorithms. You learned in Chapter 12 that symmetric or private-key encryption is much less CPU intensive than pure public-key encryption. That derived Diffie-Hellman key can now:

▼   Be used as a "session key" for all subsequent exchanges for the duration of the current communication session

▲   Encrypt other randomly generated keys to easily pass them safely over the secure connection that was created using this procedure at the initiation of the current communication session

In Chapter 13, we also noted that in and of itself, Diffie-Hellman is not entirely safe from man-in-the-middle attacks. A man-in-the-middle attack against Diffie-Hellman is certainly not easy, but it can be done. So, as we mentioned earlier, we need authentication like the Interlock protocol to protect against such attacks.

Diffie-Hellman allows you to generate newly shared keys that are independent of older keys to use for symmetric encryption, thereby providing Perfect Forward Secrecy. To be able to use different combinations of algorithms with Diffie-Hellman, or alone for that matter, you will need a way for the peers to agree upon such an arrangement beforehand.

That is what the Diffie-Hellman *parameter* in the ISAKMP/IKE SA is for. The parameter contains information on a *group* to perform the Diffie-Hellman exchange. The group consists of generation material used for creating new keys.

THE PSEUDO-RANDOM FUNCTION

The *pseudo-random function* (PRF) is really just another name for a hash function. In IKE, you can use the PRF both for authentication purposes and to generate additional key material as a randomizer.

DOMAIN OF INTERPRETATION (DOI)

The DOI for IPSec specifies all the parameters associated with the AH and ESP protocols and assigns them unique identifiers. In effect, the DOI serves as a database of values to be referenced during the IPSec SA negotiation. For example, as discussed in Chapter 16, ESP is one of the IPSec security protocols. It consists of one mandatory encryption *transform* (algorithm) and several optional transforms.

For instance, the IPSec DOI will identify the ESP protocol with a *Protocol ID* of 3, and assign each ESP encryption transform a unique *Transform ID* value between 0 and 10.

Each security protocol or service that uses the ISAKMP protocol will require its own DOI in order to specify all relevant parameters. For this reason, the IANA registers DOIs and also is responsible for assigning them unique identifiers.

INTERNET SECURITY ASSOCIATION AND KEY MANAGEMENT PROTOCOL (ISAKMP)

The Internet Security Association and Key Management Protocol (ISAKMP) is based on the central concept of a security association (SA). Unlike the IPSec SA, the ISAKMP/IKE SA is a bidirectional security contract between communicating systems that specifies how a secure connection will be established. Both ISAKMP/IKE and IPSec use SAs, but the SAs are independent of one another.

IPSec SAs are unidirectional, and they are unique for each security protocol. A set of SAs is needed for a protected communications channel:

▼ One SA per direction
▲ One SA per protocol

For example, if you have a pipe that supports ESP between peers, one ESP SA is required for each direction. SAs are uniquely identified by destination (IPSec endpoint) address, security protocol (AH or ESP), and Security Parameter Index (SPI). A user can also establish IPSec SAs manually.

The SA consists of all the parameters needed to fully specify the security contract. This includes items such as the authentication and encryption algorithms to be used, keying material, key lengths, and key lifetimes. All SA parameters are organized into a structure called the *Security Parameter Index* (SPI). An SA is negotiated for each security protocol used between communicating systems and can have multiple SAs established.

ISAKMP/IKE negotiates and establishes SAs on behalf of IPSec. An ISAKMP/IKE SA is used by ISAKMP/IKE only, and unlike the IPSec SA, it is bidirectional.

The establishment of SAs between systems takes place in two phases. First, an ISAKMP bidirectional SA is negotiated between the systems (such as between two ISAKMP servers). The ISAKMP bidirectional SA is then used as a secure channel to protect traffic between the systems for the second phase of negotiation, the non-ISAMP unidirectional SAs. All non-ISAKMP SAs are considered to be protocol SAs, such as an AH SA or ESP SA in the case of IPSec.

ISAKMP is not limited to IPSec. The ISAKMP protocol was designed to support security at all layers of the protocol stack, and it is possible for other protocols (SSL, TLS, OSPF, and so on) to use ISAKMP to establish their own SAs as well.

Each subsequent security protocol or service used by the communicating systems will have its own SA and corresponding SPI. The IANA has assigned UDP Port 500 for use by the ISAKMP protocol.

ISAKMP Message Structure

ISAKMP defines a very flexible method of building messages that can be adapted to almost any type of service, not just IPSec. ISAKMP messages are very modular. Building these messages is accomplished using various types of *payloads*.

Currently, there are 14 payload types defined:

- ▼ Security Association payload
- ■ Proposal payload
- ■ Transform payload
- ■ Key Exchange payload
- ■ Identification payload
- ■ Certificate payload
- ■ Certificate Request payload
- ■ Hash payload
- ■ Signature payload
- ■ Nonce payload
- ■ Notification payload
- ■ Notify Message payload
- ■ Delete payload
- ▲ Vendor ID payload

These payloads are the basic building blocks of an ISAKMP message. Each payload has a generic header indicating what the next payload is and the length of the payload. This enables chaining payloads together and nesting payloads within another payload.

Cookies

In addition to the payloads, the ISAKMP Header has the cookies to protect against de-nial-of-service attacks, the exchange type to indicate what type of flow is occurring (Aggressive mode, for example), and a message ID to uniquely identify the message against an SA negotiation.

Payloads may also require certain attributes to be defined, and ISAKMP defines how data attributes are to be formatted within the payload.

How these payloads are coded or formatted depends on the services using ISAKMP. These definitions are known as the Domain of Interpretation (DOI), which we described earlier in this chapter. The DOI also contains the exchange types and naming conventions. This means that if IPSec is the service being used in the IPSec DOI, then ISAKMP defines how the payloads are coded.

Along with the DOI, is the concept of a *situation*. A situation allows a device to make policy decisions with regard to security services that are being negotiated. For example, the IPSec DOI defines three situations:

▼ Identity only

■ Secrecy

▲ Integrity

IPSEC IKE

IKE is a combination of several existing key-management protocols that are combined to provide a specific key-management system. IKE is very complex and several different methods can be used in the establishment of trust and in providing keying material.

IKE is IPSec's answer to protocol negotiation, key management, and key exchange. Recall that IKE is actually a hybrid protocol that integrates the ISAKMP, SKEME, Photuris, and a subset of the Oakley key-exchange scheme. IKE provides the means to:

▼ Negotiate which protocols, algorithms, and keys to use

■ Ensure from the beginning to the end of the exchange that you're talking to whom you think you're talking and to authenticate the origin

■ Manage those keys after they've been agreed upon

▲ Exchange material for generating those keys safely

IKE SA Negotiation

To establish an IKE SA, the initiating node needs to negotiate the following:

▼   The encryption algorithms used to protect the data.

■   The hash algorithms used to create a message digest of the data to be signed.

■   The authentication algorithms used to create the digital signature for the data.

■   The group information needed for the Diffie-Hellman exchange.

■   A Pseudo-Random Function (PRF) used to hash certain values during the key exchange for verification purposes. This procedure is optional. If you want, you can just use the hash algorithm by itself.

▲   The Security Protocol to use, such as ESP or AH in IPSec's case.

Oakley and ISAKMP protocols, which are included in IKE, each define separate methods of establishing an authenticated key exchange between systems. Oakley defines *modes* of operation to build a secure relationship path, and ISAKMP defines *phases* to accomplish much the same process in a hierarchical format. The relationship between these two is represented by IKE employing a combination of ISAKMP's *exchanges* and Oakley's *modes* together using ISAKMP's *two phases* concept.

Phases and Modes

Phase 1 is the process that allows the two ISAKMP peers to establish a secure, authenticated communication channel. Each system is verified and authenticated against its peer to allow for future communications. *Phase 2* exists to provide keying information and material to assist in the establishment of SAs for an IPSec communication.

Within Phase 1 there are two modes of operation defined in IKE:

▼   Main mode

▲   Aggressive mode

Each of these modes is used in the first phase to set up secure exchange of ISAKMP SAs and can only exist in Phase 1. After the first phase is complete, the second phase makes use of two other modes:

▼   Quick mode

▲   New Group mode

Quick mode is used to establish SAs on behalf of the underlying security protocol. *New Group* mode is designated as a Phase 2 mode because it can only execute after Phase 1 is complete. However, the real purpose of the service provided by New Group mode is to benefit Phase 1 operations. One of the advantages of a two-phased approach is that the second phase can provide additional ISAKMP SAs, which eliminates the reauthorization of the peers.

Phase 1 is initiated using ISAKMP-defined *anti-clogging tokens* (derived from Photuris) that are most commonly referred to as "cookies." (Arrr…mmmm…cooookieees—Easy there, Homer!) No, not those cookies. The purpose of these cookies is to guard against denial of service (DoS) attacks. The *initiator* cookie and *responder* cookie are used to establish an ISAKMP SA, which provides end-to-end authenticated communications. Remember, ISAKMP communications are bidirectional. Once established, either peer may initiate a Phase 2 Quick mode function to establish SA communications for the security protocol.

The order of the cookies is crucial for future second phase operations. A single ISAKMP SA can be used for many second phase operations, and each second phase operation can be used for several IPSec SAs or SA bundles.

Main mode and Aggressive mode each use Diffie-Hellman keying material to provide authentication services. In *Main* mode the first two messages determine a communication policy, the next two messages exchange Diffie-Hellman public data, and the last two messages authenticate the Diffie-Hellman exchange.

Aggressive mode is an option available to vendors and developers that provides much more information with fewer messages and acknowledgements. The first two messages in Aggressive mode determine a communication policy and exchange Diffie-Hellman public data. In addition, a second message authenticates the responder, thus completing the negotiation.

Both phases may use different ISAKMP exchange types called *modes* to negotiate security associations. Phase 1 may use either Main mode or Aggressive mode. Main mode is required by IKE specification, while support for Aggressive mode is recommended but not required from the implementation. The difference between the two modes is that Aggressive mode uses fewer transactions to establish the ISAKMP SA, but at the expense of not providing Perfect Forward Secrecy (PFS) for all authentication methods.

Phase 2 is much simpler in nature because it provides keying material for the initiation of SAs for the security protocol. This is the point where the key management is utilized to maintain the SAs for IPSec communications. After the ISAKMP SA has been established, negotiation of the IPSec SA may be initiated by setting the *Message ID* of the ISAKMP header.

The second phase has one mode designed to support IPSec and that is Quick mode. Quick mode verifies and establishes the keying process for the creation of IPSec SAs. The communication is used to derive key material and negotiate shared policies for non-ISAKMP SAs—in our case, one or more IPSec SAs. The Phase 2 negotiation is protected by the ISAKMP SA that was established in Phase 1. The Quick mode negotiation is done by exchanging nonces that provide replay protection by frequently generating fresh key material. This is to protect against attackers generating bogus SAs. The key-refresh interval is agreed upon in Phase 1.

Not related directly to IPSec SAs is the *New Group* mode of operation. New Group mode provides services for Phase 1 to create additional ISAKMP SAs. Although often classified as a Phase 2 function, New Group mode is not really a Phase 1 or Phase 2 mode, but may be used after Phase 1 to establish a new Diffie-Hellman group to be used in future SA negotiations using Diffie-Hellman exchanges. The *mode* term is taken from the Oakley protocol. An IKE implementation may not switch modes in the middle of an exchange.

Table 17-1 gives a brief summary of the different modes and their relation to the two phases.

System Trust Establishment

The first step in establishing communications is verification of the remote system. There are three primary forms of authenticating a remote system:

▼ Shared secret

■ Public/private key

▲ Certificate

You learned about these methods in Part IV, so this should be familiar to you.

Phase	Mode	Required	Description
1	Main mode	Yes	Guaranties Perfect Forward Secrecy for all authentication methods but extensive communication to establish ISAKMP SAs
1	Aggressive mode	No	Reduced number of round-trips to establish ISAKMP SA. Does not offer Perfect Forward Secrecy for all authentication methods
2	Quick mode	Yes	Derives keying material for IPSec SAs from nonces using the established ISAKMP SAs
After 1 Before 2	New Group mode	No	An optional step after Phase 1 but before Phase 2, between the establishment of an ISAKMP SA and IPSec SA. Used to negotiate further SAs

Table 17-1. Relationship Between Phases and Modes

Shared secret authentication is currently used widely due to the relatively slow integration of certificate authority (CA) systems and the ease of implementation. However, shared secret is not scalable and can become unmanageable very quickly due to the fact that there can be a separate secret for each communication.

Public and private key use is employed in combination with Diffie-Hellman to authenticate and provide keying material. During the system authentication process, hashing algorithms are utilized to protect the authenticating shared secret as it is forwarded over untrusted networks. This process of using hashing to authenticate is nearly identical to the authentication process of an AH security protocol. However, the message, in this case a password, is not sent with the digest. The map previously shared or configured with participating systems will contain the necessary data to be compared to the hash.

Certificates are a different process of trust establishment. Each device is issued a certificate from a CA. When a remote system requests communication establishment, it will present its certificate. The recipient will query the CA to validate the certificate. The trust is established between the two systems by means of an ultimate trust relationship with the CA and the authenticating system. Because certificates can be made public and are centrally controlled, there is no need to attempt to hash or encrypt the certificate.

Key Establishment

Key exchange is very similar to IPSec's SA management. When you need to create an SA, you need to exchange keys. So IKE wraps both together and delivers them as an integrated package automatically. IKE provides a very elegant and safe solution for negotiating SAs and exchanging keys over public networks and can scale easily to any size, even globally.

IPSec standard mandates that key management must support two forms of key establishment:

▼ Manual
▲ Automatic

The other IPSec protocols (AH and ESP) don't really care what kind of key management is used. However, there may be issues with implementing antireplay options, and the level of authentication can be related to the key-management process supported, so key management is a factor in the ultimate security of the communication. If the key is compromised, the communication can be vulnerable to attack. To guard against such attacks, rekeying mechanisms ensure that even if a key is compromised, its validity is limited either by time, amount of data encrypted, or a combination of both.

Manual Key Management

Keys can also be exchanged manually. That means if you wish to use manual, or "face-to-face" key exchange, you can, and IPSec specifies that manual key exchange be supported.

Manual key management requires that an administrator provide the keying material and necessary security association information for communications. Manual techniques are practical for small environments with a limited number of gateways and hosts. Manual key management does not scale to include many sites in a meshed or partially meshed environment. The complexities involved in the addition of more participants to a large IPSec environment increases exponentially.

Let's say that you have an organization that wants to implement a VPN using the Internet as its medium. Let's assume that they have five locations around the world, and they want the ability for each office to communicate directly with any other office, so each VPN relationship will need a unique key.

The number of keys can be calculated by the formula $n(n-1)/2$ where n is the number of office locations. In the preceding situation, the number of keys would be ten. That's not so bad, right? Now let's apply this formula to 25 sites, five times the number of sites in our previous situation. The number of keys is not five times ten, or 50 keys. Using the formula, you will see that the number of keys grows exponentially to 300.

The complexity doesn't just stop at the number of keys. Each device must be configured, and the keys must be shared with all corresponding systems. The use of manual keying reduces the flexibility and number of options that make IPSec so attractive. Antireplay, on-demand rekeying, and session-specific key management are not available in manual key creation.

Automatic Key Management

Automatic key management addresses the limited manual process and provides for widespread, automated deployment of keys. The goal of IPSec is to build on existing Internet standards to accommodate a fluid approach to interoperability. As described earlier in this chapter, the IPSec default automated key management is our IKE four-protocol hybrid. However, based on the structure of the standard, any automatic key management may be employed. Automated key management, when instituted, may create several keys for a single SA. There are various reasons for creating multiple keys:

▼   Encryption algorithm requires more than one key

■   Authentication algorithm requires more than one key

■   Encryption and authentication are used for a single SA

▲   Rekeying

The encryption and authentication algorithms use multiple keys, and if both algorithms are used, multiple keys will need to be generated for the SA. An example of this would be if Triple DES were used to encrypt the data. We talked about several versions of Triple DES in Chapter 13—DES-EEE3, DES-EDE3, and DES-EEE2—and each of those versions uses more than one key. DES-EEE2 uses two keys, one of which is used twice.

The process of rekeying protects future data transmissions in the event a key is compromised. This process requires the rebuilding of an existing SA. The whole purpose of rekeying during data transmission is to provide unpredictable communication flow. When the keys are changing faster than an attacker can compromise them, it really does-n't make a difference if the algorithm is ridiculously hard to crack (large keys) or rela-tively easy to crack (shorter keys). Even if the attackers can crack them in a short time, it won't help them at all if the key has changed before they can apply their solution to a key that just went by.

Automatic key management may provide two primary methods of key provisioning:

▼ Multiple string

▲ Single string

Multiple strings are passed to the corresponding system in the SA for each key and for each type. For instance, the use of Triple DES for the ESP will require more than one key to be generated for a single type of algorithm; in this case, it's the encryption algorithm. The recipient will receive a string of data representing a single key, and once the transfer has been acknowledged, the next string representing another key will be transmitted.

In contrast, the *single string* method sends all of the required keys in a single string. This requires a strict set of rules for management. It is very important for each of the sys-tems involved to ensure that the corresponding bits be properly mapped to the same key strings for the SA being established. To ensure that IPSec-compliant systems properly map the bit to keys, the string is read from the left, with the highest bit order listed first for the encryption key(s) and the remaining string used for authentication. The number of bits that are used is determined by the specific encryption algorithm as well as the num-ber of keys required for the encryption being utilized for that SA.

SUMMARY

In this chapter, we discussed protocols and mechanisms used for key exchange and key management. We then gave you a basic overview of all the elements involved in key ex-change and key management and how they work together. In the next chapter, we will discuss key exchange and key management protocols in more detail.

CHAPTER 18

Key Management/Exchange Protocols

In Chapter 17, we provided an overview of the main concepts and components of key management and key exchange. In this chapter, we go deeper into the specifics and mechanics of key management and exchange protocols.

The key-management protocol is an essential and important issue in IP security. Several protocols have been suggested for the IPSec key-management standard, and they all had some impressive capabilities. Some of the more common proposals have included ISAKMP, Oakley, SKIP, SKEME, and Photuris.

The number of proposals gave rise to a dilemma for the people who had to create a standard. The IPSec Working Group (WG) found itself in a position similar to that of the forefathers of the United States when faced with the decision of where to put the nation's new capital. The IPSec WG used the same wisdom in the creation of IKE as the forefathers used in the creation of Washington, D.C.; they placed it in between all of the candidates being considered. Fortunately, by doing this, they took advantage of the best features from each protocol. They combined them in a way that would allow the flexibility of using the features of the current protocols as well as features from protocols designed in the future.

So, in this chapter we discuss some of these individual protocols and explain the techniques and procedures that were borrowed to create the final implementation IPSec chose: the Internet Key Exchange (IKE) protocol.

Key management and distribution is one of the most critical services for cryptography. Defining secure cryptographic algorithms and protocols is not easy, but generating, exchanging, and keeping secret keys can be even tougher. Many successful attacks exploit flaws in the key-management procedures used.

Much of the overall security for a cryptosystem resides in the strength of key generation and distribution procedures. There are two different types of keys:

▼ **Master keys**   Encryption keys used to encrypt and establish other keys

▲ **Session keys**   Encryption keys used to encrypt and protect the actual data

IN-BAND AND OUT-OF-BAND KEY EXCHANGES

Any key-management scheme requiring significant scalability must be based on an authenticated public-key infrastructure. Public keys can be authenticated using a variety of mechanisms, such as public-key certificates, secure directory servers like LDAP, and so on. ITU standard X.509 provides an example of a public-key infrastructure, built using certificates. Preshared master keys and other out-of-band key distribution techniques may also be used when the system must be upheld on a limited scale. Private and public keys are usually not used to encrypt data traffic directly.

They are mostly used to authenticate some other key-exchange mechanisms able to establish session keys. The latter will then be effectively used for data encryption. Two different kinds of key exchange enable you to use public keys to establish session keys and eventually provide datagram protection.

Authenticated Session Key

The first alternative is to establish an authenticated session key, using some out-of-band session-key establishment protocols. This session key is then used to encrypt or authenticate IP data traffic. IKE/ISAKMP protocol types are the most common of these. Such a scheme has the disadvantage of having to establish and securely manage a pseudo-session layer underneath IP, which is a stateless protocol.

The IP source would need to first communicate with the IP destination to acquire this session key. Also, if and when the session key needs to be changed, the IP source and the IP destination need to communicate again to make this happen. Each such communication could involve the use of a computationally expensive public-key and Diffie-Hellman operation.

Secure management of pseudo-session states is further complicated by crash recovery considerations. If one side crashes and loses all session state, then mechanisms need to be in place that can securely remove the half-opened session state on the side that did not crash. These mechanisms need to be secure because insecure unauthenticated removal of half-opened sessions opens the door to a trivial denial-of-service attack.

Stateless Key Management

The second alternative is to utilize public keys in a stateless key-management scheme. This is how it is done in the Simple Key-Management for Internet Protocols (SKIP). This method works using in-band signaling of the packet encryption key, where the packet encryption key is encrypted in the recipient's public key. This is also the way Privacy Enhanced Mail (PEM) and other secure mail programs perform message encryption.

Although this scheme avoids the session-state establishment requirement and prior out-of-band communication to set up and change encryption keys, it has the disadvantage of having to carry the actual encryption key in each IP packet, encrypted in the recipient's public key. This kind of scheme has a high overhead in the most common implementations of 64 to 128 bytes of keying information in every packet. On top of that, whenever the encryption key changes, a public-key operation needs to be performed in order to derive the new encryption key making the computational overhead of such a scheme even higher.

DIFFIE-HELLMAN KEY EXCHANGE

The Diffie-Hellman key exchange is a well-known public-key technique used to distribute keys but not to encrypt data. The Diffie-Hellman exchange allows two entities to agree on a shared secret key, even when they can only exchange messages in public.

To begin the Diffie-Hellman exchange, it is necessary to define two numbers:

▼ p, a prime number, which is usually about 512 bits
▲ g, a number less than p

Both p and g can be known publicly beforehand. Once p and g are mutually agreed upon, both entities choose a 512-bit number at random and keep it secret. They elevate g to this randomly chosen value and send the result modulo p to the other entity, even through a public channel. Then, each entity uses its own generated value, exchanged only in a derived form, together with the other party's precomputed value to generate a common secret. Even if the channel were not protected and an enemy were to intercept the exchanged values, it would be extremely difficult to retrieve the final shared secret.

It is important to note that Diffie-Hellman by itself is not foolproof. As we discussed throughout Part IV, if both p and g are known beforehand, and an attacker knows these numbers, you can have problems. When you know p and g and can listen to the conversation during the resultant modulo exchange, you can pose as either side at will as long as you are able to spoof the IP address for a side. When dealing with security, you should never reveal anything you don't have to reveal.

SIMPLE KEY-MANAGEMENT FOR INTERNET PROTOCOLS (SKIP)

The Simple Key-Management for Internet Protocols (SKIP) is a stateless and sessionless key-management protocol used to uphold session keys protected with a public-key cryptosystem. In SKIP, each IP-based source and destination has an authenticated Diffie-Hellman public value. This public value can be authenticated in numerous ways.

Some possibilities for authenticating DH public values are the use of X.509 certificates, Secure DNS, or PGP certificates. These certificates can be signed using any signature algorithm, such as RSA or DSA.

Each IP entity in SKIP has a secret value and a public value, as in any public-key cryptosystem. A shared secret is derived from these values, and it is used as the basic-key encrypting key that provides IP packet-based authentication and encryption. This secret is called the *long-term* secret, and it is used as key for a block symmetric cryptosystem like DES, RC2, IDEA, and so on.

This master key is used to encrypt a transient key, called *session key*, which is used for data packet encryption. This is done to limit the actual amount of data encrypted using the long-term key, since it is desirable to keep the latter for a relatively long period. Transient keys are only encrypted in this long-term key and use the transient keys to encrypt or authenticate IP data traffic.

This limits the amount of data protected using the long-term key to a relatively small amount even over a long period, since session keys represent a relatively small amount of data.

If a node wants to change the session key, the receiver can discover this fact without having to perform a public-key operation. It uses the cached long time value to decrypt the new session key. Thus, without requiring communication between transmitting and receiving ends, and without necessitating the use of a computationally expensive public-key operation, the packet encrypting or authenticating keys can be changed by the transmitting side and discovered by the receiving side.

SKIP is simple to implement and provides straightforward and scalable solutions to permit dynamic rerouting of protected IP traffic through alternate encrypting intermediate nodes for crash-recovery, fail-over, and load-balancing scenarios. It is also especially suited for multicast and broadcast traffic. It may instead not be the best choice when bulk data transmission is done between two stable and well-connected hosts. In the last case, session-oriented protocols, such as IKE/ISAKMP, should be preferred.

PHOTURIS

Developed since 1995 by Phil Karn and William Simpson, Photuris uses the same principle as the Station-To-Station (STS) protocol created by Diffie, Van Oorschot, and Wiener. Photuris was described, independently from any working group, in several Internet drafts and RFCs, and is designed for use with IPSec. A few implementations use it, although IKE is much more common.

Contrary to SKIP, Photuris is a connection-oriented protocol in the sense that it is composed of a certain number of exchanges for the negotiation of options and the generation of keys that take place prior to any exchange of encrypted messages. UDP port 468 was assigned to Photuris by the IANA.

Photuris is based on the generation of a shared secret using Diffie-Hellman. This shared secret has a short life span because it is used to generate the session keys necessary to protect the rest of the communication. To counter the man-in-the-middle attack to which Diffie-Hellman is vulnerable, the exchange of the values being used to generate the shared secret is followed by an authentication of these values thanks to long-term secrets. As these long-term secrets are used only for authentication, Photuris provides perfect forward secrecy.

A problem with Diffie-Hellman is that it requires operations with a high cost in system resources, which makes it vulnerable to denial of service attacks called "flooding attacks." To make that type of attack more difficult, Photuris uses a *cookie exchange* before carrying out the Diffie-Hellman values exchange. The value of the cookies depends on the involved peers based on their IP addresses and on the UDP port number. Using these criteria aims at preventing an attacker from obtaining a valid cookie and using it to flood the victim with requests coming from arbitrary IP addresses and/or ports. In addition, it should not be possible for an attacker to generate cookies that will be accepted by an entity as generated by itself. This implies that the transmitting entity uses some local secret information in the generation of its cookies and in their later checking.

Photuris encompasses the following stages:

1. A *cookies exchange* to counter some simple denial of service attacks. Each involved party generates a cookie, and the cookies are transmitted with each message.

2. A *public values exchange* for the generation of a shared secret.

3. An *identities exchange* so that the parties are identified to one another and can check the authenticity of the values exchanged during the preceding stage. This exchange is protected in confidentiality thanks to private keys derived, among others, from the shared secret and the cookies.

Other messages can later be exchanged to modify the session keys or the security parameters. These messages will be protected in confidentiality in the same way as the messages from exchange 3.

Simultaneously with the other exchanges, the communicating peers agree on the method for shared secret generation and on some security parameters useful for the security association.

SKEME

Developed specifically for IPSec, SKEME is an extension of Photuris suggested in 1996 by Hugo Krawczyk. Contrary to Photuris, which imposes a precise course for the protocol, SKEME provides various modes of key exchange.

Like STS and Photuris, SKEME's basic mode is based on the use of public keys and a Diffie-Hellman shared-secret generation. However, SKEME is not restricted to the use of public keys, but also allows the use of a preshared key. This key can be obtained by manual distribution or by the intermediary of a key distribution center (KDC) such as Kerberos. The KDC enables the communicating peers to share a secret by the intermediary of a trusted third party. The use of this secret for the authentication of the Diffie-Hellman secret and not directly as a session key decreases the necessity of trust in the KDC. Lastly, SKEME also makes it possible to carry out faster exchanges by not using Diffie-Hellman when Perfect Forward Secrecy (PFS) is not required.

In short, SKEME contains four distinct modes:

▼ Basic mode, which provides a key exchange based on public keys and ensures PFS thanks to Diffie-Hellman.

■ A key exchange based on the use of public keys, but without Diffie-Hellman.

■ A key exchange based on the use of a preshared key and on Diffie-Hellman.

▲ A mechanism of fast rekeying based only on symmetrical algorithms.

In addition, SKEME is composed of three phases: SHARE, EXCH, and AUTH:

▼ During the SHARE phase, the peers exchange half keys, encrypted with their respective public keys. These two half-keys are used to compute a secret key K. If anonymity is wanted, the identities of the two peers are also encrypted. If a shared secret already exists, this phase is skipped.

■ The exchange phase (EXCH) is used, depending on the selected mode, to exchange either Diffie-Hellman public values or nonces. The Diffie-Hellman shared secret will only be computed after the end of the exchanges.

▲ The public values or nonces are authenticated during the authentication phase (AUTH), using the secret key established during the SHARE phase.

The messages from these three phases do not necessarily follow the order described; in actual practice they are combined to minimize the number of exchanged messages.

Another phase, known as the *cookies phase*, can be added before the share phase to provide protection against denial of service attacks, thanks to the cookies mechanism introduced by Photuris.

OAKLEY

Initially proposed by Hilarie Orman from the computer department of the University of Arizona, Oakley has been the subject of several Internet drafts within the IPSec group and is, along with ISAKMP and SKEME, an integral part of IPSec's key exchange.

Oakley is a key-exchange protocol that is closely related to SKEME in that it, too, has several modes, uses cookies, and does not require the computation of the Diffie-Hellman shared secret before the end of the protocol.

Where it differs comes from the fact that it explicitly enables the peers to agree on the key exchange, encryption, and authentication mechanisms to use.

Actually, Oakley's goal is to allow the peers to securely share a set of parameters for the protection of the exchanges. Those parameters include:

▼ Name of the key

■ Secret key

■ Identities of the peers

■ Encryption algorithms

■ Authentication

▲ Hash functions

Several options are available in Oakley. In addition to the traditional Diffie-Hellman key exchange, Oakley can be used to derive a new key from an old one or to distribute a key by encrypting it. These options result in the existence of several modes.

The main principle of Oakley is that the initiator of the exchanges starts by specifying as much information as they wish in the first message. The recipient also answers by providing as much information as they wish. The conversation goes on until the required state is reached. The choice of the quantity of information to include in each message depends on the selected options. The options are things like the use of stateless cookies, identity protection, PFS, nonrepudiation, and so on.

The three components of the protocol are

▼ Cookies exchange (optionally stateless)

■ Diffie-Hellman public values exchange (optional)

▲ Authentication (options: anonymity, PFS on the identities, nonrepudiation)

Establishing the first phase ISAKMP SA involves authenticated key exchange between the communicating systems in order to protect subsequent traffic and negotiation of protocol SAs.

The Oakley Key Determination Protocol is the mechanism used to accomplish this secure exchange. Oakley uses the Diffie-Hellman key-exchange algorithm, a public-key algorithm that allows systems to independently generate a unique secret key based on the knowledge of each other's public key.

The ISAKMP standard requires that the public keys used to establish the ISAKMP SA be authenticated using digital signatures. ISAKMP does not specify a specific signature algorithm or certificate authority (CA), but instead allows systems to indicate which CAs they support. Oakley provides a mechanism for the identification of certificate types, supporting CAs, and the exchange of certificates between systems.

Once Diffie-Hellman public keys are exchanged and authenticated, and a secret key generated, the communicating systems can establish an IPSec SA that characterizes the subsequent use of the AH and ESP security protocols.

ISAKMP

As we stated earlier in this chapter, the Internet Security Association and Key Management Protocol (ISAKMP) is a framework that defines standards and procedures for establishing, negotiating, modifying, and deleting different security parameters between two or more communicating entities. You can think of ISAKMP as a high-level protocol within which actual key management and other security protocols may operate.

In 1994, the IETF started work on finding a key management/exchange protocol that would have the following six fundamental properties:

▼ Confidentiality

■ Authentication

■ Integrity

■ Nonrepudiation

■ Authorization

▲ Accessibility

Along with ISAKMP, other protocols were also being considered, such as SKIP, SKEME, Oakley, Photuris, and so on. ISAKMP outperformed the others in terms of flexibility and generality and, in 1996, was adopted as the key-management protocol for IPSec.

ISAKMP satisfied the first two of the six fundamental security properties stated by IETF. The latter four properties could be satisfied within ISAKMP based on the specific needs at hand by having ISAKMP negotiate the additional security services with the other communicating parties.

The Internet Security Association and Key Management Protocol (ISAKMP) provides a framework for authentication, security association management, and key generation and distribution. It does not specify any key-exchange protocols because it is supposed to be key-exchange protocol independent.

ISAKMP is a powerful and flexible protocol and supports the negotiation of the security policies under which the secure communication channel has to be established.

However, ISAKMP's flexibility is provided at the cost of complexity, increased latency in association establishment, and increased packet traffic.

ISAKMP uses UDP as a transport protocol, without relying on any error checking information such as checksums for its processing. ISAKMP can negotiate SAs for security protocols operating at any OSI layer.

ISAKMP Concepts

We mentioned that many of the concepts in this chapter would be discussed several times due to the fact that many of the terms intersect with, and in some cases are identical to, the terms used by other protocols. Though related, those definitions are not always identical, and it is important that you once again understand this. So, that being said (…again), we will explain some of the terms as they relate to ISAKMP specifically.

Security Protocol

A *security protocol* is a security service for network communication performed at a single point in the network stack. In the case of IPSec, for example, the Encapsulating Security Payload (ESP) and Authentication Header (AH) protocols are the security protocols operating in the Network layer.

ISAKMP Security Association (SA)

A *security association* (SA) defines a relationship between two or more communicating entities. This relationship describes how the entities will use security services, such as encryption, to communicate securely. In cleartext, a security association tells what the commonly agreed security parameters and services are between the communicating parties. Examples of such parameters are authentication algorithms, authentication keys, encryption algorithms, encryption keys, and the time of validity for the SA itself.

Depending on what kind of traffic is in question, different kinds of SAs are set up. Each type of security protocol requires its own set of SAs. AH and ESP, for example, require one set of SAs, while Secure Socket Layer (SSL) traffic requires another. SAs are often named according to the associated security protocol, for example, ESP SA. Note that ISAKMP doesn't define *what the contents* of the SAs are, but rather *how the communicating parties agree* on what the contents of the SAs are.

The Internet Security Association and Key Management Protocol (ISAKMP) is used to establish an agreement on the encryption algorithm, the authentication method, and the hash algorithm, which will have to be used in AH and ESP extensions. An instance of this set of parameters is named a security association (SA).

The ISAKMP protocol defines two different kinds of security associations and establishes them in two distinct phases:

▼ **ISAKMP SA (bidirectional)**   This defines the policy used to establish a secure channel for further communication. This policy specifies the security parameters used, such as authentication method, encryption algorithm, Diffie-Hellman group, and hash algorithm.

▲ **Non-ISAKMP SA (unidirectional)**   Once a channel has been secured using an ISAKMP SA, a *protocol SA* or *non-ISAKMP SA* can be established. The non-ISAKMP SA is the set of the security parameters used for the safe communication of data such as IPSec parameters for AH and ESP. A non-ISAKMP SA is also known as a *Session SA*.

ISAKMP is also used to exchange keys in a secure way, but it is not intended to define the specific way to do this. ISAKMP is intended to provide a consistent framework within which virtually any key-exchange technique, present or future, can work. This is done in order to separate SAs and key management from the details of the key exchange itself. This should be starting to sound very familiar to you right about now (the OSI Layering principle…).

ISAKMP Phases

ISAKMP protocol negotiation consists of two separate phases. In the first phase, ISAKMP *Phase 1*, the peers establish a secure channel for further communication by negotiating ISAKMP SAs. In the second phase, ISAKMP *Phase 2*, the peers negotiate non-ISAKMP SAs, which will be used to protect real communication like AH and ESP for IPSec.

In ISAKMP Phase 1, the initiator sends some proposals, and the responder either chooses one of these proposals or rejects the communication. Each proposal contains all the security parameters used to establish an ISAKMP SA. ISAKMP Phase 2 is done only if Phase 1 has been successful. It is similar to Phase 1 and can even be much simpler if no Perfect Forward Secrecy (PFS) is required. Phase 2 can indeed be very fast and simple if the correspondent Phase 1 guarantees a very secure channel.

Should the ISAKMP Phase 2 also be successful, the peers can start to communicate using IPSec with the non-ISAKMP SA they agreed on. ISAKMP SAs are bidirectional in nature. This means that once the SA is established, each of the two entities can begin a Phase 2 negotiation.

This two-phase approach has a higher startup cost than a single-phase approach, but there are also several advantages. The cost of Phase 1 can be shared among several Phase 2 negotiations. Multiple SAs can be established with a single Phase 1 exchange.

Domain of Interpretation (DOI)

ISAKMP defines a default set of payload formats and message exchanges. However, in most cases it will be necessary to define more. A *Domain of Interpretation* (DOI) defines these additional payload formats and exchange types. A DOI also defines naming conventions for security information such as cryptographic algorithms and modes. In other words, with the help of a DOI, the contents of all ISAKMP payloads can be correctly interpreted and all message exchanges correctly completed. To be more specific, a DOI defines:

▼ An interpretation for the contents of the Situation field

■ The set of mandatory, recommended, and optional security policies

■ A naming scheme for encryption algorithms, key-exchange algorithms, security-policy attributes, and Certificate Authorities

■ The syntax for DOI-specific SA attributes (for Phase 2 negotiations)

■ The syntax for DOI-specific payload contents

■ Additional key-exchange types (should the existing ISAKMP types not suffice)

▲ Additional notification message types (should they be needed)

ISAKMP Situation

A *situation* is a collection of information that is relevant when deciding what security services and parameters are needed for a session. In other words, a situation defines the outer conditions that must be taken into account when establishing a communications session.

An example situation can contain information such as source and destination IP addresses, desired security classification (public, confidential), and so on. This information is sent between ISAKMP servers so that according to each one, the servers can decide precisely what security services can be chosen for the SAs from a set of proposed services.

For example, the IPSEC DOI defines three situations:

▼ SIT_IDENTITY_ONLY

■ SIT_SECRECY

▲ SIT_INTEGRITY

Security Parameter Index (SPI)

A *Security Parameter Index* (SPI) is an identifier for an established SA in the scope of a security protocol. With the help of an SPI value and the Destination Address of the communicating party, you can uniquely identify a SA. This unique ID allows security protocols to access the correct SAs as well as the security mechanisms and parameters found within, to provide protection for the communication process.

ISAKMP Cookies

ISAKMP uses cookies to protect the protocol from denial of service (DoS) attacks. Some basic secret material is used to generate cookies, which requires very little computational cost. In this way, the cookie generation is secure and unique. One way of generating cookies is to use Karn's method. This is where a fast hash is performed over the IP Source and Destination Addresses, the UDP Source and Destination Ports, and a locally generated secret random value. The time and date are then added, so the cookie is unique for each SA establishment.

ISAKMP Exchanges

ISAKMP defines a default set of exchange types for ISAKMP Phases 1 and 2, such as establishment of both ISAKMP SAs and non-ISAKMP SAs. The payload format of these types is also defined. Exchange types define the content and ordering of ISAKMP messages during communications between peers. The primary difference between exchange types is the ordering of the messages and the payload ordering within each message. These exchanges provide different security protection for the exchange itself and information exchanged.

Five default exchange types are currently defined for ISAKMP:

▼ **Base exchange** Identity protection is not provided. Identity identifiers are exchanged before encryption.

■ **Identity Protection exchange** Separating the key-exchange information from the identity and authentication information provides identity protection. The identities are encrypted before being sent.

■ **Authentication Only exchange** The information is in plaintext. There is authentication without the computational expense of computing keys.

■ **Aggressive exchange** This is faster since it has fewer round-trips, but it does not provide identity protection.

▲ **Informational exchange** This is a one-way exchange with Error Notification or Deletion.

These exchange types can be used in either phase of negotiation. However, they may provide different security properties in each of the phases. With each of these exchanges, the combination of cookies and SPI fields identifies whether this exchange is being used in the first or second phase of a negotiation.

ISAKMP Features

Since ISAKMP is used to establish a secure channel for applications or network protocols like IPSec AH and ESP, it is important to provide authentication of the entities involved in the communication.

The authentication method is part of the ISAKMP SA exchange regardless of the key-exchange protocol being used. ISAKMP defines four possible authentication methods:

▼ Signature

■ Public-key encryption

■ Encrypted nonces

▲ Preshared keys

ISAKMP exchanges are linked together to avoid insertion of bogus messages. The authentication, key, and SA exchanges are linked together to avoid an enemy jumping in and impersonating one of the peers. Time-sensitive information is inserted in the cookies for protection against replay attacks.

ISAKMP Architecture

ISAKMP services are usually offered by a dedicated ISAKMP server using port 500. According to specification, ISAKMP must offer services with the UDP protocol. However, the specification says nothing about not supporting other transport protocols as well, such as TCP or even IP directly.

Take a look at Figure 18-1 to see the ISAKMP architecture. Even though the figure doesn't explicitly show that ISAKMP offers security service negotiation for all layers of the protocol stack, that is what ISAKMP has been designed for.

ISAKMP Messages

When ISAKMP starts negotiating security services for the communicating entities, it does so in two distinct phases.

Phase 1 is designed to establish an initial secure communications channel by creating security associations between the ISAKMP servers. This SA, also called the *ISAKMP SA,* provides the initial security parameters that are used for protecting subsequent communication between ISAKMP servers. After this, any number of *Phase 2* negotiations may take place.

Phase 2 negotiations are used for establishing the security protocol SAs needed for the actual communication between the communicating parties. That would be AH and ESP in the case of IPSec.

The messages exchanged by the ISAKMP servers are made up of different kinds of *payloads*. Each payload defines some security information that is needed to establish and maintain shared SAs and keys. Exchanges define what combinations of payloads must be sent and in what order. They can be seen as protocol definitions for ISAKMP servers.

Take a look at Figure 18-2 to see what an ISAKMP Protocol Header looks like. The descriptions for the individual fields follow. On the protocol level, every ISAKMP message starts with the ISAKMP Protocol header.

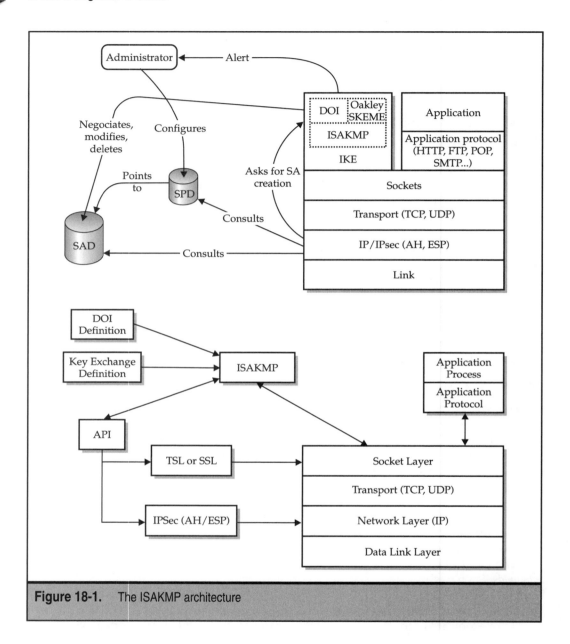

Figure 18-1.   The ISAKMP architecture

▼ **Initiator Cookie**   The Initiator Cookie field contains pseudo-random nonces that help prevent replay and denial of service attacks.

■ **Responder Cookie**   The Responder Cookie field contains pseudo-random nonces that help prevent replay and denial of service attacks.

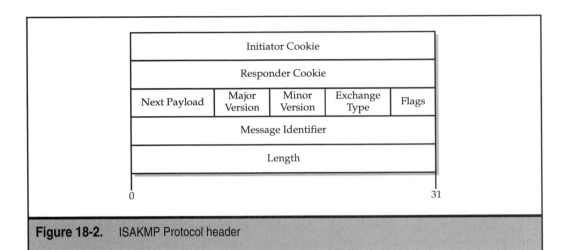

Figure 18-2.   ISAKMP Protocol header

- **Next Payload**   The Next Payload field is used to indicate the type of the next payload and allows the chaining of many payloads into one ISAKMP message.
- **Version**   The Version field indicates the version of the ISAKMP protocol implementation being used.
- **Exchange Type**   The Exchange Type field expresses the type of message exchange. This is an identifier for stating what payload combinations are exchanged between the negotiating ISAKMP servers and in what order.
- **Flags**   The Flags field contains the simple options for the ISAKMP exchange itself. They include:
    - The Encryption bit
    - The Commit bit for signaling key-exchange synchronization
    - The Authentication Only bit to be used with a subset of exchanges
- **Message Identifier**   The Message Identifier field is a randomly generated number that is used to describe the state of the ISAKMP Phase 2 negotiation.
- **Length**   The Length field contains the value that represents the length of the payload.

When ISAKMP messages are received, they are carefully inspected. The length of the payloads is checked as well as the initiator and responder cookies, the exchange type and flags (to check the validity of the next payload pointer), message identifier, and finally the version information. Should any of these tests fail, with the exception of the version, the ISAKMP packet will be dropped. If there should be a mismatch found in the version numbers, the packet should be processed. This is because each new ISAKMP implementation should be designed to be backward compatible. Checks like these improve the reliability of the ISAKMP protocol as well as protect against denial of service and replay attacks.

Payloads

ISAKMP payloads are the building blocks from which all ISAKMP messages are created. With the help of payloads, ISAKMP servers share information about common security associations and keys.

Physically, a payload is constructed with a generic header and the information that follows. That subsequent information can be completely unintelligible by ISAKMP.

The idea is that ISAKMP should know the overall syntax of its own default payloads and the extension payloads introduced by the DOI specification and should leave the exact interpretation of their actual contents to the DOI. This is what allows ISAKMP to accommodate any new and/or enhanced protocols.

Figure 18-3 shows what a Generic Payload header looks like, followed by a description of the individual fields within.

▼ **Next Payload**   The Next Payload field tells what type of payload follows the current one in an ISAKMP message:

Number	Payload
0	None
1	Security Association
2	Proposal
3	Transform
4	Key Exchange
5	Identification
6	Certificate
7	Certificate Request
8	Hash
9	Signature
10	Nonce
11	Notification
12	Delete
13	Vendor Identifier
14–127	Reserved
128–255	Private Use

■ **Reserved**   The Reserved field is currently unused and fixed to 0.

▲ **Payload Length**   The Payload Length field indicates the length of the current payload's contents that follow immediately after the header.

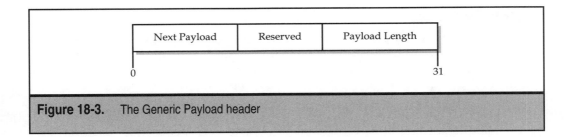

Figure 18-3.   The Generic Payload header

Security Association Payload

This payload is used to negotiate security parameters for establishing SAs and to define the domain of interpretation for negotiation. It also defines the outer requirements that the services to be negotiated must meet; in other words, it defines the situation. The length of the Situation field is defined in the DOI. Take a look at Figure 18-4 to see the Security Association Payload header.

The IPSec DOI, for example, has been assigned the DOI identifier value of 1. The contents of the Situation field for the IPSec DOI can be one of the three mentioned earlier:

▼ SIT_IDENTITY_ONLY

■ SIT_SECRECY

▲ SIT_INTEGRITY

A number of other parameters are also in the remainder of the IPSec DOI's SA Payload. Things like *secrecy level, secrecy category, integrity level,* and so on, describe the situation in more detail. The IPSec DOI's definition of the SA Payload is one example of how ISAKMP's default payload definitions can be extended with DOI-specific definitions.

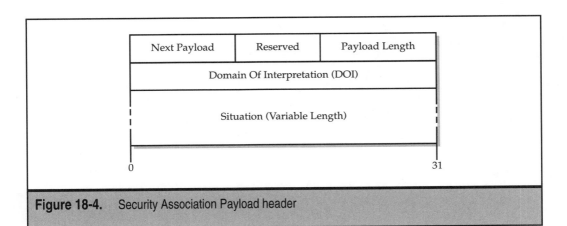

Figure 18-4.   Security Association Payload header

Proposal Payload

Proposal Payloads describe what security services are applicable when trying to select security services for creating a new SA. In other words, they are security service suggestions sent from the initiating ISAKMP server to the recipient in descending order of preference. That means the proposal that best meets the mutually agreed security requirements is used to set the security parameters in an SA. Take a look at Figure 18-5 to see the Proposal Payload header.

It is important to note that a Proposal Payload works on a per-security-protocol basis. This means that every proposal is related to the security protocol set in the Protocol ID field. Examples of protocols here would be AH and ESP in the case of IPSec.

The SPI identifies the future SA should this proposal be accepted. A Proposal Payload also tells how many *Transform Payloads* (the number of transforms), the individual suggestions for security parameters, are contained in the proposal. These transforms follow in separate Transform Payloads immediately after the Proposal Payload.

Transform Payload

A *Transform Payload* can be seen as one detailed security parameter suggestion in a proposal, as described earlier. Such a suggestion consists of a specific security mechanism and the associated attributes that will be included in the SA if this suggestion gets accepted. It is strongly recommended that the SA attributes be defined according to the Data Attribute syntax. (You will be seeing this shortly.) Take a look at Figure 18-6 to see the Transform Payload header.

Key Exchange Payload

This payload is for exchanging key information between ISAKMP servers. The data field carries key material for the generation of the actual encryption keys. The exact interpretation of its contents is left to the key-exchange protocols and the DOI in question. Many asymmetric key generation algorithms are supported, such as Oakley (Diffie-Hellman) and RSA.

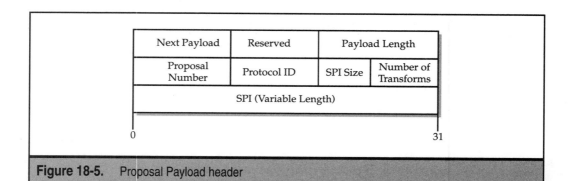

Figure 18-5.   Proposal Payload header

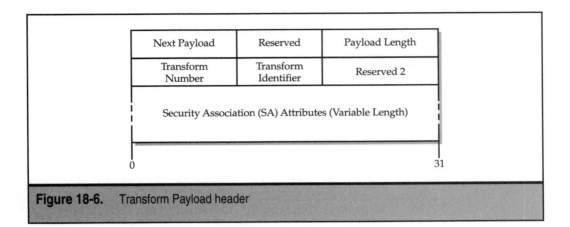

Figure 18-6.   Transform Payload header

Identification Payload

This payload is needed for exchanging information about the identities of the communicating parties. It contains information about the type of identification being used and identity information of the party identifying itself. The exact interpretation of the contents depends on the DOI.

Certificate Payload and Certificate Request Payload

These payloads are used for exchanging information about the certificates with ISAKMP in case the normal certificate distribution services such as secure DNS are unavailable. Ten types of certificates are recognized, the most common being X.509, SPKI, PGP, and PKCS.

The *Certification Type* or *Certification Encoding* defines what type of certificate is transported within the certificate (authority or data) field. Take a look at Figure 18-7 to see the Certificate Payload header.

Hash Payload

The hash data that can be inserted into the Hash Payload has two uses:

▼ Ensuring the integrity of ISAKMP messages

▲ Authenticating the communicating entities

The hash algorithm used, such as MD5, SHA, and so on, is selected during establishment of the SA.

Signature Payload

The Signature Payload is similar to the Hash Payload in that it can be used for integrity verification of ISAKMP messages as well as nonrepudiation. The Signature Payload

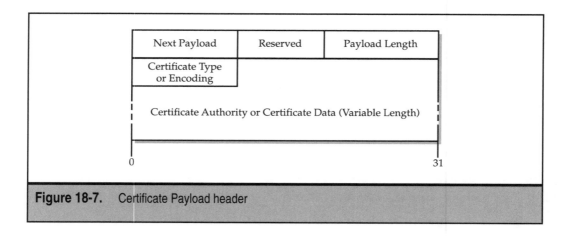

Figure 18-7. Certificate Payload header

carries the data generated by a digital signature algorithm that, like the Hash Payload, has been selected during establishment of the SA.

Nonce Payload

A *nonce* is randomly chosen data that is time-sensitive during an exchange and is designed to protect against replay attacks. Nonces may also be sent as part of the key-exchange protocols.

Notification Payload

The *Notification Payload* is used for sending codes, error messages, and other similar status information. Take a look at Figure 18-8 to see the Notification Payload header.

There are 30 predefined error codes, but more can be defined in the DOI or based on private needs. Examples of such codes (and their values) are

▼ Invalid-payload-type (1)

■ DOI-not-supported (2)

▲ Authentication-failed (24)

Delete Payload

The Delete Payload is intended for informing the communicating party that an SA is no longer valid and has been deleted. Upon receipt of this payload, the recipient should delete its own instance of the SA. To uniquely identify the SA, the security protocol's identifier and the SPI are included in the payload.

The modification of already established SAs in ISAKMP is actually achieved with the help of this payload. Updating of SAs is done by establishing a new SA with the desired properties and then deleting the old one, the one that had to be modified.

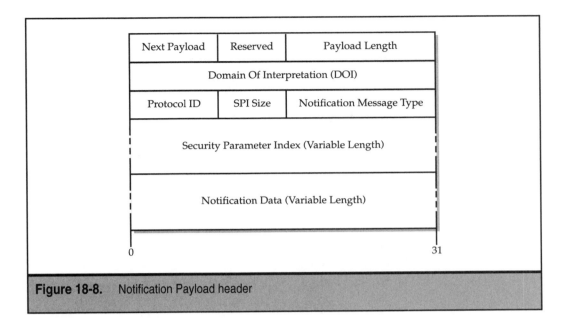

Next Payload	Reserved	Payload Length
Domain Of Interpretation (DOI)		
Protocol ID	SPI Size	Notification Message Type
Security Parameter Index (Variable Length)		
Notification Data (Variable Length)		

0   31

Figure 18-8.   Notification Payload header

Vendor Identification Payload

The vendor payload contains vendor-specific information. The contents of this payload have no practical importance from the point of view of ISAKMP and security. The Vendor Identification payload is solely a product development mechanism for vendors to use without affecting the rest of ISAKMP.

Data Attributes

Data attributes are parameters that are contained within other payloads. One payload may contain several data attributes. A data attribute allows including an additional parameter in key-exchange protocols if such are needed. Data attributes are particularly useful in Transform Payloads for passing security association parameters. Take a look at Figure 18-9 to see the Data Attributes Payload header.

What the additional attributes are and how they are interpreted is defined in the key-exchange definitions that are fed to the ISAKMP server at startup. ISAKMP only defines the generic structure of data attributes.

AF or Attribute Flag bit   The AF or attribute flag bit indicates how the third and fourth field should be interpreted. In other words, AF indicates whether *field three* and *field four* are

Attribute Length for *field three* and Attribute Value for *field four*

or

Attribute Value for *field three* and Empty for *field four*

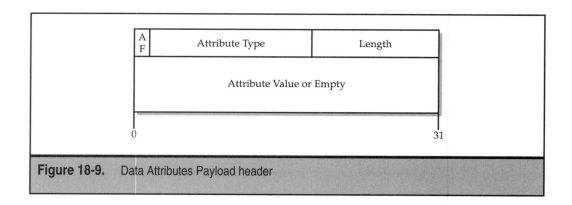

Figure 18-9. Data Attributes Payload header

Exchanges

Payloads, as just described, define "atomic" pieces of information that were contained within ISAKMP messages. ISAKMP *exchanges* define what combinations of payloads must be sent and in what order for establishing and maintaining SAs and keys. In other words, exchanges define protocol steps for ISAKMP servers that they in turn follow to negotiate security services and parameters.

Values zero through five are the ISAKMP default values. The rest of the values are reserved and can be used at the discretion of the following three groups:

▼ ISAKMP future use

■ DOI specific use

▲ Private use

The following table lists the possible exchange values:

Number	Exchange
0	None
1	Base
2	Identity Protection
3	Authentication Only
4	Aggressive
5	Informational
6–31	Reserved for ISAKMP Future Use
32–239	DOI-Specific Use
240–255	Reserved for Private Use

Base Exchange

A *Base exchange* is made of four separate ISAKMP messages. It provides for encapsulating authentication and key-exchange information into one ISAKMP message for transmission to the receiving ISAKMP server. It's worth noting that this exchange doesn't protect the identities of the communicating parties since no shared encryption keys have yet been established between them, and therefore no traffic encryption is possible.

This exchange is used primarily for establishing SAs. During the Base exchange, the initiator and responder cookies (nonces) shown in the ISAKMP header are generated. These prevent denial-of-service attacks and replay attacks against ISAKMP servers.

When this exchange is successfully completed, you will have

▼ Selection of one of the proposed security protocols consisting of one proposed set of security services

■ The generation of the necessary shared keys

▲ The verification of the communicating entities' identification

In other words, we have just described what happens when a new SA has been established. So, a Base exchange is essentially an exchange that is made of a small number of individual ISAKMP messages sent between the ISAKMP servers.

Identity Protection Exchange

The Base exchange offers no secrecy or protection of the identities of the communicating entities. The solution to that problem can be addressed by using the Identity Protection exchange. This solution is not without cost, however. The Identity Protection exchange adds two more ISAKMP messages that must be sent, unlike the Base exchange.

The protection of the identities is achieved by separating the key-exchange information from the authentication and identity information. The way this works is the exchange starts with the generation of key material that both ISAKMP servers use to calculate the shared encryption key.

When this key has been calculated, the exchange proceeds, sending the identities of the communicating entities to each other. These identities are encrypted with the shared key prior to sending. Once all of this has taken place, the new SA is then established.

Authentication Only Exchange

The Authentication Only exchange is used for interchanging authentication information only, meaning that keys for protecting further communication will not be generated. This also means that no shared encryption keys need to be computed.

Encryption can be involved in the authentication phase, however, if the keys generated in the Phase 1 negotiation of the ISAKMP SAs are used to protect the authentication information of this exchange.

Aggressive Exchange

The Aggressive exchange is intended to minimize network traffic by forcing all information to be transferred within a single ISAKMP message. That information includes the SA Payload, the Key Exchange Payload, and any authentication-related payloads.

We should also point out that the Aggressive exchange doesn't provide secrecy of the identities of the communicating entities for the same reasons the Base exchange doesn't. This also means that there can only be one Proposal Payload including only one Transform Payload, due to the restriction of one message. The reason for this is that the initiator would have to "guess" which proposal the recipient had chosen in case there were several proposals. This is, of course, not possible.

Informational Exchange

This exchange is the only mandatory exchange type that must be offered by all ISAKMP. It is used for sending information about errors and SA deletion requests to the communicating party. (See the "Notification Payload" and "Delete Payload" sections earlier.) Another point to recognize is that in contrast to the other exchanges, the Informational exchange is a one-way exchange, meaning that no acknowledgement message needs to be returned by the recipient.

ISAKMP Exchanges Compared with IKE Modes

The Internet Key Exchange (IKE) is the next key-exchange protocol we will be discussing. Before we get started on that, let's reinforce some of the terms. IKE provides for the negotiation of SAs and authenticated keying material and derives its key-exchange functionality from ISAKMP, with the difference being it includes methods that are a subset of Oakley and SKEME.

IKE defines two methods for performing an authenticated key exchange:

▼ Main mode

▲ Aggressive mode

These modes comply fully with the ISAKMP payload formats, attribute encoding, and so on, and can be implemented with the default exchanges of ISAKMP.

▼ IKE's Main mode corresponds to the Identity Protection exchange of ISAKMP.

▲ IKE's Aggressive mode corresponds to the Aggressive exchange of ISAKMP.

IKE's modes have four different authentication variants:

▼ Authentication with digital signatures

■ Normal Authentication with public-key encryption

■ Revised Authentication with public-key encryption

▲ Preshared key

THE INTERNET KEY EXCHANGE (IKE)
OR ISAKMP/OAKLEY

The Internet Key Exchange (IKE) protocol combines ISAKMP, part of Oakley Key Determination Protocol, and part of SKEME Key Exchange Mechanism to provide authenticated keying material for use with ISAKMP, IPSec AH/ESP, and other security associations. IKE uses two separated phases:

▼ IKE Phase 1

▲ IKE Phase 2

IKE Phases and Security Associations

We briefly covered phases and SAs in Chapter 17, and we will go over them in much more detail very soon in the mechanics section ahead, but we will recap here briefly. In IKE Phase 1, the two entities negotiate a bidirectional security association (SA) for protection of further ISAKMP message exchange (ISAKMP SA). They negotiate the authentication method (digital signature, two forms of authentication with public-key encryption, or preshared key), the Diffie-Hellman group, hash algorithm, encryption algorithm, the ISAKMP SA lifetime, and so on. With Diffie-Hellman key exchange, the peers agree on a common secret without exposing it to any third party.

This shared secret, together with other information like nonces, cookies, and so on, is then used to derive keying material for ISAKMP message encryption and authentication and for further derivation of keys for other services such as IPSec, during IKE Phase 2 exchanges.

In IKE Phase 2, IPSec SAs are negotiated under the protection offered by ISAKMP SA. It is also possible to make an additional Diffie-Hellman key exchange to provide Perfect Forward Secrecy (PFS). Otherwise, the master keying material generated in Phase 1 is used to derive keys for IP packet encryption and authentication.

As already pointed out, IKE is used to define ISAKMP SAs consisting of a hash algorithm, an encryption algorithm, an authentication method, and a Diffie-Hellman group. Some possible values for the latter are defined as mandatory in the IKE specification, while others are defined as optional. Any possible values may be supported by IKE.

SA values are defined in the IKE specification as follows:

▼ **Hash algorithm**   MD5 and SHA algorithms as mandatory and the Tiger algorithm as optional

■ **Encryption algorithm**   DES-CBC algorithm as mandatory and the 3DES algorithm as optional

■ **Authentication method**   Preshared key method as mandatory and the DSS and RSA algorithms as optional

▲ **Diffie-Hellman group**   MODP Oakley Group 1 as mandatory and MODP Oakley Group 2 as optional

IKE implementations may support any additional encryption algorithm like Blowfish, MD5, RIPEMD, and so on, and any other Diffie-Hellman groups like elliptic-curve cryptosystems. IKE also defined four different operation modes, based on the ISAKMP exchange types:

- ▼ **Main mode**   This mode is performed in IKE Phase 1, and it is a part of the ISAKMP Identity Protection exchange. (This feature is required for all IKE implementations.)

- ■ **Aggressive mode**   This mode is performed in IKE Phase 1 as an alternative to Main mode, and it is a part of the ISAKMP Aggressive exchange. (This is an optional feature and is not required by IKE.)

- ■ **Quick mode**   This mode is used in IKE Phase 2 such as a non-ISAKMP SA (AH or ESP).

- ▲ **New Group mode**   This mode is optional and it is used after IKE Phase 1 to define a new group for Diffie-Hellman key exchange.

There is also a fifth mode, named Informational Exchange, which is directly derived from ISAKMP, and it is used when anomalous situations need to be audited.

IKE Main Mode and Aggressive Mode

Main mode and Aggressive mode are the two IKE modes used during ISAKMP Phase 1 exchange. They both establish an SA and provide an authenticated Diffie-Hellman key exchange to create a shared secret key to be used in IKE Phase 2.

Main mode requires six one-way messages between the peers, whereas *Aggressive* mode requires three one-way messages. This means that Aggressive mode is a little faster than Main mode. However, it does not provide identity protection, because the peers' identities are transmitted before a secure channel has been established. With Aggressive mode, an enemy can get information about the involved entities. Moreover, SA negotiation is limited in Aggressive mode, and the Diffie-Hellman group cannot be negotiated.

Main mode provides the richest negotiation support. In Main mode, the first two messages are used for policy negotiation, the third and the fourth exchange Diffie-Hellman material and nonces, while the last two are used to authenticate previously exchanged data.

In Aggressive mode, the first two messages are used for negotiating policy, for Diffie-Hellman exchange, and to provide additional data and identification. With the second message, the responder also authenticates itself. The third message is instead used for the initiator's authentication.

The keys used in IKE Phase 2 to generate IPSec keys are computed on some of the exchanged or derived values. In particular, the first key-seed that the entities have to calculate is SKEYID, and the peers may derive it in different ways according to the selected authentication method. Other keys are in various ways derived from SKEYID.

IKE Authentication Methods

The result of Phase 1 is the establishment of a secure channel through an authenticated key exchange. Four different authentication methods can be applied with either Main mode or Aggressive mode. These four authentication methods are digital signature, the two public-key encryption schemes, and the preshared key method. Along with the use of a preshared key, a hash value is calculated to provide authentication.

This value includes the two Diffie-Hellman public values, the two cookies, the SA Payload originally sent by the initiator, and the Identity Payload of the peer the hash is referred to. The hash algorithm is negotiated during the first IKE exchanges, and the SKEYID is used as a key. The hash value has to be signed and verified if authentication by digital signature is used. For the other three authentication methods—public-key, revised public-key, and preshared key—the hash directly authenticates the negotiation. Some of the fields included in the definition of the hashes are exchanged under encryption.

To be able to calculate the right hash value, each peer has to correctly decrypt each field, to provide proof that it possesses the decrypting keys. When the opposite peer receives the hash, it calculates the hash on its own and checks if the calculated value and the one received match. If they do match, the sender is authenticated as being the possessor of the right secret keys. If Authentication is done using digital signatures, after the Diffie-Hellman key exchange in Phase 1, each entity involved will compute a hash value over the previously exchanged data. Then each peer signs (encrypts) this hash with its private key and sends the signature together with its identifier to the other peer. After receiving the signed data, a peer calculates the hash in the same way as the other entity has done. It also decrypts the signed data with the other peer's public key to recover the original hash value. Finally, it compares the two values and, if they match, authentication is provided.

The Internet Key Exchange (IKE) or ISAKMP/Oakley Mechanics

We have mentioned in previous chapters that the crucial elements of IPSec are security associations (SA) and the information that they provide in regard to identifying the partners of a secure communications channel, the cryptographic algorithms, and the keys to be used.

As we have pointed out throughout this chapter, many of the terms you see are shared by the different key management/exchange protocols. In some cases, the terms that are shared may be identical, but their definitions may not always be identical.

A good example of this is IKE, ISAKMP, and Oakley. ISAKMP and Oakley are distinct protocols unto themselves and are *not* identical to IKE.

Keeping that in mind, you should know that IKE and ISAKMP/Oakley *are* identical and are often used interchangeably. There was a time when the IPSec Working Group (WG) decided to use ISAKMP and Oakley together as the key management/exchange protocol for the IPSec standard, and since there wasn't an official name for the time being, it was referred to as ISAKMP/Oakley. It wasn't until later that they decided to name the combination the Internet Key Exchange (IKE) protocol.

We will be using "ISAKMP/Oakley" throughout the discussions and descriptions that follow in order to show the points where IKE uses each of these protocols.

Another good example of terms being used interchangeably is the security association (SA). You should always remember that there is a difference between IPSec SAs and IKE/ISAKMP SAs. IPSec SAs are unidirectional, whereas IKE/ISAKMP SAs are bidirectional.

The Internet Key Exchange (IKE) framework supports automated negotiation of SAs as well as automated generation and refreshment of cryptographic keys. The ability to perform these functions with little or no manual configuration of machines will be a critical element as a VPN grows in size.

As a reminder, we would also like to reiterate that the secure exchange of keys is by far the most critical factor in establishing a secure communications environment. It doesn't matter how strong your authentication and encryption are; they are worthless if your key is compromised.

IKE or ISAKMP/Oakley Exchanges

ISAKMP requires that all information exchanges must be both encrypted and authenticated so that no one can eavesdrop on the keying material, and the keying material will be exchanged only among authenticated parties. This is required because the ISAKMP procedures deal with initializing the keys, so they must be capable of running over links where no security can be assumed to exist. ISAKMP protocols use the most complex and processor-intensive operations in the IPSec protocol suite. In addition, the ISAKMP methods have been designed with the explicit goals of providing protection against several well-known exposures:

▼ **Denial-of-service** The messages are constructed with unique cookies that can be used to quickly identify and reject invalid messages without the need to execute processor-intensive cryptographic operations.

■ **Man-in-the-middle** Protection is provided against the common attacks such as deletion of messages, modification of messages, reflecting messages back to the sender, replaying of old messages, and redirection of messages to unintended recipients.

▲ **Perfect Forward Secrecy (PFS)** Each refreshed key will be derived without any dependence on predecessor keys. This means that compromise of past keys provides no useful clues for breaking any other key, whether it occurred before or after the key was compromised.

The Two Phases of ISAKMP/Oakley

The effectiveness of any cryptography-based solution depends much more on keeping the keys secret than it does on the actual details of the cryptographic algorithms chosen. The IETF IPSec WG has prescribed an extremely robust exchange protocol known as ISAKMP/Oakley. ISAKMP/Oakley uses two phases:

▼ **Phase 1** This set of negotiations establishes a master key from which all subsequent cryptographic keys will be derived for protecting the data itself. In the most general case, public-key cryptography is used to establish an ISAKMP SA between systems, and to establish the keys that will be used to protect the ISAKMP messages that will flow in the subsequent Phase 2 negotiations. Phase 1 is concerned only with establishing the security protocol for the ISAKMP messages, but it does not establish any SAs or keys for protecting the data. In Phase 1, the cryptographic operations are the most processor-intensive, but need not be done often. This is because a single Phase 1 exchange can be used to support multiple subsequent Phase 2 exchanges.

▲ **Phase 2** Phase 2 exchanges are less complex. This is because they are used only after the secure communications channel negotiated in Phase 1 has been activated. A set of communicating systems negotiates the SAs and keys that will protect the actual data exchanges. The ISAKMP SA generated in Phase 1 protects the subsequent Phase 2 ISAKMP messages. Phase 2 negotiations generally occur more frequently than Phase 1. For example, a typical application of a Phase 2 negotiation is to refresh the cryptographic keys once every two to three minutes. You will see how ISAKMP/Oakley is used to initially establish SAs and exchange keys between two systems in the "Initializing Security Associations with ISAKMP/Oakley" section. However, the ISAKMP protocol offers a solution even when the remote host's IP address is not known in advance. ISAKMP allows a remote host to identify itself by a permanent identifier, such as a fully qualified domain name (FQDN) or an e-mail address. The ISAKMP Phase 1 exchanges will then authenticate the remote host's permanent identity using public-key cryptography:

■ Certificates create a binding between the permanent identifier and a public key. ISAKMP's certificate-based Phase 1 message exchanges can authenticate the remote host's permanent identity.

■ Since the ISAKMP's messages are carried within IP datagrams, the ISAKMP partner (for example, a firewall, router, or destination host) can associate the remote host's dynamic IP address with its authenticated permanent identity.

In the upcoming section, "Initializing Security Associations with ISAKMP/Oakley," you will see how ISAKMP/Oakley exchanges authenticate the remote host to its peer and set up the SAs.

The Main Mode Exchange of ISAKMP/Oakley Phase 1

Now let's outline what happens during a Main Mode exchange. As we've stated, Main mode has six message flows. In the first two messages, a proposal is offered by the *initiator* with one or more transforms, and the *responder* accepts the proposal with the chosen transform.

Additionally, cookies are generated to protect against denial of service attacks. The initiator cookie and responder cookie pair identify the ISAKMP SA.

During the next two messages, an exchange occurs as a Diffie-Hellman key exchange along with some nonces. After these two messages, each party now has the keying material to generate keys for encryption and authentication of subsequent ISAKMP messages. Keying material is also derived that will be used to generate keys for other non-ISAKMP SAs in Phase 2. All ISAKMP messages from this point are then encrypted.

In the last two messages of Phase 1, authentication occurs. Depending on the chosen authentication method, the appropriate messages and identities are exchanged here so that each party can authenticate the other. To see a typical Main Mode exchange, take a look at Figure 18-10.

The Aggressive Mode Exchange of ISAKMP/Oakley Phase 1

Only three messages are required to establish the SA in Aggressive mode. The difference here, though, is that the identities of the parties involved are revealed.

In the first message, the initiator sends the proposal, Diffie-Hellman key exchange, nonce, and ID to the responder.

At this point the responder can generate its nonce and Diffie-Hellman key exchange. In conjunction with what the responder has just received from the initiator, the responder has all the components to generate the keying material.

The responder on the second messages sends the same information to the initiator so that the keying material can be generated. The responder also attaches the information necessary to authenticate itself.

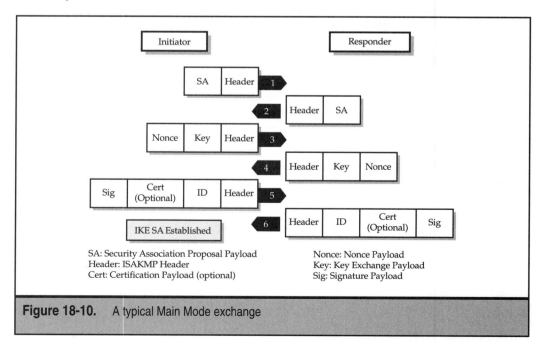

Figure 18-10. A typical Main Mode exchange

When the initiator receives the second message, it is able to generate the keying material and authenticate the responder. All that is required now is for the responder to authenticate the initiator. To do this, the initiator sends to the responder the last Phase 1 message containing information that will enable the responder to authenticate the initiator.

As with Main mode, the information that is exchanged to facilitate authentication is dependent on the negotiated authentication method. To see a typical Aggressive Mode exchange, take a look at Figure 18-11.

Initializing Security Associations with ISAKMP/Oakley

This section outlines in detail how ISAKMP/Oakley protocols initially establish SAs and exchange keys between two systems that wish to communicate securely. To provide a concrete example, we describe a message sequence with the following characteristics:

▼ ISAKMP messages themselves will be carried as a UDP payload.

■ ISAKMP Phase 1 exchanges will be authenticated with digital signatures based on certificates obtained from a valid certificate authority.

■ Parties participating in the ISAKMP Phase 1 exchanges will be identified by user-based certificates (by name) rather than by host-based certificates (by IP addresses). This is a more general solution, since it can be used even when a host receives a dynamically assigned IP address.

▲ ISAKMP Phase 2 exchanges will be used to negotiate the protection of user traffic with the ESP protocol, making use of its optional authentication function.

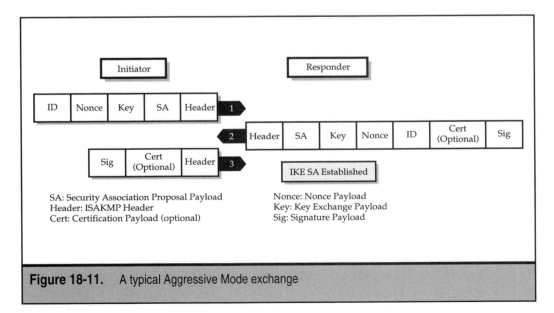

Figure 18-11.   A typical Aggressive Mode exchange

Other message sequences are possible within the ISAKMP framework, but they are not described in this chapter.

In the remainder of this section, we assume that the parties involved are named Host-A and Host-B. Host-A will be the initiator of the ISAKMP Phase 1 exchanges, and Host-B will be the responder. If needed for clarity, subscripts $_A$ or $_B$ will be used to identify the source of various fields in the message exchanges. Subscripts $_i$ and $_r$ will be used to describe initiator and responder, respectively. Superscripts ^{x} and ^{y} will represent the private and public values of a public-key pair, respectively.

Phase 1—Setting Up ISAKMP/Oakley Security Associations

The SAs that protect the ISAKMP messages are set up during the Phase 1 exchanges. Since we are starting fresh (no previous keys or SAs have been negotiated between Host-A and Host-B), the Phase 1 exchanges will use the IKE Main mode that was derived from the Oakley protocol. (This exchange is equivalent to ISAKMP's Identity Protect exchange.)

Six messages are needed to complete the exchange:

▼ *Messages 1 and 2* negotiate the characteristics of the SAs. Messages 1 and 2 are in cleartext for the initial Phase 1 exchange, and they are unauthenticated.

■ *Messages 3 and 4* exchange nonces and execute a Diffie-Hellman exchange to establish a master key (SKEYID). Messages 3 and 4 are in cleartext for the initial Phase 1 exchange, and they are unauthenticated.

▲ *Messages 5 and 6* exchange digital signatures and optionally the pertinent user-based certificates for mutually authenticating the party's identities. The encryption algorithm and keying material established in Messages 1 through 4 will be used to protect the payloads of Messages 5 and 6.

ISAKMP Phase 1, Message 1

Since Host-A is the initiating party, it will construct a cleartext ISAKMP message (Message 1) and send it to Host-B. The ISAKMP message itself is carried as the payload of a UDP packet, which in turn is carried as the payload of a normal IP datagram. Message 1 of an ISAKMP Phase 1 exchange can be seen in Figure 18-12.

The Source Address and Destination Address to be placed in the IP header are those of Host-A (initiator) and Host-B (responder), respectively. The UDP header will identify that the destination port is 500, which has been assigned for use by the ISAKMP protocol. The payload of the UDP packet carries the ISAKMP message itself.

In Message 1, Host-A, the initiator, proposes a set of one or more security protocols for consideration by Host-B, the responder. The ISAKMP Message contains at least the following fields in its payload:

▼ The ISAKMP header in Message 1 will indicate an Exchange Type of Main Mode and will contain a Message ID of 0. Host-A will set the Responder Cookie field to 0 and will fill in a random value of its choice for the Initiator Cookie, which we will denote as Cookie$_A$.

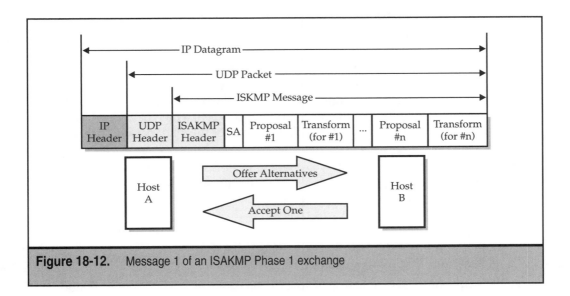

Figure 18-12.   Message 1 of an ISAKMP Phase 1 exchange

- The Security Association field identifies the Domain of Interpretation (DOI). Since the hosts plan to run IPSec protocols between themselves, the DOI is simply IP.

- Host-A's Proposal Payload will specify the protocol PROTO_ISAKMP and will set the SPI value to 0.

NOTE: For ISAKMP Phase 1 messages, the actual SPI field within the Proposal Payload is not used to identify the ISAKMP SA. During Phase 1, the pair of values <*Initiator Cookie, Responder Cookie*>, both of which must be nonzero values, identifies the ISAKMP SA instead. Since the *Responder Cookie* has not yet been generated by Host-B, the ISAKMP SA is not yet identified.

- ▲ The Transform Payload will specify KEY_OAKLEY. For the KEY_OAKLEY transform, Host-A must also specify the relevant attributes, namely, the authentication method to be used, the pseudo-random function to be used, and the encryption algorithm to be used. Host-A will specify:
  - Authentication using digital signatures
  - Pseudo-random function of HMAC-MD5
  - Encryption algorithm of DES-CBC

ISAKMP Phase 1, Message 2

In Message 1, Host-A proposed one or more candidate security protocols to be used to protect the ISAKMP exchanges. Host-B uses Message 2 to indicate which one, if any, it will support. Note that in our example, Host-A proposed just a single option, so Host-B merely needs to acknowledge that the proposal is acceptable.

The message contents will be as follows:

▼ The Source Address and Destination Address to be placed in the IP header are those of Host-B (responder) and Host-A (initiator), respectively. The UDP header will identify that the destination port is 500, which has been assigned for use by the ISAKMP protocol. The payload of the UDP packet carries the ISAKMP message itself.

■ The ISAKMP Header in Message 2 will indicate an exchange type of Main Mode and will contain a Message ID of 0. Host-B will set the Responder Cookie field to a random value, which we will call $Cookie_B$, and will copy into the Initiator Cookie field the value that was received in the $Cookie_A$ Cookie field of Message 1. The value pair $<Cookie_A, Cookie_B>$ will serve as the SPI for the ISAKMP SA.

■ The Security Association field identifies the DOI. Since the hosts plan to run IPSec protocols between themselves, the DOI is simply IP.

■ Host-B's Proposal Payload will specify the protocol PROTO_ISAKMP and will set the SPI value to 0.

NOTE: For ISAKMP Phase 1 messages, the actual SPI field within the Proposal Payload is not used to identify the ISAKMP SA. During Phase 1, the pair of values *<Initiator Cookie, Responder Cookie>*, both of which must be nonzero values, identifies the ISAKMP SA instead.

■ The Transform Payload will specify KEY_OAKLEY. For the KEY_OAKLEY transform, the attributes that were accepted from the proposal offered by Host-A are copied into the appropriate fields. That is, Host-B will confirm that it will use

■ Authentication using digital signatures

■ Pseudo-random function of HMAC-MD5

▲ Encryption algorithm of DES-CBC

Properties of ISAKMP SA Are Established   At this point, the properties of the ISAKMP SA have been agreed to by Host-A and Host-B. The identity of the ISAKMP SA has been set equal to the pair $<Cookie_A, Cookie_B>$. However, the identities of the parties claiming to be Host-A and Host-B have not yet been authoritatively verified.

ISAKMP Phase 1, Message 3

The third message of the Phase 1 ISAKMP exchange begins the exchange of the information from which the cryptographic keys will eventually be derived. Message 3 of an ISAKMP Phase 1 exchange can be seen in Figure 18-13. All information is exchanged in

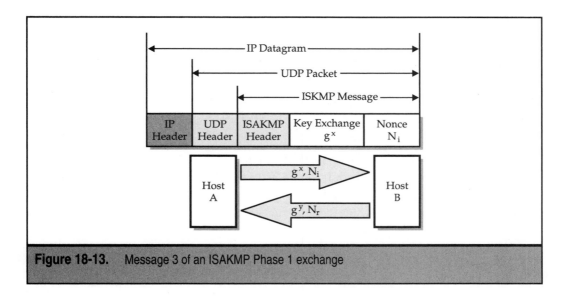

Figure 18-13. Message 3 of an ISAKMP Phase 1 exchange

the clear. None of the messages carries the actual cryptographic keys. Instead, they carry inputs that will be used by Host-A and Host-B to derive the keys locally. The ISAKMP payload will be used to exchange two types of information:

▼ **Diffie-Hellman public value** This is g^x from the initiator. The exponent x in the public value is the private value that must be kept secret.

▲ **Nonce** This is N_i from the initiator. (*Nonce* is a fancy name for a value that is considered to be random according to some very strict mathematical guidelines.)

These values are carried in the Key Exchange field and the Nonce field.

ISAKMP Phase 1, Message 4

After receiving a Diffie-Hellman public value and a nonce from Host-A, Host-B will respond by sending to Host-A its own Diffie-Hellman public value (g^y from the responder) and its nonce (N_r from the responder).

Generating the Keys (Phase 1)

At this point, each host knows the values of the two nonces (N_i and N_r). Each host also knows its own private Diffie-Hellman value (x and y) and knows its partner's public value (g^x or g^y). Each side can construct the composite value g^{xy}. Finally, each side knows the values of the initiator cookie and the responder cookie.

Given all these bits of information, each side can then independently compute identical values for the following quantities:

▼ **SKEYID** This collection of bits is sometimes referred to as *keying material,* since it provides the raw input from which actual cryptographic keys will be derived later in the process. It is obtained by applying the agreed-to pseudo-random function, in our example, HMAC-MD5, to the known inputs:

$$SKEYID = HMAC\text{-}MD5 (N_i, N_r, g^{xy})$$

■ Having computed the value SKEYID, each side then proceeds to generate two cryptographic keys and some additional keying material

■ SKEYID_d is keying material that will be subsequently used in Phase 2 to derive the keys that will be used in non-ISAKMP SAs for protecting user traffic:

$$SKEYID\_d = HMAC\text{-}MD5 (SKEYID, g^{xy}, Cookie_A, Cookie_B, 0)$$

■ SKEYID_a is the key used for authenticating ISAKMP messages:

$$SKEYID\_a = HMAC\text{-}MD5 (SKEYID, SKEYID\_d, g^{xy}, Cookie_A, Cookie_B, 1)$$

▲ SKEYID_e is the key used for encrypting ISAKMP exchanges:

$$SKEYID\_e = HMAC\text{-}MD5 (SKEYID, SKEYID\_a, g^{xy}, Cookie_A, Cookie_B, 2)$$

Keys Are Now Available At this point in the protocol, both Host-A and Host-B have derived identical authentication and encryption keys that they will use to protect the ISAKMP exchanges. In addition, they have derived identical keying material from which they will derive keys to protect the actual data during Phase 2 of the ISAKMP negotiations. However, at this point, the two parties' identities still have not been authenticated to one another.

ISAKMP Phase 1, Message 5

At this point in the Phase 1 flows, the two hosts will exchange identity information with each other, using the Digital Signature Algorithm (DSA) to authenticate themselves. Message 5 of an ISAKMP Phase 1 exchange can be seen in Figure 18-14. Here, the ISAKMP message will carry an Identity Payload, a Signature Payload, and an optional Certificate Payload. Host-A uses Message 5 to send information to Host-B that will allow Host-B to authenticate Host-A.

When an actual certificate is present in the Certificate Payload field, the receiver can use the information directly, after verifying that it has been signed with a valid signature of a trusted certificate authority (CA). If there is no certificate in the message, then it is the responsibility of the receiver to obtain a certificate using some other method.

For example, a query may be sent to a trusted CA using a protocol like LDAP. Other possibilities include sending a query to a secure DNS server by maintaining a secure local

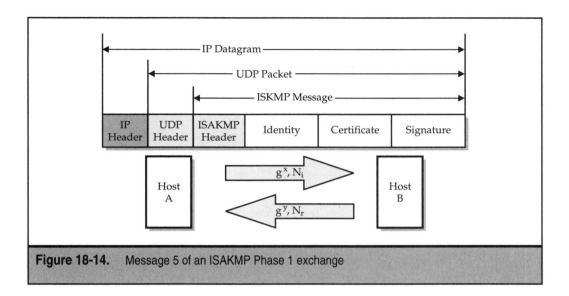

Figure 18-14. Message 5 of an ISAKMP Phase 1 exchange

cache that maps previously used certificates to their respective ID values or by sending an ISAKMP Certificate Request message to its peer, which, in turn, must then immediately send its certificate to the requester.

NOTE: The method for obtaining a certificate is a local option and is not defined as part of ISAKMP/Oakley. In particular, it is a local responsibility of the receiver to check that the certificate in question is still valid and has not been revoked.

Bear in mind several points:

▼ At this stage of the process, all ISAKMP payloads, whether in Phase 1 or Phase 2, are now encrypted, using the encryption algorithm negotiated in Messages 1 and 2 and the keys derived from the information in Messages 3 and 4. The ISAKMP header itself, however, is still transmitted in the clear.

■ In Phase 1, IPSec's ESP protocol is not used; that is, there is no ESP header. The recipient uses the Encryption Bit in the Flags field of the ISAKMP header to determine if encryption has been applied to the message. The pair of values $<Cookie_A, Cookie_B>$, which serves as an SPI for Phase 1 exchanges, provides a pointer to the correct algorithm and key to be used to decrypt the message.

■ The digital signature is not applied to the ISAKMP message itself. Instead, it is applied to a hash of information that is available to both Host-A and Host-B.

▲ The identity carried in the Identity Payload does not necessarily bear any relationship to the source IP address; however, the identity carried in the Identity Payload must be the identity to which the certificate applies.

Because the pseudo-random function for the ISAKMP SA is HMAC-MD5 in our example, Host-A (the initiator) will generate the following hash function, sign it using the Digital Signature Algorithm (DSA), and then place the result in the Signature Payload field:

$$\text{HASH\_I} = \text{HMAC-MD5 (SKEYID, } g^x, g^y, \text{Cookie}_A, \text{Cookie}_B, \text{SA}_p, \text{ID}_A)$$

ID_A is Host-A's identity information that was transmitted in the Identity Payload of this message, and SA_p is the entire body of the Security Association Payload that was sent by Host-A in Message 1, including all proposals and all transforms proposed by Host-A. The cookies, public Diffie-Hellman values, and SKEYID were explicitly carried in Messages 1 through 4 or were derived from their contents.

ISAKMP Phase 1, Message 6

After receiving Message 5 from Host-A, Host-B will verify the identity of Host-A by validating the digital signature. If the signature is valid, then Host-B will send Message 6 to Host-A to allow Host-A to verify the identity of Host-B. The structure of Message 6 is the same as that of Message 5, with the obvious changes that the Identity Payload and the Certificate Payload now pertain to Host-B. A less obvious difference is that the hash that is signed by Host-B is different from the one previously signed by Host-A:

$$\text{HASH\_R} = \text{HMAC-MD5 (SKEYID, } g^y, g^x, \text{Cookie}_B, \text{Cookie}_A, \text{SA}_p, \text{ID}_B)$$

Notice that the order in which Diffie-Hellman public values and the cookies appear has been changed, and the final term now is the Identity Payload that Host-B has included in Message 6.

Phase-1 Is Complete   When Host-A receives Message 6 and verifies the digital signature, the Phase 1 exchanges are then complete. At this point, each participant has authenticated itself to its peer, both have agreed on the characteristics of the ISAKMP SAs, and both have derived the same set of keys or keying material.

Additional Phase 1 Facts

There are several additional Phase 1 facts worth noting:

▼ The preceding Phase 1 message flows pertain to the case where the ISAKMP/Oakley messages will be authenticated by the Digital Signature Standard (DSS). There are other permissible ways to authenticate ISAKMP messages. For example, preshared keys or public-key encryption could also be used.

Regardless of the specific authentication mechanism that is used, there will be six messages exchanged for IKE (derived from Oakley) Main mode. However, the content of the individual messages will differ, depending on the authentication method. The key calculation formula for SKEYID will also differ depending on the authentication method.

■ Although ISAKMP/Oakley exchanges make use of both encryption and authentication, they do not use either IPSec's ESP or AH protocol. ISAKMP exchanges are protected with Application layer security mechanisms, not with Network layer security mechanisms.

■ ISAKMP messages are sent using UDP. There is no guaranteed delivery for them.

▲ The only way to identify that an ISAKMP message is part of a Phase 1 flow rather than a Phase 2 flow is to check the Message ID field in the ISAKMP header. For Phase 1 flows, this must be zero. Although not explicitly stated in the ISAKMP protocol specification for Phase 2, the Phase 2 flows must be nonzero.

Phase 2—Setting Up Non-ISAKMP Security Associations

After having completed the Phase 1 negotiation process to set up the ISAKMP SAs, Host-A's next step is to initiate the ISAKMP/Oakley or IKE Phase 2 message exchanges known as Quick Mode (derived from Oakley) to define the SAs and keys that will be used to protect IP datagrams exchanged between the pair of users. These are referred to as "non-ISAKMP SAs."

Because the purpose of the Phase 1 negotiations was to agree on how to protect ISAKMP messages, all ISAKMP Phase 2 payloads, but not the ISAKMP header itself, must be encrypted using the algorithm agreed to by the Phase 1 negotiations.

When IKE's (derived from Oakley's) Quick mode is used in Phase 2, authentication is achieved via the use of several cryptographically based hash functions. The input to the hash functions comes partly from Phase 1 information (SKEYID) and partly from information exchanged in Phase 2. Phase 2 authentications are based on certificates, but the Phase 2 process itself does not use certificates directly. Instead, it uses the SKEYID_a material from Phase 1, which itself was authenticated via certificates.

IKE's (derived from Oakley's) Quick mode comes in two forms:

▼ **Without a Key Exchange attribute**   Quick mode can be used to refresh the cryptographic keys, but does not provide the property of PFS.

▲ **With a Key Exchange attribute**   Quick mode can be used to refresh the cryptographic keys in a way that provides PFS. This is accomplished by including an exchange of public Diffie-Hellman values within messages 1 and 2.

NOTE: Apparently, PFS is a very important feature that should always be used, but for some reason the specification treats the PFS property as "Optional." ISAKMP standards do require that a system must be capable of handling the Key Exchange field when it is present in a Quick Mode message; however, they do not require a system to include the Key Exchange field within the message.

An overview of the three messages within Quick Mode follows. In our example, we have Host-A propose two security protocols, and then show how Host-B will choose just one of them:

▼ **Proposal 1**   This will offer protection using ESP with DES-CBC for encryption and HMAC-MD5 for authentication.

▲ **Proposal 2**   This will offer protection using AH with HMAC-MD5 for authentication.

ISAKMP Phase 2, Message 1

Message 1 of a Quick Mode exchange allows Host-A to:

▼ Authenticate itself

■ Select a nonce

■ Propose SA(s) to Host-B

■ Execute an exchange of public Diffie-Hellman values

▲ Indicate if it is acting on its own behalf or as a proxy negotiator for another entity

An overview of the format of Message 1 in Figure 18-15 shows what happens in Message 1 of an ISAKMP Phase 2 Quick Mode exchange.

NOTE: Inclusion of a Key Exchange field is optional. However, when PFS is desired, the field must be present.

Because we have assumed that Host-A and Host-B are each acting on their own behalf, the User Identity fields shown in Figure 18-15 will not be present. The message will consist of the following:

▼ **ISAKMP Header**   This will

■ Indicate an Exchange Type of Quick Mode

■ Include a nonzero Message ID chosen by Host-A

■ Include the initiator and responder cookie values chosen in Phase 1 ($Cookie_A$ and $Cookie_B$)

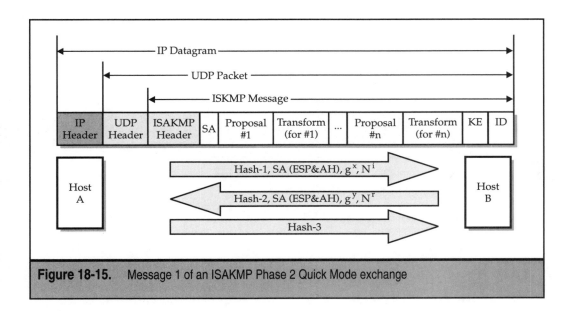

Figure 18-15.   Message 1 of an ISAKMP Phase 2 Quick Mode exchange

- Turn on the Encryption Flag to indicate that the payloads of the ISAKMP message are encrypted according to the algorithm and key negotiated during Phase 1

- **HASH_1**   A Hash Payload must immediately follow the ISAKMP header. HASH_1 uses the pseudo-random function that was negotiated during the Phase 1 exchanges. In our example, this is HMAC-MD5, and so:

    HASH_1 = HMAC-MD5 (SKEYID_a, M-ID, SA, N_i, KE) HASH_1

is derived from the following information:

- *SKEYID_a* was derived from the Phase 1 exchanges.
- *M-ID* is the Message ID of this message.
- *SA* is the Security Association Payload carried in this message, including all proposals that were offered.
- *KE* is the public Diffie-Hellman value carried in this message. This quantity is chosen by Host-A, and is denoted as g_{qm}^{x}.

NOTE: This is not the same quantity as g^x that was used in the Phase 1 exchanges.

- **Security Association Payload**   Indicate IP as the DOI.
- **Proposal, Transform Pairs**   There will be two of these pairs in this example:
  - **First Security protocol**   The first Proposal Payload will be numbered 1, will identify ESP as the protocol to be used, and will include an SPI value that is randomly chosen by Host-A for use with the ESP protocol. The Proposal Payload will be followed by a single Transform Payload that indicates ESP_DES as the algorithm.

NOTE: In the current DOI document, the *code point* ESP_DES calls for both encryption with DES-CBC and authentication with HMAC-MD5.

  - **Second Security protocol**   The second Proposal Payload will be numbered 2, will identify AH as the protocol to be used, and will include an SPI value that is randomly chosen by Host-A for use with the AH protocol. The Proposal Payload will be followed by a single Transform Payload that names HMAC-MD5 as the algorithm.
- **Nonce**   This contains the nonce N_i that was chosen randomly by Host-A.
- ▲ **KE**   This is the Key Exchange Payload that will carry the public Diffie-Hellman value chosen by Host-A, $g_{qm}{}^x$. There is also a Group field that indicates the prime number and generator used in the Diffie-Hellman exchange.

ISAKMP Phase 2, Message 2

After Host-B receives Message 1 from Host-A and successfully authenticates it using HASH_1, it constructs a reply, Message 2, to be sent back to Host-A. The Message ID of the reply will be the same one that Host-A used in Message 1. Host-B will choose new values for the following:

- ▼ Nonce Payload now carries N_r, a random value chosen by Host-B.
- Key Exchange Payload now carries Host-B's public Diffie-Hellman value, $g_{qm}{}^y$.
- Hash Payload now carries the value HASH_2, which is defined as:

    HASH_2 = HMAC-MD5 (SKEYID_a, N_i, M-ID, SA, N_r, KE)

- ▲ Security Payload only describes the single chosen proposal and its associated transforms, not all of the security protocols offered by Host-A. In this example, Host-B will select Proposal #1, which will use ESP_DES (encryption plus authentication) as the protocol. Host-B also chooses an SPI value for the ESP_DES, the selected protocol; Host-B's SPI does not depend in any way on the SPI that Host-A assigned to that protocol when it offered the proposal. That is, it is not necessary that SPI_A be the same as SPI_B; it is only necessary that they each be nonzero and that they each be randomly chosen.

Keys for Non-ISAKMP Can Be Derived   At this point, Host-A and Host-B have exchanged nonces and public Diffie-Hellman values. Each one can use these values and information to derive a pair of keys, one for each direction of transmission.

Generating the Non-ISAKMP Keys

Using the nonces, public Diffie-Hellman values, SPIs, protocol code points exchanged in Messages 1 and 2 of Phase 2 Quick mode, and the SKEYID value from Phase 1, each host now has enough information to derive two sets of keying material:

▼ For data generated by Host-A and received by Host-B, the keying material is

$$KEYMAT_{BA} = HMAC\text{-}MD5(SKEYID, g_{qm}{}^{xy}, protocol, SPI_B, N_i, N_r)$$

▲ For data generated by Host-B and received by Host-A, the keying material is

$$KEYMAT_{BA} = MAC\text{-}MD5(SKEYID, g_{qm}{}^{xy}, protocol, SPI_A N_i, N_r)$$

In this example, Host-A needs to derive four keys:

1. Key for generating the *Integrity Check Value* (ICV) for transmitted datagrams
2. Key for validating the ICV of the received datagrams
3. Key for encrypting the transmitted datagrams
4. Key for decrypting the received datagrams

Like Host-A, Host-B needs to derive the "mirror image" of the same four keys. For example, the key that Host-B uses to encrypt its outbound messages is the same key that Host-A uses to decrypt its inbound messages, and so on.

Non-ISAKMP Security Associations (Phase 2, Message 3)

At this point, Host-A and Host-B have exchanged all the information necessary for them to derive the necessary keying material. The third message in the Quick Mode exchange is used by Host-A to prove its active existence by producing a hash function that covers the Message ID and both nonces that were exchanged in Messages 1 and 2. Message 3 consists only of the ISAKMP Header and a Hash Payload that carries:

$$HASH\_3 = HMAC\text{-}MD5(SKEYID\_a, 0, M\text{-}ID, N_i, N_r)$$

When Host-B receives this message and verifies the hash, then both systems can begin to use the negotiated security protocols to protect their user data streams.

NEGOTIATING MULTIPLE SECURITY ASSOCIATIONS

The examples covered in the "ISAKMP Phase 2, Message 1" through "ISAKMP Phase 2, Message 3" sections outlined a case where a single non-ISAKMP SA was negotiated by means of a Quick Mode message exchange. However, it is also possible to negotiate multiple SAs within a single three-message Quick Mode exchange. Each SA will have its own set of keying material.

The message formats are very similar to the formats discussed previously, so only the differences will be outlined:

▼ Message 1 will carry multiple SA Payloads with each offering a range of security protocols.

■ HASH_1 will cover the entire set of all offered SAs carried in Message 1. This means that each SA and all of its offered proposals are included.

■ In Message 2, for each offered SA, Host-B will select a single security protocol. This means that if n SAs are open for negotiation, then Host-B will choose n security protocols, one from each proposal.

■ As was the case for HASH_1, HASH_2 will now cover the entire set of all offered SAs carried in Message 1. This means that each SA and all of its offered proposals are included.

■ After Messages 1 and 2 have been exchanged, then Host-A and Host-B will be able to generate the keying material for each of the accepted security protocols, using the same formulas as in the "Generating the Keys" section, applied individually for each accepted SA. Even though the nonces and the public Diffie-Hellman values are the same for all selected security protocol, the keying material derived for each selected security protocol will be different because each proposal will have a different SPI.

▲ Since multiple SAs have been negotiated, it becomes a matter of local choice as to which SA is used to protect a given datagram. The receiving system must be capable of processing a datagram that is protected by any SA that has been negotiated. That is, it would be legal for a given source host to send two consecutive datagrams to a destination system, where each datagram was protected by a different SA.

USING ISAKMP/OAKLEY WITH REMOTE ACCESS

The critical element in the remote access scenario is the use of ISAKMP/Oakley to identify the remote host by name, rather than by its dynamically assigned IP address. Once the remote host's identity has been authenticated and the mapping to its dynamically assigned IP address has been determined, the rest of the steps are the same as we have described for the other scenarios.

For example, if the corporate intranet is trusted, then the remote host needs to establish a single SA between itself and the firewall. However, if the corporate intranet is not trusted, then it may be necessary for the remote host to set up two SAs:

▼   One between itself and the firewall

▲   A second one between itself and the destination host

Recall that a single ISAKMP Phase 1 negotiation can protect several subsequent Phase 2 negotiations. Phase 1 ISAKMP negotiations use computationally intensive public-key algorithms, while Phase 2 negotiations use the less CPU-intensive symmetric-key algorithms. This means that the heavy CPU loads will only occur in Phase 1, which is only executed once when the dial-up connection is first initiated.

The main points to remember with respect to remote access case are

▼   The remote host's dynamically assigned address is the one that is placed in the IP header of all ISAKMP messages.

■   The remote host's permanent identifier (such as an e-mail address or FQDN, for example) is the quantity that is placed in the ID field of the ISAKMP Phase 1 messages.

■   The remote host's certificate used in the ISAKMP exchange must be associated with the remote host's permanent identifier.

▲   In traffic-bearing datagrams, the remote host's dynamically assigned IP address will be used. This is necessary since the Destination IP Address that appears in the datagram's IP header is used in conjunction with the SPI and Protocol Type to identify the relevant IPSec SA for processing the inbound datagram.

SUMMARY

In this chapter we talked about the specific protocols and protocol mechanics used in key exchange and key management. In the next chapter we will tie all of this together with IPSec to provide you with a comprehensive understanding of how these protocols work together to provide for secure communications.

CHAPTER 19

IPSec Architecture and Implementation

In this chapter, we will put it all together and talk about the different IPSec architectures and how they are implemented. We will start with the overall architecture and then break out some of the specifics.

IPSEC ARCHITECTURE AND IMPLEMENTATION

As we stated in Chapter 17, IPSec has two basic working environments:

▼ Host

▲ Security gateway

A *security gateway* (SG) is usually a router or IPSec-compliant firewall. A *host* is always a final endpoint of a connection. IPSec can be deployed in three different models using two kinds of modes (Tunnel and Transport):

▼ Between a host and another host (host-to-host)

■ Between security gateways (SGs) and another SG (SG-to-SG)

▲ Between a host and an SG (host-to-SG)

Host-to-host is between two endpoints, SG-to-SG is the LAN-to-LAN model, and the host to SG is the mobile host or "road warrior" model. Take a look at Figures 19-1, 19-2, and 19-3 to get a basic idea of what we mean. The figures use the following conventions:

▼ A *monitor sitting to the side* of a CPU tower represents a host.

■ A *monitor sitting above* a CPU tower represents a security gateway.

■ The *padlock* indicates a device with IPSec implemented.

■ The *solid black line* represents a physical connection.

▲ The *dual gray line (pipe)* represents a logical connection using one or more security associations (AH or ESP, Transport or Tunnel).

Figure 19-1 shows a host-to-host model.
Figure 19-2 shows a security gateway to security gateway model.
Figure 19-3 shows a host-to–security gateway model.

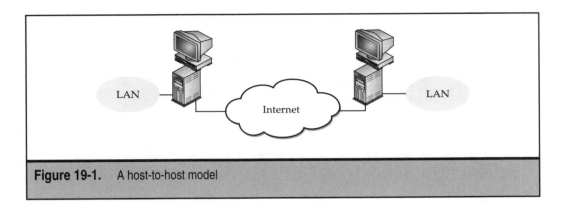

Figure 19-1.   A host-to-host model

Implementing IPSec

Different approaches can be taken on how to implement IPSec in a host, a security gateway, or a network system. These approaches depend on the environment IPSec is deployed in. Three common solutions are provided next:

▼   IPSec can be implemented as an integral part of the IP stack. This requires
access to the source code and can be used in both hosts and security gateways.
For most hosts, this will probably be the best solution, because IPSec will then

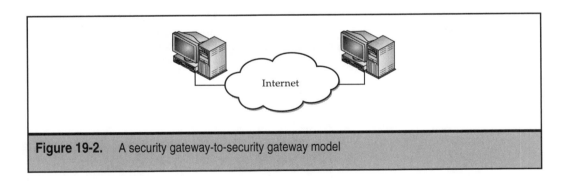

Figure 19-2.   A security gateway-to-security gateway model

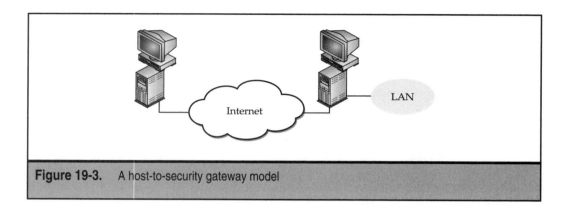

Figure 19-3.   A host-to-security gateway model

be built into the operating system. This way IPSec will be more streamlined for better performance. Also consider that software upgrades on hosts happen quite frequently compared with upgrades of security gateways/routers. To see an example of IPSec integrated into the IP stack, take a look at Figure 19-4.

■   IPSec can be implemented as a security layer "underneath" the existing IP protocol. This is also called the *bump-in-the-stack* (BITS) approach. It does not require source code access, but does require a clearly defined interface between the IP layer and the underlying network driver. This approach might be a little less efficient than IP stack integration, but it is easier to implement since it only affects the individual machine. This does increase the workload to the machine, because it is now responsible for the IPSec processing. This might not be the best solution if bandwidth is scarce. To see an example of a bump-in-the-stack (BITS) IPSec implementation, take a look at Figure 19-5.

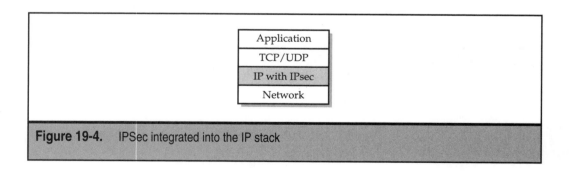

Figure 19-4.   IPSec integrated into the IP stack

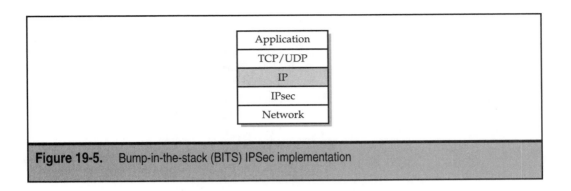

Figure 19-5. Bump-in-the-stack (BITS) IPSec implementation

▲ In some instances, it may be difficult to upgrade the machine on which you wish to install IPSec using the BITS method. Even if it can be upgraded, the added workload of processing IPSec can make it ineffective. If this is the case, it may be more desirable to use an outbound cryptographic processor. This approach is what is often referred to as bump-in-the-wire (BITW). The cryptographic processor machine is usually IP addressable and can support either a host or router. In the case of a router, it is forced to function as a security gateway. To see an example of a *bump-in-the-wire* (BITW) IPSec implementation, take a look at Figure 19-6.

Committees and Standards

The Internet Engineering Task Force (IETF) is an open, international community of network designers, operators, vendors, and researchers concerned with the evolution of the Internet architecture standards and the smooth operation of the Internet. IPSec is one of those standards.

Several groups have volunteered to perform industry standard testing to assure present and future interoperability amongst IPSec-compliant products. Two of these

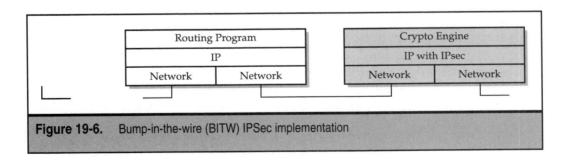

Figure 19-6. Bump-in-the-wire (BITW) IPSec implementation

organizations are the National Institute of Standards and Technology (NIST) and the International Computer Security Association (ICSA).

NIST is primarily concerned with development of new technologies and establishing standards to which industry should comply. Based upon current IPSec standards, NIST provides a web-based application that can test vendors' implementation of IPSec for compliance with the components and functionality. Among the NIST certifications is the FIPS-140-1(2), which defines requirements for development of cryptographic mechanisms. Although this certification applies to IPSec-compliant products, it is seldom sought for nongovernment application. Standards such as FIPS and Common Criteria (CC) were developed specifically for evaluation of products intended for use by the government. Due to the high cost of developing and evaluating these products, commercial industry tends to accept products bearing the less costly certification of ICSA.

The government's architectural and security requirements for cryptographic products are typically higher than that of the normal business community. To provide industry with acceptable security and engineering standards with which to develop secure products, the National Information Assurance Partnership (NIAP), an organization based upon a NIST and National Security Agency (NSA) partnership, was formed. In cooperation with other international agencies, NIAP came up with the Common Criteria (CC) with which commercial products are evaluated and assigned an evaluation assurance level based upon compliance with CC requirements.

Although the certification and testing that NIST offers may seem the ideal testing mechanism for assuring IPSec compliance, ICSA certification is far less costly for both the developer and the consumer than something that is FIPS compliant or that has a CC evaluation assurance level. The ICSA IPSec certification is widely recognized and is pretty much considered the de facto IPSec interoperability standard. However, ICSA competitors are quick to mention that the IPSec standards that ICSA uses are based more on industry demand rather than a strict adherence to that defined by the IETF.

Industry Application

As soon as the IPSec suite was made available to industry, vendors began integration of IPSec encryption mechanisms inside their common networking products. The problems that surfaced almost immediately were vendor noncompliance with IPSec standards and federal encryption exportation legislation. Even though legislation was subsequently relaxed to allow more liberal exportation of encryption, vendor compliance with IPSec standards was slow to mature.

One organization that contributed to improved interoperability standards was the Virtual Private Network Consortium (VPNC). They published technical documentation on interoperability and hosted conferences intended to allow vendors to share and discuss their VPN products.

Commercial companies were not the only developers of IPSec-compliant products. FreeS/WAN (Secure Wide Area Network) was released as a free, Linux-based IPSec VPN. Also, a NIST implementation called Cerebrus was intended to provide industry with a true, IPSec-compliant implementation with which to test products.

Encryption Past and Present

ESP is designed to support a variety of symmetric encryption algorithms. By default, ESP uses a basic DES (Data Encryption Standard) to guarantee minimal interoperability among IPSec networks, but it is up to the two negotiating parties to decide which algorithm to use.

DES, the current official encryption standard, is being phased out in favor of stronger algorithms such as 3DES. Although 3DES is not the chosen replacement for DES, it is the most common symmetric-encryption algorithm in use due to its strength and low overhead requirements. However, NIST has recently announced its selection for DES' successor, the Advanced Encryption Standard (AES). From a selection competition involving several prospective encryption algorithms, RIJNDAEL (often referred to and pronounced as "rain doll") was selected as the AES standard.

The AES will be the government's designated encryption cipher, but is expected to be adopted by the commercial world as well. This means a great deal of integration with IPSec-compliant products, because the AES standard may be the dominant mechanism for encryption.

IPSec and NAT

We talked about Network Address Translation (NAT). (Recall, NAT is often used loosely and is technically referred to as Network Port Address Translation [NPAT] in Chapter 7, but we will review it briefly.) Sometimes globally unique IP addresses are a scarce resource. In Europe, for example, it is especially hard to obtain a globally unique IPv4 address. Other times, a company simply wishes to keep secret the IP addresses of the machines in its intranet. (An unlisted address similar in concept to an unlisted phone number.) Both of these situations can be addressed with NAT.

NAT is usually implemented in a machine that resides at the boundary of a company's intranet, at the point where there is a link to the public Internet. NAT sets up and maintains a mapping between internal IP addresses and external public (globally unique) IP addresses. Since the internal addresses are not advertised outside of the intranet, NAT can be used when they are private (globally ambiguous) addresses or when they are public (globally unique) addresses that a company wishes to keep secret.

The weakness of NAT is that by definition the NAT-enabled machine will change some or all of the address information in an IP packet. When IPSec authentication is used, a packet whose address has been changed will always fail its integrity check under the AH protocol, since any change to any bit in the datagram will invalidate the Integrity Check Value that was generated by the source.

Within the IETF, a working group is looking at the deployment issues surrounding NAT. This group has been advised by the Internet Engineering Steering Group (IESG) that the IETF will not endorse any deployment of NAT that would lead to less security than can be obtained when NAT is not used. Since NAT makes it impossible to authenticate a packet using IPSec's AH protocol, NAT should be considered as a temporary measure at best, but should not be pursued as a long-term solution to the addressing problem.

IPSec protocols offer some solutions to the addressing issues that were previously handled with NAT. We will see in later scenarios that there is no need for NAT when all the hosts that constitute a given Virtual Private Network use globally unique (public) IP addresses; address hiding can be achieved by IPSec's Tunnel mode. If a company uses private addresses within its intranet, IPSec's Tunnel mode can keep the addresses from appearing in cleartext form in the public Internet.

AH, ESP, and NAT

ESP and AH are different in design, but there is some overlap within the functionality of both. ESP, AH, or a combination of both may be used for production, but a particular application will require either one or the other. Because of the frequent conflicts between NAT and AH's checksum verification mechanism, using ESP is typically favored.

The AH provides strong integrity and authentication for IP datagrams by tying data in each packet to a verifiable signature. This allows verification of the identity of the sender and assurance that the data has not been altered. Because AH does not encrypt the data being sent, it is typically used when integrity and authenticity of the data is more important than confidentiality.

An AH is normally inserted after an IP header and before the other information being authenticated. Because some of the fields in the IP header change in transit, AH cannot protect everything. However, AH works around these fields by providing three basic types of protection for what it has access to:

▼ A checksum of the immutable fields is generated to assure data integrity.

■ A shared secret key is included in the authentication data to assure data origin.

▲ A sequence number field within the AH header is used to prevent replay attacks.

AH processing is applied only to nonfragmented IP packets. However, an IP packet with AH applied can be fragmented en route to its destination. In this case, the destination first reassembles the packet and then applies AH processing to it.

Packets that fail authentication are discarded and never delivered to upper layers. This mode of operation greatly reduces the chances of successful denial of service attacks, which aim to block the communication of a host or security gateway by flooding it with forged packets.

However, the increased usage of NAT has the same effect, as do forged packets. AH digitally signs the data payload and headers of the outbound packet with a hash value appended to the packet. While this would not normally cause disruption, a NAT device in between the IPSec endpoints will rewrite either the source or destination address with one of its own choosing. Upon receipt, the packet will fail the checksum verification and will be dropped. The VPN device at the receiving end doesn't know about the NAT in the middle and assumes that the data has been tampered with.

This issue has resulted in the more frequent use of ESP for authentication. ESP is compatible with NAT because integrity checks are performed on portions of the header that are not altered by NAT devices.

In some implementations of IPSec using ESP, corruption of the calculated checksum is still possible. Vendors have responded to this problem by using User Datagram Protocol (UDP) encapsulation to circumvent the damage done by the NAT device.

Tunnel and Transport Mode

As we discussed in Chapter 16, IPSec protocols can implement security associations in two modes, *Transport mode* and *Tunnel mode*.

IPSec Transport Mode

In Transport mode, the original IP datagram is taken, and the IPSec header is inserted right after the IP header. In the case of ESP, the trailer and the optional authentication data are appended at the end of the original payload. If the datagram already has IPSec header(s), the new header would be inserted before any of those. That is hardly ever the case, and it would be better to use Tunnel mode. Take a look at Figure 19-7 to see the IPSec Transport mode headers.

The Transport mode is used by hosts only, not by security gateways. Security gateways are not even required to support Transport mode. The advantage of the Transport mode is its lower overhead requiring fewer CPU cycles. One disadvantage is that the mutable fields are not authenticated. ESP in Transport mode provides neither authentication nor encryption for the IP header.

This is a disadvantage, since false packets (spoofing attack) might be delivered for ESP processing. Another disadvantage of Transport mode is that the addresses of the original IP datagram must be used for delivery. This can be a problem when using private

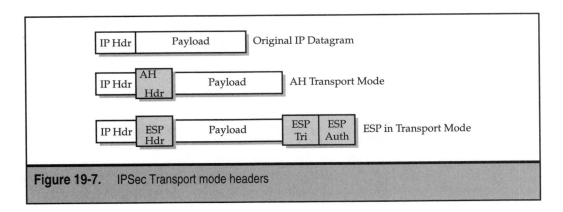

Figure 19-7.   IPSec Transport mode headers

address schemes or when internal addressing structures need to be hidden from the public network or Internet.

IPSec Tunnel Mode

In the case of Tunnel mode, a new IP datagram is created that encapsulates the original IP datagram in its payload. Then, IPSec in Transport mode is applied to the resulting datagram. Take a look at Figure 19-8 to see IPSec Tunnel mode headers. In the case of ESP, the original datagram becomes the payload data for the new ESP packet, and therefore its protection is total if both encryption and authentication are selected. However, the new IP header is still not protected.

Tunnel mode is used whenever either end of an SA is a security gateway. This means that between two firewalls Tunnel mode is always used for traffic that is passing through the firewalls between the secure networks through an IPSec tunnel. Although security gateways are supposed to support Tunnel mode only, sometimes they can also work in Transport mode. This mode is allowed when the security gateway acts as a host, that is, in cases when traffic is destined to itself. Why would that ever happen? Examples are maintenance-related protocols such SNMP commands or ICMP echo requests.

In Tunnel mode, the outer headers' IP addresses do not need to be the same as the inner headers' addresses. For example, two security gateways may operate an AH tunnel that is used to authenticate all traffic between the networks they connect together. This is a very typical mode of operation. Hosts are not required to support Tunnel mode, but they often do, and they have to support it for some remote access scenarios.

The advantages of the Tunnel mode are total protection of the encapsulated IP datagram and the possibility of using private addresses. However, there is an extra processing overhead associated with this mode.

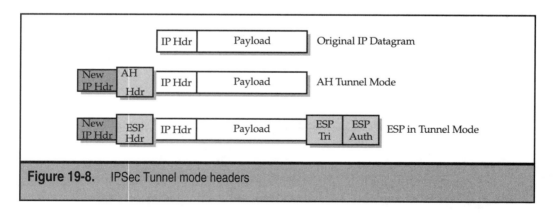

Figure 19-8. IPSec Tunnel mode headers

Tunnel Versus Transport

With the introduction of commercial encryption products, Tunnel modes have become an important consideration for many types of encryption architecture. The introduction of such concepts as *secure enclaves* and *virtual private networking* (VPN) has defined various strategies of establishing end-to-end encryption. One of these strategies involves connection between single *hosts* and *security gateways. Security Gateways* protect internal hosts, or secure enclaves.

The ideal secure tunnel typically involves at least a pair of security gateways. The dedicated VPN security gateways provide a secure tunnel to protect the transmissions of their protected enclaves. The tunnel endpoints are protected on at least one end by a security gateway protecting the identity of the inside hosts. This mode is commonly referred to as Tunnel mode because at least one of the machines is *not* an endpoint of the connection, thereby encapsulating the IP data of the protected machines. The machine(s) that is not an endpoint acts as a tunnel entrance, encrypting and decrypting the data on behalf of the machines it is protecting.

Transport mode is used between two endpoints of a connection, such as a host-to-host (client-to-client) scenario. As we stated earlier, the primary difference between the two is that Tunnel mode encapsulates the IP headers of the protected machines and Transport mode doesn't. The actual mechanics of the process are more involved, but the distinction primarily serves as an abstraction to define various encryption architectures.

MANAGING THE SECURITY ASSOCIATIONS

We talked about IPSec security associations in Chapter 16. As you will recall, IPSec SAs are a set of security services supplied by using one of the IPSec protocols. SAs can be formed in both the Tunnel mode or Transport modes we spoke of in the previous sections.

A Tunnel mode SA has an "outer" as well as an "inner" IP header. The outer header specifies the system that must process the IPSec information, whereas the inner header specifies the actual and final destination of the packet. The AH and ESP protocols support different sets of services, so an SA can only offer the services supported by the IPSec protocol. This means that things like header protection can only be offered by the AH protocol, and data encryption can only be handled by the ESP protocol. This is where SA bundling comes in, because SA bundles are suited for implementing complex security policies.

SA Bundles

SA bundles can be made in Transport mode, Tunnel mode, and as a combination of both. Tunnel mode implementations offer a wider array of implementation choices and would be the natural choice among VPN implementations that might require services demanding these combined SAs to accomplish the necessary transport. SA bundles basically have two formats:

▼ **Fine granularity** This is when you assign an SA to each communication process. Data transmitted over a single SA is protected by a single security protocol, meaning the data can be protected by AH or ESP, but not both, since SAs can have only one security protocol.

▲ **Coarse granularity** This is when you combine services from several applications into an SA bundle. This allows for communication with two levels of protection using multiple SAs.

Let's say you have some computer out on the Internet that wants to establish a Tunnel mode SA with your corporate router and a Transport mode SA to the actual final destination of a host on the internal network behind the router. This method provides secure communications over an untrusted medium like the Internet and then continues that security once it is on the internal network for a secure point-to-point connection. This would require an SA bundle that would terminate at two different destinations.

Security Association Combinations

Security associations can be combined in two ways:

▼ **Transport adjacency** This is applying more than one security protocol to the same IP datagram without implementing Tunnel mode for communications. Using both AH and ESP provides a single level of protection and no nesting of communications, since the endpoint of the communication is the final destination. This application of *transport adjacency* is applied when Transport mode is implemented for communication between two hosts, each behind a security gateway.

▲ **Iterated tunneling** This is the application of multiple layers of security protocols within Tunnel mode SAs. This allows for multiple layers of nesting since each SA can originate or terminate at different points in the communication stream. There are three occurrences of iterated tunneling:

■ The endpoints of each SA are the same.

■ One of the endpoints of the SAs is the same.

■ Neither endpoint of the SAs is the same.

Identical endpoints can refer to Tunnel mode communications between two hosts behind a set of security gateways, where SAs terminate at the hosts, and AH and/or ESP is contained in an ESP providing the tunnel. You can see an example of identical endpoints using iterated tunneling in Figure 19-9.

With only one of the endpoints identical, an SA can be established between the host and security gateway and between the host and an internal host behind the security gateway. We used this model earlier as an example of one of the applications of SA bundling. In Figure 19-10, you can see an example using iterated tunneling in which one of the endpoints is identical.

In the event that neither SA terminates at the same point, an SA can be established between two security gateways and between two hosts behind the security gateways. This application provides multilayered nesting and communication protection. An example of this application is a VPN between two security gateways that provide Tunnel mode operations for their corresponding networks to communicate. Hosts on each network are provided secured communication based on client-to-client SAs. This provides for several layers of authentication and data protection. In Figure 19-11, you can see an example using iterated tunneling in which neither of the endpoints is identical.

Establishing a VPN

Now that we have defined the components of a VPN, we will discuss the form that they create when combined. To be IPSec compliant, four implementation types are required of any VPN implementation. Each type is merely a combination of options and protocols

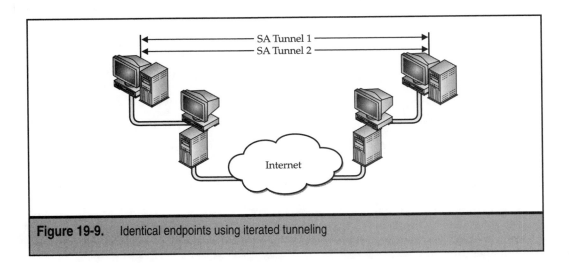

Figure 19-9.   Identical endpoints using iterated tunneling

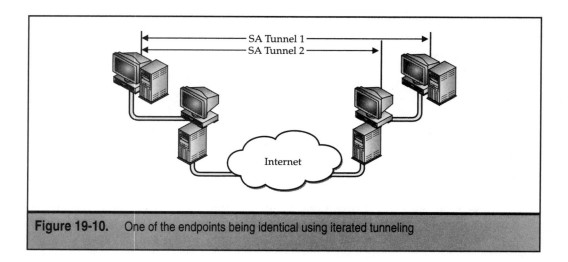

Figure 19-10. One of the endpoints being identical using iterated tunneling

with varying SA control. The four detailed here are only the minimum required formats, and vendors are encouraged to build on the four basic models.

The VPNs can use either or both the AH and ESP security protocols. The mode of operation is defined by the role of the endpoint, except in client-to-client communications, which can be Transport mode or Tunnel mode. Let's take a look at the four cases:

▼ **Case 1** Providing end-to-end security between two hosts across the Internet/intranet. This is the simplest case, because the host is not only acting on behalf

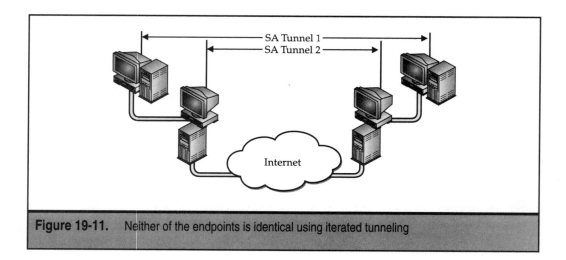

Figure 19-11. Neither of the endpoints is identical using iterated tunneling

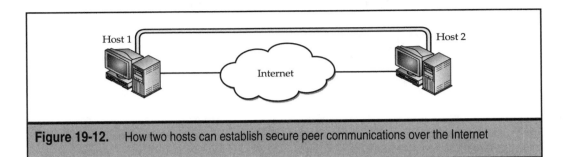

Figure 19-12. How two hosts can establish secure peer communications over the Internet

of itself, but also because this case can facilitate most protocol combinations. It is worth noting that even though it is an end-to-end communication, Tunnel mode *may* be applied. Take a look at Figure 19-12 to see how two hosts can establish secure peer communications over the Internet.

■ **Case 2** This case illustrates a simple Virtual Private Network, where two LANs are interconnected by using two security gateways in Tunnel mode acting on behalf of each LAN. Take a look at Figure 19-13 to see a typical security gateway–to–security gateway VPN. In this case, neither host needs to have the IPSec protocol stack installed in order to communicate securely. Also, when using Tunnel mode, there is no requirement to support both AH and ESP at the same tunnel. If it is desirable to both authenticate and encrypt the packet, ESP also supports authentication as an option.

■ **Case 3** This case combines the two previous cases. It does not add any new requirements to the hosts and security gateways, except that the security gateways must allow IPSec traffic to pass. The protocol combinations can

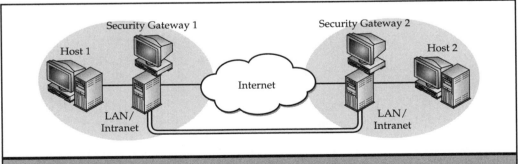

Figure 19-13. A typical security gateway-to-security gateway VPN

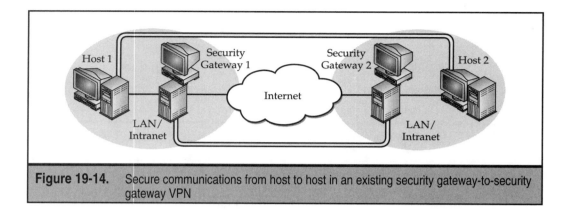

Figure 19-14. Secure communications from host to host in an existing security gateway-to-security gateway VPN

also, therefore, be any of the combinations from the two preceding cases. Figure 19-14 shows how this case combines Case 1 and 2 to allow secure communications from host to host in an existing security gateway-to-security gateway VPN.

■ **Case 4** This is the case where a remote host uses the Internet to reach the security gateway of the corporate network. It then connects to a host on the inside LAN. A typical situation is where the remote host first uses a PPP connection to an ISP to access the Internet. It then traverses the Internet in order to reach the security gateway. Tunnel mode is the only mode required between Host 1 and the corporate security gateway, so it can use any protocol combinations just as in Case 1. Take a look at Figure 19-15 to see a remote host connected to Corporate network. This is also known as the "road warrior" model.

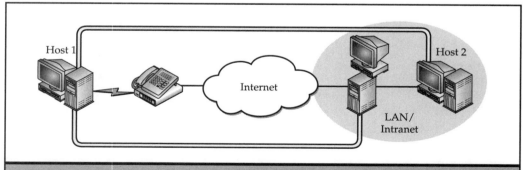

Figure 19-15. A remote host connected to Corporate network. Also called the Road Warrior model

SECURITY ASSOCIATION DATABASES

You were given a general description of these in databases in Chapter 16. Security associations require two different kinds of databases:

▼ **Security Policy Database (SPD)**   The Security Policy Database specifies the security services that will be provided for IP packets. This database contains an ordered list of policy entries. Each entry includes information about the type of packets the policy should apply to, such as source and destination address. An IPSec policy entry also would include an SA specification, listing the IPSec protocols, if it is Tunnel mode or Transport mode, and the security algorithms to be used.

▲ **Security Association Database (SAD)**   The Security Association Database defines the parameters associated with individual SAs. Each SA has an entry in the SAD. For outbound packets, entries are pointed to by entries in the SPD. For inbound processing, each entry in the SAD is indexed by a destination IP address, protocol type, and SPI. Each SA specification in the SPD points to an SA, or a bundle of SAs, in the SAD.

Here is a quick overview of what happens:

1.   When an IPSec system sends a packet, it first matches the packet against the entries in the Security Policy Database to see if there is an SA for the packet in the SAD.

2.   If an SA does not exist, the IPSec system creates one. It then applies the processing specified and puts the SPI from the security association into the IPSec header.

3.   When the IPSec peer receives the packet, it looks up the security association in its database by destination address, SPI, and security protocol and then processes the packet as required.

Security Policy Database (SPD)

The SPD is a security association management database designed to enforce policies in an IPSec environment. One of the most important parts of SA processing is an underlying security policy that specifies what services are offered to IP datagrams and in what fashion they are implemented.

The SPD is consulted for all IP and IPSec communications. This applies to inbound and outbound connections, which is why it is always associated with an interface. There are two kinds of interfaces as far as IPSec is concerned:

▼ **Black interface**   An interface that performs IPSec is referred to as a *Black interface*.

▲ **Red interface** An interface where IPSec is not being performed is referred to as a *Red interface,* meaning that no data will be encrypted over that interface.

The number of SPDs and SADs is directly related to the number of Black and Red interfaces being supported by the router. The SPD must control *both* traffic that is IPSec based *and* traffic that is not IPSec related. When a packet is received on an interface, the SPD can do one of three things:

▼ Forward it and not apply IPSec.

■ Discard the packet.

▲ Forward it and apply IPSec.

The SPD can be configured so that only IPSec traffic gets forwarded. This would provide a basic firewall function by allowing only IPSec protocol packets into the Black interface. If IPSec is to be applied to the packet, the SPD policy entry will specify an SA or SA bundle to be employed. Within the specification are the IPSec protocols, mode of operation, encryption algorithms, and any nesting requirements.

SPD SA Selectors

The SPD controls all traffic through an IPSec system. To choose which packets a policy should act upon, the SPD enables the administrator to specify one or several traffic or packet *selectors.* Selectors constitute the field or portion of the SPD entry that selects which traffic will be affected by the database policy, which in turn can specify one or many SAs. Some selectors may include things such as:

▼ Any host

■ A specific IP source address

■ A destination address

■ A specific user ID

■ A network IP range

■ The specific protocol used

▲ The high-level port number specified in the connection

Almost any criteria that can be used to identify a packet can be used as an SA selector for the SPD. A selector is used to apply traffic to a policy. A security policy may determine that several SAs be applied for an application in a defined order, and the parameters of this bundled operation must be detailed in the SPD. An example policy entry may specify that all matching traffic be protected by an ESP using DES, nested inside an AH using

SHA-1. Each selector is used to associate the policy to SAD entries. The SPD is policy driven and is concerned with system relationships.

Security Association Database (SAD)

Just like the IPSec standard that specifies a database in order to keep track of the services available and how policies are applied, the IPSec standard also specifies a different database that tracks every active SA. This database entry contains a record of all the values negotiated at the time each SA was initially created. This database is used by the SPD to quickly link packets to an existing security association in order to process them according to existing policies and conforming to the information contained in the SPI. This database is called the *Security Association Database* (SAD). The SAD is responsible for each SA in the communications defined by the SPD. Each SA has an entry in the SAD. The SA entries in the SAD are indexed by the three SA properties:

▼ Destination IP address

■ IPSec protocol

▲ Security Parameter Index (SPI)

The SAD database contains the following parameters for processing IPSec protocols and the associated SA:

▼ Sequence number counter for outbound communications.

■ Sequence number overflow counter that sets an option flag to prevent further communications utilizing the specific SA.

■ A 32-bit antireplay window that is used to identify the packet for that point in time traversing the SA and that provides the means to identify that packet for future reference.

■ Lifetime of the SA that is determined by a byte count or time frame, or a combination of the two.

■ The algorithm used in the AH.

■ The algorithm used in the authenticating ESP.

■ The algorithm used in the encryption of the ESP.

■ IPSec mode of operation, Transport or Tunnel mode.

▲ Path MTU (PMTU). This is data that is required for ICMP data over an SA.

Each of these parameters is referenced in the SPD for assignment to policies and applications. The SAD maintains a record of all existing security associations for as long as they live, and it is always used in conjunction with the SPD to correctly sort and process all IPSec traffic. Any of the core components of an SA can become a selector parameter for the SAD, including destination IP, IPSec protocol (AH or ESP), and SPI.

IPSec Database and System Processing

It is important to understand how systems process datagrams when it comes to using IPSec and IKE. When implementing IPSec in place, datagrams can no longer be simply processed, forwarded, or discarded, but must be subject to a security policy to determine if additional IPSec processing is required and when it has to occur. Though there are slight differences among platforms as to how they implement IPSec on their particular IP stacks, the general principle of IPSec processing for *host* and *security gateway* systems can be summarized by the four subsections that follow.

Outbound IPSec Processing for Host Systems

With IPSec active, any outbound packet is subject to the Security Policy Database (SPD) to determine if IPSec processing is required or what else to do with the packet.

If IPSec is required, the Security Association Database (SAD) is searched for an existing security association (SA) for which the packet matches the profile.

If that is not the case and IKE as well as on-demand outbound SAs are supported, a new IKE negotiation is started that ultimately results in the establishment of the desired SA(s) for this packet.

Finally, IPSec is applied to the packet as required by the SA, and the packet is delivered. This process is illustrated in Figure 19-16.

NOTE: In general, the routing table is consulted to determine if the packet can be delivered at all. If no route is found, IPSec processing should not be performed, but instead the user should be informed of this problem. We are assuming, however, that *host* systems usually have a default route defined so that packets are sent in any case.

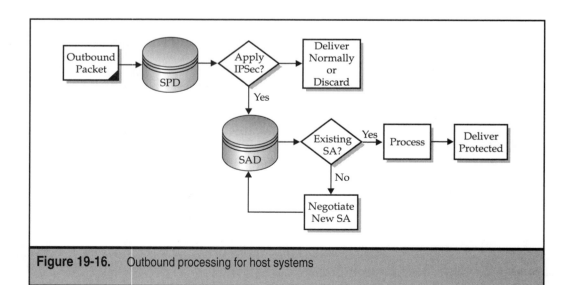

Figure 19-16. Outbound processing for host systems

Inbound Processing for Host Systems

With IPSec active, any inbound packet is subject to the SPD to determine if IPSec processing is required or what else to do with the packet. If IPSec is required, the SAD is searched for an existing Security Parameter Index (SPI) to match the SPI value contained in the packet. If that is not the case, there are essentially two options:

▼ Silently discard the packet. (Do not inform the sender, but log the event if configured to do so.) This is the default action performed by most IPSec implementations.

▲ If IKE as well as on-demand inbound SAs are supported, a new IKE negotiation is started that ultimately results in the establishment of SA(s) to the sender of the original packet. In this case, it does not matter if the original packet was IPSec protected or in the clear, which would only depend upon the local policy. However, it requires that the sender of the original packet respond to the IKE negotiations, and it would mean that packets were discarded until an SA is established.

Finally, IPSec is applied to the packet as required by the SA, and the payload is delivered to the local process. This process is illustrated in Figure 19-17.

Outbound Processing for Security Gateway Systems

On a *security gateway* system, any outbound packet is usually subject to the SPD of the secure (Black) interface to determine what to do with it.

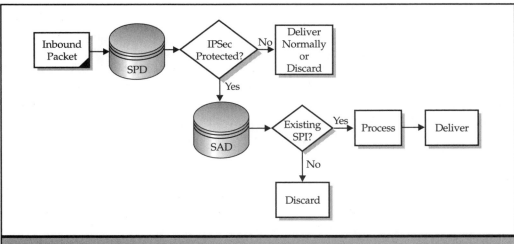

Figure 19-17.   Inbound processing for host systems

If the decision is to route the packet, the routing table is consulted to determine if the packet can be delivered at all.

If no route is found, IPSec processing should not be performed, but instead the original sender may be informed of this problem using ICMP network-unreachable messages.

We are assuming, however, that security gateway systems either employ routing protocols or have a default router defined so that a successful routing decision can be made. From this point on, processing is essentially the same as it is on host systems.

The packet is then forwarded to the SPD of a nonsecure (Red) interface to determine if IPSec processing is required or what else to do with the packet.

If IPSec is required, the SAD is searched for an existing SA for which the packet matches the profile.

If that is not the case and IKE as well as on-demand outbound SAs are supported, a new IKE negotiation is started that ultimately results in the establishment of the desired SA(s) for this packet.

Finally, IPSec is applied to the packet as required by the SA, and the packet is delivered. This process is illustrated in Figure 19-18.

Inbound Processing for Security Gateway Systems

On a security gateway system with IPSec active, any inbound packet is subject to the SPD to determine if IPSec processing is required or what else to do with the packet.

If IPSec is required, the SAD is searched for an existing SPI to match the SPI value contained in the packet. If that is not the case, there are essentially two options:

▼ Silently discard the packet. (Do not inform the sender, but log the event if configured to do so.) This is the default action performed by most IPSec implementations.

▲ If IKE as well as on-demand inbound SAs are supported, a new IKE negotiation is started that ultimately results in the establishment of SA(s) to the sender of the original packet. In this case, it does not matter if the original packet was IPSec protected or in the clear, which would only depend upon the local policy. However, it requires that the sender of the original packet respond to the IKE negotiations, and it would mean that packets were discarded until an SA is established.

Once the packet has been successfully processed by IPSec, which may be an iterative process for SA bundles, a routing decision has to be made as to what to do with the packet next.

If the packet is destined to another *host*, it is delivered over the appropriate interface according to the routing tables. If the packet is destined to the *security gateway* itself, the payload is delivered to the local process. This is illustrated in Figure 19-19.

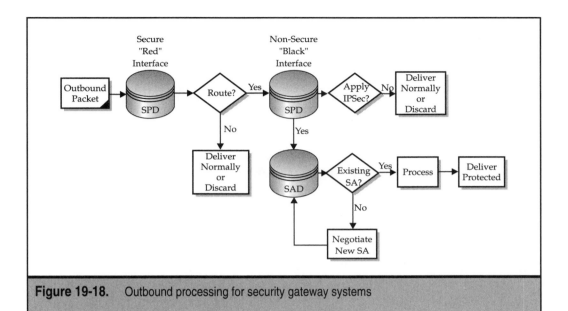

Figure 19-18. Outbound processing for security gateway systems

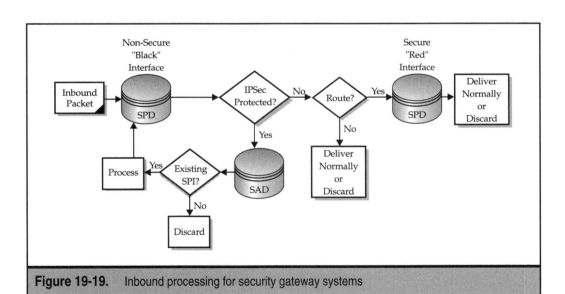

Figure 19-19. Inbound processing for security gateway systems

NOTE: If the *security gateway* receives a packet with IPSec applied in an iterated tunnel SA bundle, it only has to process the outer SA that is carried in a datagram destined to the *security gateway*. Since the inner SA is carried in a datagram destined to another *host,* it will be forwarded by the security gateway (provided the policy permits it), regardless of its IPSec protection.

SUMMARY

This is the last of the four chapters in Section V. IPSec is an amalgam of many different kinds of protocols that work carefully together to provide security over a public medium. In this chapter, we talked about the architecture and implementation of the IPSec protocol using all of the subjects we talked about in the first three chapters of this section. Figure 19-20 shows how it all comes together.

Now we will move on to the final part of the book, Section VI. Multi-Protocol Label Switching (MPLS) is technology of the future with respect to enterprise and Service Provider VPN implementations.

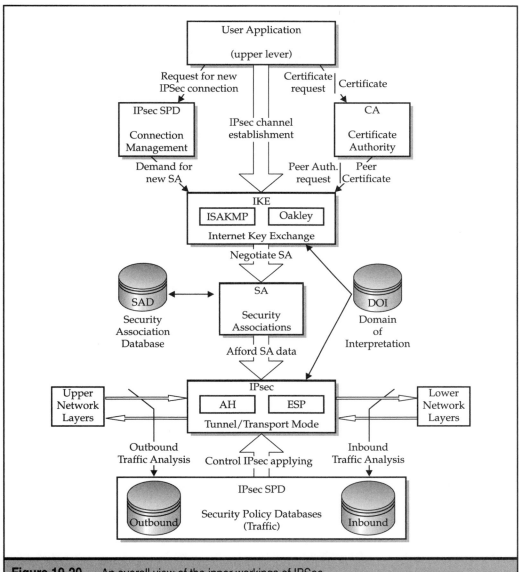

Figure 19-20.   An overall view of the inner workings of IPSec

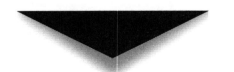

PART VI

MPLS

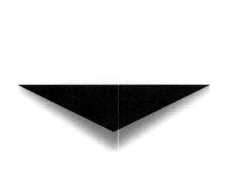

CHAPTER 20

Quality of Service (QoS)

This brings us to the last section of the book and easily the most exciting, MPLS. MPLS is the future of VPNs. But it is much more than that, it is the future of the Internet itself because of the many new services it will allow ISPs and other carriers to offer.

Until just recently, the nature of these services all required there own technologies and infrastructure to operate successfully. MPLS will be the technology that allows all of these services to work together using one common infrastructure, the Internet.

Before we get into the specifics of MPLS, however, it is important that you have a very solid grasp of a concept that you have seen mentioned throughout this book from the very beginning and one that has become as common as VPNs. That concept is *Quality of Service* (QoS). QoS is the concept used to address the different nature of the new Internet services we mentioned above. Different QoS protocols govern how data travels over a network or any number of networks up to and including the global Internet. In this chapter, we will talk about these concepts and protocols to give you a better understanding of why they are needed and how they work.

BASIC TERMS

There are a number of terms and topics that are very hot in the computer industry right now and if you haven't heard them yet, you will soon. These terms are being thrown about in a marketing frenzy to promote what are often called "killer applications." We are going to introduce you to a handful of these terms now so that you will recognize them as we talk about the nature and specific requirements of these killer applications.

Probably the most notable of all is IP Telephony or Voice over IP (VoIP). This is the Holy Grail of killer applications. Many trillions of dollars are spent every year worldwide on voice communication. When you consider the financial implications of making it possible to use the Internet to "reach out and touch somebody," you should not be left wondering why. Internet TV and radio have already started and will soon be upon us. Distance learning, videoconferencing, and personal closed circuit communications are just another drop in the bucket when you see what the possibilities of the future Internet are. You have already seen what a profound effect e-mail and surfing the web have had upon all of us.

Two of the terms that go along with QoS and the services we have just mentioned are *Convergence* and *Service Level Agreements* (SLA).

Convergence

Convergence refers to combining all applications into one medium or infrastructure; in this case, the Internet. Currently, most large organizations separate their data, phone, and mission critical mainframe communication needs over different service providers, telephone, and long distance telephone companies. The nature of these technologies meant that these different networks needed to be physically separate. Each could be optimized to provide an operating environment that best suited the nature of the specific type of traffic.

Having a physically separate network infrastructure removes many of the administrative complexities. Unfortunately, it is also much more expensive in terms of resources. Not only would you have to pay high monthly fees for the extra communications link, you would also need to maintain separate equipment, operations, and technical support personnel.

To integrate all of those services onto one physical infrastructure, because of the different nature of these services, you need to specify different classes of service in order to distinguish which packets are being transported and when. The goal is to make sure each type of data gets handled in a way that meets, but preferably does not exceed, the specific application's requirements. If each type of traffic gets just what it needs and no more, then (theoretically) the capacity of this one physical link could be optimized. The cost savings will be a mere fraction of the previous situation. *Convergence* offers many benefits such as:

▼ Technology and equipment consolidation reduces hardware and software complexity.

■ Parallel physical networks and equipment can be eliminated, not to mention the reduction in the number of vendors involved.

■ Native compatibility among standardized network components obviates the need for custom gateways and conversion of application data.

■ Enterprise networks can be more easily integrated with the service provider's infrastructure and the Internet.

■ Higher performance is possible if QoS-based optimizations can be invoked at both the local and long distance levels.

■ Global deployment of new multimedia applications can be faster without tedious conversions.

■ Lower costs and higher productivity by training employees on just one network.

▲ Network management systems complexities and operating procedures can be more fully integrated.

Service Level Agreements

Service Level Agreements, most commonly referred to as *SLAs,* are contracts that companies maintain with those carriers for a certain guaranteed level of performance, such as the amount of bandwidth in the case of ISPs and number of simultaneous telephone line connections in the case of telephone companies.

An SLA is a written, unmistakable outline that defines end-to-end service specifications and, in a generic sense, usually consists of the following:

▼ *Availability* This generally equates to guaranteed uptime. It can also include service latency or the minimum delay accessing the network.

- *Services Offered* A detailed specification of the service levels being offered.

- *Service Guarantees* This consists of guarantees for things like each class of service, throughput, packet loss rates, delay (latency), delay variation (jitter), and so on.

- *Responsibilities* Clearly defined responsibilities and/or consequences for not meeting agreed SLAs, which equipment and the hours of support and possible financial penalties.

▲ *Auditing* Monitoring activity on the infrastructure and the services used for appropriate billing.

Now that we have covered some of the different terms and applications you are going to someday encounter, we can move on to the topic of QoS itself.

WHAT IS QUALITY OF SERVICE (QoS)?

We will start off by giving you a simplified definition of QoS (mainly because a precise official definition doesn't exist). In a nutshell, QoS involves the capability of a network to provide a set of characteristics that tailors the delivery of data to meet the specific requirements of a user or application. That is really only a "twenty-thousand-foot view" of the subject, and you will soon see why.

The simplest "real world" analogy would be vehicle traffic and freeways. Imagine a four-lane freeway with exits and onramps every mile or so. The freeway's policy is "first come first served," meaning that anyone can pull on or off the freeway at any time they wish. This would be referred to as a *"best effort"* approach. This approach could work forever theoretically regardless of how much congestion resulted. Of course, in this scenario everyone has a job where it doesn't matter what time they get to work or which path they choose to get there, as long as they put in their eight hours. This example is the vehicle traffic equivalent of the Internet. Unfortunately, most people's lives don't work this way.

The best example of vehicle traffic QoS are the High Occupancy Vehicle (HOV) lanes, or as they are more affectionately known, "the damn diamond lanes" (what with all its three-passenger automobiles and their smug little drivers that get to arrogantly travel to and fro at a brisk 60 miles per hour while you sit and read the license plate frame in front of you and try to guess whether the fumes you are breathing are regular or super unleaded....oh sorry, I got little carried away there).

HOV lanes are a great example of using a *"differentiation"* technique to solve automobile traffic congestion while making use of the existing four-lane highway. In this case, you differentiate the cars that can share the high performance diamond lane based on a given criteria or "classification." In this case, the classification is "three or more people," as opposed to the regular traffic with the "less than three" classification. The latter group's choice consists of the three remaining low performance lanes (notice that they use a "diamond" icon to mark that lane as opposed to, say, a "cat litter" icon).

Another technique (albeit not very efficient) using the existing four lanes would be an *"integrated"* approach. In this case, you would have all cars with three people line up at their respective onramps and, at different intervals, let one car at a time pull onto the diamond lane and drive all the way to their exit, individually driving as fast as their car can carry them. They could drive at this speed without worry because they would be the only car in that lane during that interval. The very instance that car exits, the lane would open up and the next car in line would pull onto the diamond lane and zip down to their exit at blazing speed and so on until there were no cars left. At this point, the lane could be opened up for use by anyone (usually by 10:00 A.M. or so).

If you are anything like me, "Johnny solo" (I have a foot odor problem…) you commute everyday cursing under your breath wondering, "why don't they just add ten more lanes?" Then they could avoid all of this blatant favoritism and everyone could drive as fast as possible regardless of how many people are in the vehicle. This would be known as a *"Bandwidth"* approach and tends to be the approach most people would choose (if the money was available) since it is the least complicated and requires no defining of rules, posting of signs, scheduling, or need for enforcement. Currently, this also tends to be the solution most ISPs choose when it comes to the Internet.

So much for the simple definition of QoS. The reality is that QoS is extremely complex and involves many different characteristics that depend on certain specific qualities that make it difficult to achieve from an end-to-end perspective. Another major reason this concept is so complicated is uncertainty.

The Internet is a global network of networks. All of these networks have different capabilities, architectures, capacities, and number of other anomalies. When you take those anomalies into consideration and add the biggest challenge of all, the fact that they are out of your control from an administrative standpoint, it becomes evident that guaranteeing some semblance of assured quality is a monumental task, indeed.

After you read this chapter, you will hopefully come away with a whole new and real appreciation for what the term QoS really means.

WHY DO WE NEED QoS?

When you ask, "Why do we need QoS?," you first need to consider the nature of the network infrastructure upon which your data runs; in this case, the Internet itself. Along with the nature of the network infrastructure, you need to consider the nature of the application using the network.

Nature of the Network Infrastructure

Let's start with our network infrastructure, the Internet. The main reason for the Internet's tremendous success is its simplicity and ability to scale. The Internet's simplicity revolves around the principle behind its underlying protocol. That protocol is IP and the principle is that responsibility for successful communication rests with the endpoints.

This concept puts "smarts" in the ends of the network, namely the source and destination hosts, leaving the network "core" dumb.

When data is transmitted between computers on the Internet, it is broken down into manageable chunks, labeled, and then sent as separate packets. Upon arrival at their destination, these packets are stripped of their labels (which contain information about their contents and routing information). The information contained in the packets is then re-assembled to reproduce the original set of data that was transmitted.

Traditional services such as e-mail, file transfer, and web pages have been transported in this manner for years, and it has worked very successfully. This method means that end-user computers do not need to remember routing information. In fact, it works in much the same way as sending a letter through the post office. Whenever you send a letter through the mail, you simply label it and place it in the nearest mailbox. You have no way of knowing the details of the route by which the letter reaches its destination, nor do you care.

In the Internet, routers pass these packets along a chain of hops until the address is reached. These hops have various kinds of routers that generally maintain a queue (or multiple queues) of outgoing packets on each outgoing physical port. If the queues of outgoing data packets become full, so that the hop is too busy to handle all of the packets it is being sent, it simply discards packets randomly to ease the build up of congestion. Internet packets adhering to the protocol contain enough information (source and destination address, sequence number, contents checksum…) for the control protocols to be able to reassemble the data correctly, and to request re-transmits should some of the data fail to arrive. In the simplest Internet environment, packets are dropped by routers on a completely random basis, without regard to their contents or destination.

In recent years, however, the Internet has seen an increase in the use of applications that relies on the timely, regular delivery of packets, and which cannot tolerate the loss of packets, or the delay caused by waiting in queues on router ports. As we stated earlier in our traffic analogy, IP provides a best effort service that is subject to unpredictable delays and data loss.

Originally, IP was designed for applications such as e-mail distribution, basic file transfer and remote terminal access, none of which are particularly sensitive to delay or bandwidth. The purpose of the Internet was to enable any kind of computer to communicate over a wide variety of network systems and devices using a robust, self-healing, and inexpensive network.

Due to the tremendous growth of the Internet, congestion has been increasing at an alarming rate. Increasing the available bandwidth to avoid congested Internet links seems like the obvious solution, but unfortunately the problem is more than a simple capacity issue. The issue is that not only has traffic increased in volume, it has also changed in nature. There are many new types of traffic, from many new IP-based applications, and they vary tremendously in their operational requirements.

Now most new Internet applications are multimedia and require significant bandwidth. Others have strict timing requirements, or function one-to-many or many-to-many (multicast). These require network services beyond the simple *best effort* service that IP delivers. In effect, they require that the once "dumb" IP networks gain some "intelligence."

Different network applications have different operational requirements that demand different network services. Increased network traffic requires increased network bandwidth capacity, but new applications like IP telephony have other requirements. Increasing the network "pipe-size" is not the only answer. More than any other, the desire to provide telephone service over the Internet is driving the convergence of the telephone and Internet industries. This development is quite interesting, since the infrastructure and technology used in modern telephone networks are almost the exact opposite of the infrastructure and technology used in the Internet's IP networks.

Whereas IP uses packet-switching (connectionless) and provides *best effort* services, telephone networks use circuit-switching (connection-oriented) to provide a pre-defined "nailed up" connection for the duration of the call. There's a good reason for this requirement as two-way, real-time telephone flow is a very demanding application for a network to satisfy.

Ultimately, this discussion takes us back to the never-ending debate of what technology is best. My father always taught me to always avoid discussing religion or politics. That turned out to be very good advice because there is no point in arguing over anything that doesn't have a finite or clearly "right" answer, and as sure as there will always be Republicans and Democrats, Catholics and Protestants, or "Less filling!" and "Tastes great!," there will be the never-ending debate over connection-oriented communication and connectionless communication.

There is no "right" answer because each has its own strengths and weaknesses. So with that point in mind, we have discussed the connectionless packet-switching nature of the Internet along with its strengths and shortcomings. Now let's discuss the nature of the connection-oriented, circuit-switched infrastructure of the telephone network.

When you pick up a telephone and call a distant party, you obtain a quality of service inherent in circuit-switching that makes a voice flow both possible and practical. The practicality of the call results from the basic design of the telephone company network infrastructure. That infrastructure digitizes voice flows into a 64 Kbps data stream and routes the digitized flow through a fixed "nailed up" path, established over the network infrastructure. For the entire path, 64 Kbps of bandwidth is allocated end-to-end for the duration of the call.

With circuit-switching, a 64 Kbps time slot is allocated from the entry (*ingress*) point into the telephone network, through the network, to an exit (*egress*) point. The 64 Kbps time slot (or channel) is commonly referred to as a Digital Signal (DS) level 0, or "DS0," and the path through which the DS0 signal is allocated occurs by switches reserving 64 Kbps channels of bandwidth.

As voice is digitized at the *ingress* point into the telephone company network, a slight *delay* of a few milliseconds occurs. As each switch performs a cross-connection operation, permitting digitized voice to flow from a DS0 contained in one circuit connected to the switch onto a DS0 channel on another circuit connected to the switch, a path is formed through the network and another *delay* occurs. Although each cross-connection introduces a slight *delay* to the flow of digitized voice, the switch *delay* is minimal, typically a fraction of a millisecond or less. Thus, the total end-to-end *delay* experienced by digitized voice as it flows through a telephone network can be considered as minimal. You can get an idea of these points of *delay* by taking a look at Figure 20-1.

Another characteristic of the digitized voice flow through the telephone network infrastructure concerns the variability or latency differences between each digitized voice sample (a sample is simply the discrete points at which measurements of an analog voice signal is taken and recorded for conversion to a digital form). Although voice digitization and circuit-switching processes add latency to each voice sample, that delay is consistent and predictable.

Unlike the Internet or IP infrastructure, there is no arbitrary dropping of voice samples during periods of traffic congestion. Instead, whenever the volume of calls exceeds the capacity of the network, such as on Mother's Day, Christmas, and, (well, it should be as far as I am concerned) My Birthday, all new call attempts are temporarily blocked and the caller encounters a "fast" busy signal when dialing. The nature of this method is anything but *best effort*. Instead, it is often referred to as *guaranteed*, *deterministic*, *predictable*, or *loss-less*.

So to sum up, the nature of the connection-oriented, circuit-switched telephone network is that of a low delay, and even more importantly, uniform or near-uniform delay transmission system. Those two qualities—low delay and uniform or near-uniform delay—represent two key QoS metrics (we will be discussing those metrics shortly).

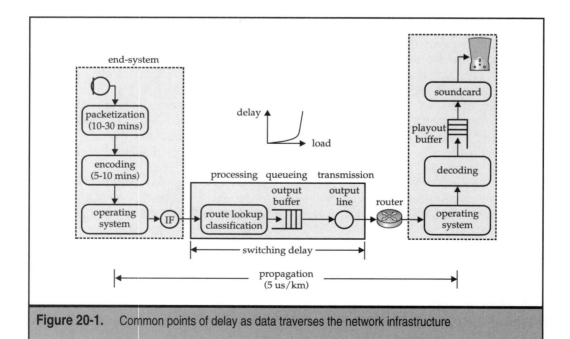

Figure 20-1.   Common points of delay as data traverses the network infrastructure

If you will recall, this situation sounds very much like our earlier vehicle traffic analogy involving the *integrated* approach to congestion. Since circuit-switched networks provide dedicated bandwidth on an end-to-end basis, you can see the similarity of our separate diamond lane that allowed the car to travel from one location to another by itself, while prohibiting other cars destined to other locations to share the highway.

On the surface, it might appear as though the *guaranteed* method of communication is much better than an unpredictable *best effort* approach. After all, everyone loves a guarantee. However, we all know that there is no such thing as a free lunch, and that *all* guaranties come at a price. Although the telephone company network infrastructure provides the QoS necessary to support real-time voice communications, its design is relatively inefficient, expensive, and difficult to scale.

These inefficiencies arise from a number of facts. The first one is that, unless the two parties are shouting at each another, flow is normally half duplex. That means that a flow can't occur in both directions at the same time. When one person is talking, the other is listening (obviously, that doesn't include marriage). This results in half of the bandwidth utilization being wasted.

Also, unless we talk non-stop, we periodically have to come up for air and pause between words as we speak. Because 64 Kbps of bandwidth is allocated for the duration of the call, bandwidth utilization is far from optimal. A good analogy is the one we used to describe Statistical Time Division Multiplexing (not STDM…no acronym here, sorry) over standard Time Division Multiplexing (TDM) back in the *WAN/Frame Relay* section of Chapter 2.

To save you a trip, we will briefly review the analogy. The best way to explain these inefficiencies is to picture a long freight train. The boxcars of this freight train are painted in groups of four all the way to the end in the colors red, green, blue, and purple, and in that same order. Now, imagine that four people are using this railroad to ship sacks of grain. Let's also assume that each boxcar can carry a maximum of three sacks of grain at a time.

The nature of the telephone infrastructure is like a situation where a different color is assigned to each of these four people. One guy has red designated for his shipments, the other guy has green, and so on. The red guy has three sacks of grain to ship, and the other three guys have 30 sacks each.

This system would be inefficient because right after the first set of four boxcars went by with three sacks loaded on them, the rest of the boxcars that follow would all have the red boxcar go by empty because the first guy only had three sacks to begin with and he already loaded those on the first red boxcar.

Another inefficiency to consider is that when the telephone infrastructure was designed, digitization of voice required a minimum of 64Kbps. With the advent of newer voice compression techniques, that same voice sample can now be digitized using only 8Kbps and, in some cases, even lower.

That example should give you some insight as to the nature of the network infrastructure itself. Now, let's talk about the other major factor in providing QoS, the nature of the applications using the network infrastructure.

Nature of the Application

Networks exist to support applications. Different network applications have different operational requirements that demand different network services. Increased network traffic requires increased network bandwidth capacity, but many of the new applications like IP telephony have other requirements. Increasing the network "pipe-size" will certainly be essential, but bandwidth by itself will not be enough to provide a solution.

The nature of network applications can be characterized in two ways:

▼ How predictable the application data rate is

▲ How tolerant of delay the application is

In Tables 20-1 and 20-2, you will see the terms used to characterize the predictability of application data rates as well as the terms used to characterize the application's sensitivity to delay, respectively.

Over the last few years, the Internet has seen increasing use of applications that rely on the timely, regular delivery of packets. These applications can't tolerate the loss of packets, or the delays caused by waiting in the queues of intermediate routers along the way. These applications tend to be those that work in real-time. Examples of such applications would be things like streaming audio and/or video, interactive audio and/or video, and application sharing, to name a few.

Rate Type	Description
Stream	This represents a predictable delivery at a relatively constant data rate. For example, even though their data rates fluctuate, audio and video data streams are considered relatively constant because they have a quantifiable upper limit.
Burst	This represents an unpredictable delivery of "blocks" of data at a variable data rate. It includes applications like file transfer. File transfers tend to move data in large blocks at a time using all available bandwidth when sending little to no bandwidth between transfers. The nature of this type means that there is no quantifiable upper limit.

Table 20-1. Application data rate predictability

Delay Tolerance	Delivery Type	Description
High	Asynchronous (Elastic)	Delivery time and delay are not important and are often referred to as *"elastic."*
Medium High	Synchronous	Data is time-sensitive, but flexible.
Medium	Interactive	Delays may be noticeable to users/applications, but do not adversely affect usability or functionality.
Medium Low	Isochronous	Time-sensitive to an extent that adversely affects usability.
Low	Mission-Critical	Data delivery delays disable functionality.

Table 20-2.   Application sensitivity to delay

All of these applications require the smooth and timely arrival of information because it needs to be played out as soon as possible after arrival. Streaming video and audio applications can buffer the data in a queue so that there is a delay between the arrival of the packets containing the data about the sound before being sent to the speakers. This buffer allows for the re-transmission of lost data, and the correct re-ordering of data that arrived in the wrong order.

In the case of interactive communications, however, although there is a buffer, it has to be small enough not to interfere with or distort the progress of communications. It is very important that the data that arrives is complete, in order, and received at regular intervals.

Unfortunately, the nature of Internet traffic is bursty and tends not to travel at regular intervals. The amount of congestion and subsequent delay along the route or path between two communicating hosts can vary enormously and unpredictably depending on the number of other applications, users, or services, taking that same path, or any part of it at the same time.

Congestion is caused by routers along the path receiving more data than they are able to process, or routers having to drop packets because the capacity of upstream links is insufficient. If network congestion is occurring during an interactive session, users will start to see a noticeable deterioration in the quality of the video or audio that they are receiving.

The audio stream may become faint, indistinct, choppy, broken, echoed, or unintelligible. The video stream may become stuttering and jerky, slow to refresh, pixilated, lose synchronization with the accompanying sound file, and so on.

For users sharing applications, the time taken to update and redraw screens can slow to the point where the application becomes unusable.

As mentioned earlier in this chapter, IP telephony is today's premier killer application. The market pressure to enable IP telephony has exposed the service deficiencies of the Internet. This pressure has upped the ante to define standards and deploy managed bandwidth on IP networks. IP telephony can be considered a multimedia (audio) application, but its bandwidth requirements are actually relatively modest (about 8Kbps). The difference is that it is a real-time, two-way transmission. In this case, bandwidth is not the issue, latency is.

For IP telephony, and any other real-time (like video conferencing) or two-way application for that matter, it is the timing requirements that are much more significant than the bandwidth requirements. With IP telephony, there is a person at each end of the flow. This arrangement means that they will have immediate and obvious evidence with respect to the quality of the connection or, more importantly, the lack of quality. Dropouts and delays are very noticeable and extremely distracting. Round-trip delivery delays above 0.5 second can make them unusable.

The current lowest acceptable standard for voice usability is the cell phone. As anyone who has used one knows, they are not perfect. Noise and dropped calls are not uncommon. However, latency has never been an issue for the simple reason that it can't be, as explained in the previous paragraph. So despite its short-comings, cell phone service is considered far better than what is typically possible using best effort IP service over the standard Internet.

Routing delays and lost packets caused by transient network congestion, from the unpredictable, bursty nature of network traffic, results in sub-optimal round-trip times on IP networks that severely limit the usability of IP-based telephone service to date.

The quality and reliability of voice traffic on the telephone network throughout its history has placed the bar very high, indeed. Of course, the telephone industry had the luxury of starting off with a single specific application (voice communication), and built a network to suit it.

The Internet, on the other hand, started out in exactly the opposite way. The Internet started with a new network technology and explored, successfully, new applications that were able to use the undefined (*best effort*) service. IP telephony is a tall order, and though the bar has been placed high, it *will* be cleared. The technology that will push it up and over will be QoS, and when that day comes, I wouldn't want to be in the telephone company's shoes (the telephone company doesn't want to be in those shoes either). This is no secret to the telephone industry. That is why AT&T commercials have gone from sappy "reach out and touch someone" slogans to "AT&T Broadband, think of the possibilities."

Increasing bandwidth capacity will improve IP service, so it is an essential first step towards solving the latency issue. However, it is not enough to satisfy telephony application requirements.

Quantifying the Nature of Networks and Application QoS Metrics

This section will detail the metrics by which QoS can be specified, and why some metrics are more suitable than others for different types of applications, as well as examine various techniques and architecture components that support QoS.

As we mentioned earlier in the chapter, it is difficult to find an adequate definition of what QoS actually is. There is a danger that because we wish to use quantitative methods, we might limit the definition of QoS to those aspects of QoS that can be measured and compared. In reality, there are many subjective and perceptual elements to QoS and a lot of time has been spent trying to map those perceptual elements to the quantifiable, especially in the telephony industry.

It is important to remember that ultimately these metrics are just simple guidelines to human encounters and many immeasurable factors can affect a person's perception of what the quality of the session is. New applications are being designed every day and without quantifying the nature of an application's characteristics in an agreed upon way, there can be no QoS standard. Interoperability is the key to successful end-to-end use of applications and we will need to have some consistent standards to accomplish our quest for the "Holy Grail" of convergence.

There is a time delay in capturing, digitizing, and compressing the data in a video frame or sample of voice. There is a delay in creation of the packet of data, examining the packets, and sending them on their way. Assuming sufficient bandwidth, even if nothing else at all was happening on the path between two machines, there would be a slight time delay in the traversal of the network. This is a speed of light delay, but it is a measurable amount of time nevertheless. There is a delay in checking, re-ordering, decompressing, and playing out the sample. In a circuit-switched network, these delays are constant and predictable.

In a packet-switched network like the Internet, it is the unpredictable and inconsistent elements of *best effort* network transmission that lead to compromises in the quality of many multimedia applications, and these are the measurable factors that are used to describe network QoS metrics. In this section, we will identify those metrics and the part they play in the overall picture of QoS.

Bandwidth

Bandwidth is the transmission capacity of a computer channel or communications line. It is usually stated in bits per second (bps), Kilobits per second (Kbps), or Megabits per second (Mbps). The bandwidth of a network is a nominal figure. Fast Ethernet has a bandwidth of 100Mbps, provided nobody is using it. In reality, as data exchange nears the maximum limit (in a shared environment), delays and collisions will mean a drop in quality. Network managers carefully monitor usage figures and available bandwidth and deliberately overprovision bandwidth so that quality of service to users is not compromised, even during bursts of high volume.

The bandwidths of all networks utilized in an end-to-end path need to be considered. Your maximum bandwidth will be the lowest common denominator of all hops to your destination. I can't tell you how many times I have overheard someone saying, "This stupid DSL connection is worthless. Whenever I download stuff, half the time it runs at lightning speeds and the other half is no better than my old modem!" That's because half the time he is downloading a patch from Microsoft and the other half he is downloading MP3 files oblivious to the fact that the server those files are on belongs to some 12-year-old kid who got his PC last Christmas, along with a shiny new 56Kbps modem so he can connect to the Internet.

Data Throughput

Bandwidth, as it compares to *throughput,* is analogous to the "this truck gets 40 miles per gallon on the highway" sticker on that big new four-wheel drive and the "your actual mileage may vary" disclaimer printed in 6-point print at the bottom of that same sticker.

Throughput is the "actual" bandwidth available to your application after factoring in collisions, congestion, etc. This is what determines the amount of data your application can get across the wire to its final destination. In other words, it is a function of the capacity of the network segments in the path between transmitter and receiver, and the number of packets currently traversing each segment, essentially something like (nominal bandwidth/current network traffic).

Delay or Latency

Delay, also known as *latency,* is the time between one device sending a packet and another device receiving it. Latency takes into account delays in a transmission path and delays from any device within the transmission path. These devices might be end stations or intermediate routers.

Some of the different types of delay are

▼ *Propagation Delay* The delay associated with signals traveling on any physical medium. In the case of fiber optics, propagation delay is somewhat more than the speed of light delay which is the theoretical minimum.

■ *Link Speed Delay* Data transfer rate is determined by the bit rate of the link. A fast link will obviously transfer a packet much faster than a slower link, so the slower link introduces a relative delay. Link speed delay is independent of propagation delay and is by far the greater of the two components.

■ *Queuing Delay* Every switch and router has queues, where packets can be stored until capacity is available to transfer them out to the link. It is that time spent in the queues that makes up the queuing delay. Queuing delay accumulates with each device it passes through.

■ *Hop Count* Each time a packet passes through a router, it is considered a hop. Queuing delay grows as hop count increases, so hop count is an important metric to control.

- ■ *Transmission Delay* This is the time it takes to put all bits of a packet onto the wire or data link.

- ▲ *Processing Delay* This is the time it takes a network devices CPU to process a packet arriving on an incoming interface to the proper outgoing interface.

End-to end *latency* is the optimum transmission time, plus the sum of all router propagation delays along the path between the two communicating devices on each end. Delay or *latency* includes the time taken to traverse network, plus propagation delays at each hop. Router *latency* is caused by processing time, plus possible queuing time. This metric is usually expressed in milliseconds. As a rule, the end-to-end *latency* should not exceed 300ms and starts to becomes serious when it is above 500ms.

Inter-stream *latency* is the gap between the arrival of sound and video packets with the same timestamp. In the case of inter-stream *latency*, ideally, audio should not arrive more than 30ms ahead of concurrent video, and no later than 40ms after, however, a variance of 100ms is considered tolerable.

Delay Variation or Jitter

The variation in end-to-end latency is referred to as *jitter*. *Jitter* can be expressed as the distortion of inter-packet arrival times when compared to the inter-packet departure times from the original sending station. Even when packets are sent out at regular intervals, they can still arrive at varying irregular intervals. *Jitter* is the variation in inter-arrival times. This may be due to congestion on the way to its destination and/or packets taking different paths to reach their destination. Extreme *jitter* can lead to packets arriving out of order.

Intra-stream *jitter* refers to the *jitter* between packets in the same media session such as the *jitter* between arriving packets containing a video stream. *Inter*-stream *jitter* is the *jitter* between different media streams such as *jitter* between the audio and video streams. *Jitter* is generally measured in milliseconds, or as a percentage of variation from the average latency of a particular connection.

Jitter is simply the difference (+ or -) of the time between the timestamp, taken at the first octet of data, of two consecutive packets, and the difference of time between the arrival of those timestamps. *Jitter* is illustrated in Figure 20-2.

Jitter can be calculated by subtracting Time_A from Time_B. Increasing *jitter* can be an indication of impending congestion.

Audio applications tend to tolerate *jitter* better than video applications because it uses smaller packets and less bandwidth. In Audio applications, jitter manifests itself as pops and clicks. In video applications, the picture can become, well, *jittery*. It can also cause pixelation and/or color loss.

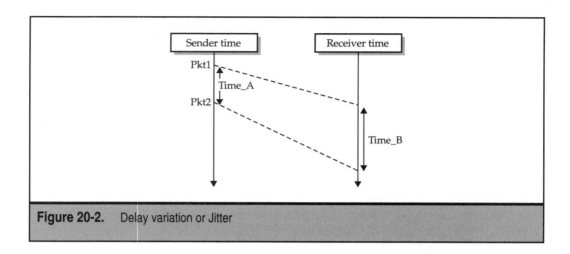

Figure 20-2. Delay variation or Jitter

Packet Loss

Lost (or dropped) packets are a product of insufficient bandwidth on at least one segment of the network path, or of congestion in at least one routing device on the network path.

Some packets may arrive but have been corrupted in transit and are, therefore, unusable. Note that loss is relative to the volume of data being sent and is usually expressed as a percentage of the total amount of data being sent.

In some cases, a high loss percentage can simply mean that the application is trying to send too much information and is overwhelming the available bandwidth. All of these factors have always existed for all data transmission over the Internet. With normal *best effort* traffic, they simply don't matter, or can be handled by re-transmissions, as they are not time sensitive.

Again, audio applications have a higher tolerance than video applications. In this case, both audio and video will experience distortion and appear choppy with a kind of stuttering playback.

The effects of packet loss will start to become noticeable on audio applications at around four to six percent. Once it reaches about ten percent and above, the audio will become unusable. Video tends to vary more because of all the anomalies based on movement, color, and compression schemes. To give you some perspective, the effects of packet loss can be seen at as little as one percent, but again, depending on the context and in some cases, far greater amounts of loss can be tolerated.

Traffic Categories

The original TCP and IP protocols were designed for applications such as e-mail distribution, basic file transfer, and remote terminal access, none of which were particularly time or bandwidth sensitive. The primary goal was to establish connectivity across a wide variety of systems and devices using a robust, inexpensive network.

The idea at the time was to assume that the network would do its best to deliver all of the data that was submitted but with no guarantees of delivery or sequencing. The original designers of the Internet never imagined that it would become the phenomenon it is today. If you had gone up to them ten years ago and told them their little project might make the might phone company a thing of the past, they would have laughed in your face. Even though the Internet is still not completely ready to accommodate voice and video traffic, I can assure you that no one in the phone company is laughing now. Once enough QoS mechanisms have been standardized and put in place, the first essential step in the transition to the new, converged network infrastructure will be complete.

Multimedia traffic is difficult to handle efficiently by, its very nature, especially in environments where resources are limited. Table 20-3 shows the different traffic categories and their sensitivities. The quantity and variation of traffic traveling over the Internet has become harder and harder to predict, making an exact matching of available capacity to actual demand impossible.

Type of Traffic	Bandwidth	Typical Style	Latency Sensitivity	Jitter Sensitivity	Loss Sensitivity
Bulk Data Transfer	10-100 Mbps	Periodic, two-party	Low	None	Low
Transaction Data	<1 Mbps	Bursty, two-party	Moderate	None	None
Voice and Facsimile	8-64 Kbps	Variable, two-party multi-party	High	High	Low
Multimedia (voice plus image)	Up to 384 Kbps for video	Variable, two-party or multi-party	High	Moderate	Low
Video on Demand & Streaming	28.8 Kbps - 1.5 Mbps	Variable, multi-party	Low	Low	Low

Table 20-3.   Traffic Categories

Performance degradation in a "best effort" network, assuming an adequate basic design, is largely the result of congestion. Congestion occurs when the output ports of a router are unable to handle all the traffic that arrives at the input ports or when the router cannot process the traffic at wire speed.

LANs are not immune to this either. In shared media networks such as Ethernet, traffic collisions can also increase delay. Any form of buffering or backoff/retry algorithms result in some amount of delay in the traffic flow. When the amount of buffering varies, then the amount of delay will also vary, leaving you with *jitter*. Also, buffer capacities are limited and packets may eventually have to be dropped, leaving you with *packet loss*. The terms *jitter* and *packet loss* are not the kind of words you want to use in the same sentence as IP telephony or video. Minimizing the effects of congestion is the main focus of QoS management.

Resource Allocation in Network Devices

Devices that provide QoS support do so by intelligently allocating resources to submitted traffic. For example, under congestion, a network device might choose to queue the traffic of applications that are more latency-tolerant instead of the traffic of applications that are less latency-tolerant.

As a result, the traffic of applications that are less latency-tolerant can be forwarded immediately to the next network device. So in this case, the network device would provide *interface capacity* as a resource for the latency-intolerant traffic, and provide *device memory* as a resource for the latency-tolerant traffic.

In order to choose which resources will be provided for these different kinds of traffic, the network device needs a way to identify the different traffic and to associate it with certain resources. This is done in the following manner:

1. Traffic arriving at network devices is identified in each device and is separated into distinct *flows* using a process called *packet classification*.
2. Traffic from each flow is then directed into a corresponding *queue*.
3. The queues are then *serviced* according to some *queue-servicing algorithm*.
4. The queue-servicing algorithm determines the rate at which traffic from each queue is to be sent onto the network.

These are the steps that determine which resources will be provided to each queue and the flows in that queue. This means that in order to provide network QoS, it is necessary to provide the following in network devices:

▼ Packet classification information by which devices separate traffic into flows.

▲ Queue-servicing algorithms that handle traffic from the separate flows.

We will refer to these jointly as *traffic handling mechanisms*. These traffic-handling mechanisms need to be configured in a way that provides useful end-to-end services across a network.

Any of the QoS technologies that we will discuss will be one of these categories:

▼   Traffic Handling Mechanism

▲   Configuration Mechanism

THE QoS FRAMEWORK

We have seen that there is more than one way to characterize Quality of Service (QoS). Generally speaking, QoS is the ability of a network (this could be an application, a host or a router) to provide some level of consistent network data delivery. We have also seen that some application characteristics are different in nature from others when it comes to their QoS requirements. To address these differences, we have two basic types of specific QoS frameworks available:

▼   Prioritization (*differentiated services*): Network traffic is classified and apportioned network resources according to bandwidth management policy criteria.

▲   Resource reservation (*integrated services*): Network resources are apportioned according to an application's QoS request and subject to bandwidth management policy.

To enable QoS, network elements will give preferential treatment to classifications identified as having more demanding requirements.

These types of QoS can be applied to individual application "flows" or to flow "aggregates." This means that there are two other ways to characterize types of QoS:

▼   Per *Flow*: A *flow* is defined as an individual, uni-directional, data stream between two applications (sender and receiver), uniquely identified by a 5-tuple, namely, Transport Protocol, Source Address, Source Port number, Destination Address, and Destination Port number).

▲   Per *Aggregate*: An *aggregate* is simply two or more flows. Generally, these flows will have something in common such as any one or more of the 5-tuple parameters, a label or a priority number, or perhaps some authentication information.

As a rule, the *integrated service* or *Intserv* mechanisms are provided on a per-flow basis. In aggregate *traffic handling mechanisms*, some set of traffic, from multiple flows, is classified to the same flow and is handled in aggregate. Aggregate classifiers generally look at some aggregate identifier in packet headers. *Differentiated Service* or *Diffserv* and 802.1p are examples of aggregate traffic handling mechanisms at Layer 3 and at Layer 2, respectively. In both these mechanisms, packets corresponding to multiple flows are marked with the same DSCP or 802.1p mark.

When traffic is handled on a per-flow basis, resources are allotted on a per-flow basis. From the application perspective, this means that the application's traffic is granted resources completely independent of the effects of traffic from other flows in the network.

Intserv and Diffserv are complementary technologies and work best when combined in the pursuit of end-to-end QoS. Together, these mechanisms can facilitate deployment of applications such as IP-telephony, video-on-demand, and various non-multimedia mission-critical applications. Intserv enables per-flow requests and Diffserv enables scalability across large networks.

While this tends to enhance the quality of the service experienced by the application, it also imposes a burden on the network equipment. Network equipment is required to keep independent state information for each flow and to apply independent processing for each flow. In the core of large networks, where there can be millions of flows simultaneously, per-flow traffic handling is probably not a good idea. When traffic is handled in aggregate, the state maintenance and processing burden on devices in the core of a large network is reduced significantly.

That brings us to the fundamental principle of "Leave the complexity at the 'edges' of the network and keep the 'core' of the network simple". That is the QoS designers credo (well, maybe not their credo…how 'bout mantra?).

QoS PROTOCOLS

A number of QoS protocols have evolved to satisfy the variety of application needs. We will describe these protocols individually, and then describe how they fit together in various architectures with the end-to-end principle in mind. Keep in mind that all of the protocols as they relate to QoS are anything but mutually exclusive. Invariably the best solution for any QoS architecture will involve the combination of one or more of these protocols. The important thing here is not to focus on the individual QoS protocols, but instead how they are used together to enable end-to-end QoS.

Traffic Handling Mechanisms

We will now discuss some of the more significant *traffic handling mechanisms*. Note that, and this is important, underlying any *traffic handling mechanism* is a set of queues and the protocols for servicing these queues.

▼ *Differentiated Service (Diffserv)*

■ *Integrated Services (Intserv)*

■ *802.1p*

▲ *ATM*

Each of these *traffic handling mechanisms* is appropriate for specific media or circumstances and is described in detail below.

802.1p

Most LANs are based on some form of IEEE 802 technology. This includes Ethernet, Token-Ring, and other variations of shared media networks. 802.1p is a *traffic handling mechanism* for supporting QoS in these networks.

802.1p defines a field in the Layer 2 header of 802 packets that can carry one of eight priority values. Typically, hosts or routers sending traffic into a LAN will mark each transmitted packet with the appropriate priority value. LAN devices, such as switches, bridges, and hubs, are expected to treat the packets accordingly (by making use of underlying queuing mechanisms). The scope of the 802.1p priority mark is limited to the LAN and only the VLAN trunks that support 802.1Q. Once packets are carried off the LAN through a Layer 3 device, the 802.1p priority is removed.

IP Precedence

Traditionally, the Internet has treated all data packets equally in that no preference was afforded one packet over another. That meant that whenever an Internet router senses that its queues were coming too close to the overflow point, it treated all of the packets equally. That treatment was applied using a Random Early Discard (RED) policy that would literally drop packets at random.

That isn't to say that there was no mechanism in place capable of prioritizing those packets, it's just that virtually all of these Internet routers simply ignored this mechanism. Ever since the first specification of the Internet Protocol (RFC 791), the mechanism available was in the Type of Service (TOS) octet in the second byte of the IP header. You can see an example of the TOS and Diffserv headers in Figure 20-3.

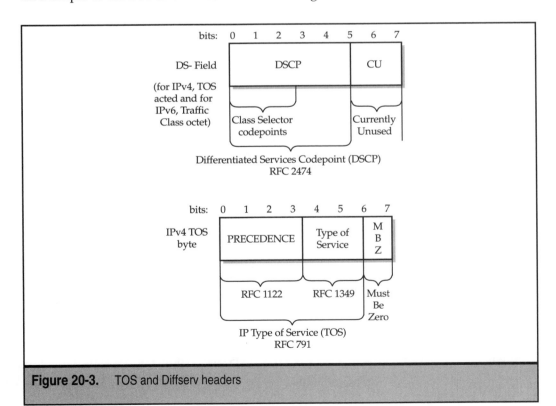

Figure 20-3.   TOS and Diffserv headers

This specification allowed for this byte to be subdivided into two fields, the first three bits (0,1,& 2) specifying IP Precedence and the next four bits specifying the treatment that the packet should be given as it traversed a network. The unused bit in this octet must be set to 0 in Ipv4. Being only three bits, the IP precedence field could specify eight discrete states. RFC 791 defines the bit patterns of the first three bits as follows:

▼ 000 – Routine

■ 001 – Priority

■ 010 – Immediate

■ 011 – Flash

■ 100 - Flash Override

■ 101 - CRITIC/ECP

■ 110 - Internetwork Control

▲ 111 - Network Control

The highest two values are reserved for network test and diagnostic use. IP Precedence uses bits in the 4-bit field to specify requests for

▼ Minimum Delay

■ Maximum Throughput

■ Maximum Reliability

▲ Minimum Cost

For many years, the IP precedence field was not used a great deal and was rarely implemented. RFC 791 specified the protocol in 1981, but this was generally not utilized either by applications, operating systems, or routers.

RFC 1812 is dated June 1995. It specified the "Requirements for IP Version 4 Routers" and included details of how routers should implement, and treat, the IP Precedence bits. It notes, "The basic mechanisms for precedence processing in a router are preferential resource allocation, including both precedence-ordered queue service and precedence-based congestion control, and selection of Link Layer priority features." It recommends that, "Routers should implement precedence-ordered queue service." Again, there was no widespread implementation.

Differentiated Services (Diffserv)

Differentiated services (DiffServ) has been proposed by the IETF with scalability as the main goal. DiffServ is a per-aggregate-class based service discrimination framework using packet tagging.

The IETF re-defined the TOS octet, and this redefinition supports the work of the Differentiated Services Work Group of the IETF. The TOS octet is now known as the Differentiated Services Field (DS Field), as defined in RFC 2474, "Definition of the Differentiated Services Field (DS Field) in the IPv4 and IPv6 Headers." This redefined the DS Field so that the first six bits (0-5) indicate the Differentiated Services Code Point (DSCP), with bits 6 and 7 currently unused (CU). You can refer to Figure 20-3 to see what we mean.

The primary goal of differentiated services is to allow different levels of service to be provided for traffic streams on a common network infrastructure. A variety of techniques may be used to achieve this goal, but the end result will be that some packets receive preferential treatment over others.

Routers within the Diffserv network use the DSCP to classify packets and apply specific queuing or scheduling behavior known as a per-hop behavior (PHB) based on the results of the classification.

The interpretation of the DSCP field is currently being standardized by the IETF. DiffServ uses DSCP to select the per-hop behavior (PHB) a packet experiences at each node. A PHB is an externally observable packet forwarding treatment which is usually specified in a relative format compared to other PHBs, such as relative weight for sharing bandwidth or relative priority for dropping. The mapping of DSCPs to PHBs at each node is not fixed. Before a packet enters a DiffServ domain, its DSCP field is marked by the end-host or the first-hop router according to the service quality the packet is required and entitled to receive.

Within the DiffServ domain, each router only needs to look at DSCP to decide the proper treatment for the packet. No complex classification or per-flow state is needed. DiffServ has two important design principles:

▼ Pushing complexity to the network boundaries

▲ Separation of policy and supporting mechanisms

The network boundary refers to application hosts, leaf (or first hop) routers, and edge routers. Since a network boundary has a relatively small number of flows, it can perform operations at a fine granularity, such as complex packet classification and traffic conditioning.

In contrast, a network core router may have a larger number of flows, it should perform fast and simple operations. The differentiation of network boundary and core routers is vital for the scalability of DiffServ. The separation of control policy and supporting mechanisms allows these to evolve independently.

DiffServ only defines several per-hop packet forwarding behaviors (PHBs) as the basic building blocks for QoS provisioning, and leaves the control policy as an issue for further work. The control policy can be changed as needed, but the supporting PHBs should be kept relatively stable. The separation of these two components is key for the flexibility of DiffServ.

A similar example is Internet routing. It has very simple and stable forwarding operations, while the construction of routing tables is complex and may be performed by a variety of different protocols. (This often reflects a software/hardware split, where PHBs are implemented in hardware, while the control policy is implemented in software.)

Currently, DiffServ provides two service models beside best effort:

▼ Premium Service

▲ Assured Service

Premium service is a guaranteed peak rate service that is optimized for very regular traffic patterns and offers little or no queuing delay. This model can provide absolute QoS assurance. One example of using it is to create "virtual leased lines," with the purpose of saving the cost of building and maintaining a separate network.

Assured service is based on statistical provisioning. It tags packets as In or Out according to their service profiles. In packets are unlikely to be dropped, while Out packets are dropped first if needed. This service provides a relative QoS assurance. It can be used to build "Olympic Service" which has gold, silver, and bronze service levels.

Integrated Services (Intserv)

Intserv is a service framework. There are two services defined within this framework. They are

▼ *Guaranteed Service*: Guaranteed Service promises to carry a certain traffic volume with a quantifiably, bounded latency.

▲ *Controlled Load Service*: Controlled Load Service agrees to carry a certain traffic volume with the "appearance of a lightly loaded network."

These are quantifiable services in the sense that they are defined to provide quantifiable QoS to a specific quantity of traffic. There are certain Diffserv services by comparison which may not be quantifiable.

Intserv services are typically (but not necessarily) associated with the RSVP signaling protocol. Each of the Intserv services define admission control algorithms which determine how much traffic can be admitted to an Intserv service class at a particular network device, without compromising the quality of the service. Intserv services do not define the underlying queuing algorithms to be used in providing the service.

ATM

You should recognize this term from Chapter 2. ATM is a Link layer technology that was tailor-made to offer every kind of QoS need your heart desires. ATM fragments packets into Link layer cells, which are then queued and serviced using queue servicing algorithms appropriate for the particular ATM service. ATM traffic is carried on virtual circuits (VC) that support one of the numerous ATM services. Some of these include:

▼ Constant Bit Rate (CBR)

■ Variable Bit Rate (VBR)

■ Real time VBR and non-real time VBR

■ Unknown Bit Rate (UBR)

▲ Available Bit Rate (ABR)

ATM actually goes beyond a strict *traffic handling mechanism* in the sense that it includes a low level signaling protocol that can be used to set up and tear down ATM VCs.

Because ATM fragments packets into relatively small cells, it can offer very low latency service. If it is necessary to transmit a packet urgently, the ATM interface can always be cleared for transmission in the time it takes to transmit one cell.

That might not sound like a big deal, but it would if you consider sending normal TCP/IP data traffic on slow modem links without the benefit of the ATM Link layer. A typical 1500-byte packet, once submitted for transmission on a 28.8 Kbps modem link, will occupy the link for about 400 msec until it is completely transmitted (preventing the transmission of any other packets on the same link).

Integrated Services Over Slow Link Layers (ISSLOW) addresses this problem. ISSLOW is a technique for fragmenting IP packets at the Link layer for transmission over slow links such that the fragments never occupy the link for longer than some threshold.

Configuration Mechanisms

In order to be effective in providing network QoS, it is necessary to effect the *configuration* of the *traffic handling mechanisms* described consistently across multiple network devices. *Configuration mechanisms* include:

▼ *Resource Reservation Protocol (RSVP)*

■ *Subnet Bandwidth Manager (SBM)*

■ *Policy mechanisms and protocols*

▲ *Management tools and protocols*

Top-Down vs. Signaled Mechanisms

It is important to note the distinction between *top-down* QoS configuration mechanisms and *signaled* QoS configuration mechanisms.

Top-down mechanisms typically "push" configuration information from a management console down to network devices.

Signaled mechanisms typically carry QoS requests and implicit configuration requests from one end of the network to the other, along the same path traversed by the data that requires QoS resources.

Top-down configuration is typically initiated on behalf of one or more applications by a network management program.

Signaled configuration is typically initiated by an application's changes in resource demands.

RSVP

RSVP is a *signaled* QoS configuration mechanism. It is a protocol by which applications can request end-to-end, per-flow QoS from the network, and can indicate QoS requirements and capabilities to peer applications. RSVP is suited primarily for use with IP traffic.

RSVP uses Intserv semantics to convey per-flow QoS requests to the network. A lot of people think (and understandably so) that RSVP is Intserv and vice versa. It is important to note that RSVP is limited to neither per-flow usage, nor to Intserv semantics.

In fact, extensions to RSVP enable it to be used to signal information regarding traffic aggregates. Other extensions enable it to be used to signal requirements for services beyond the traditional guaranteed and controlled load Intserv services.

RSVP is largely independent of the various underlying network media over which it operates and RSVP can be considered an abstraction layer between applications (or host operating systems) and media specific QoS mechanisms.

RSVP Messages   There are two significant RSVP messages:

▼ *PATH*

▲ *RESV*

You can get an idea of how these two messages work in Figure 20-4. Transmitting applications will send *PATH* messages towards the receivers. These messages describe the data that will be transmitted and follows the path that the data will take.

Receivers send *RESV* messages. The *RESV* messages follow the path seeded by the incoming *PATH* messages, back towards the senders, indicating the profile of traffic that particular receivers are interested in. In the case of multicast traffic flows, *RESV* messages from multiple receivers are "merged," making RSVP suitable for QoS with multicast traffic.

RSVP messages carry the following information:

▼ How the network can identify traffic on a flow (classification information)

■ Quantitative parameters describing the traffic on the flow (data rate, etc.)

■ The service type required from the network for the flow's traffic

▲ Policy information (identifying the user requesting resources for the traffic and the application to which it corresponds)

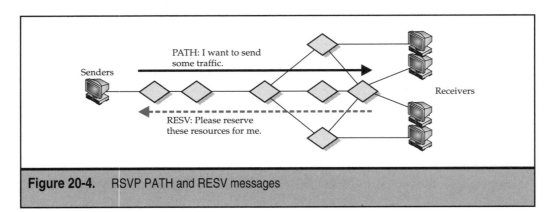

Figure 20-4.   RSVP PATH and RESV messages

Classification information is conveyed using IP source and destination addresses and ports. In the conventional Intserv use of RSVP, an Intserv service type is specified and quantitative traffic parameters are expressed using a token-bucket model. Policy information is typically a secure means for identifying the user and/or the application requesting resources. Network administrators use policy information to decide whether or not to allocate resources to a flow.

How RSVP Works   *PATH* messages wind their way through all network devices en route from sender to receivers. RSVP aware devices in the data path note the messages and establish state for the flow described by the message. (Other devices pass the messages through transparently).

When a *PATH* message arrives at a receiver, the receiver responds with a *RESV* message (if the receiving application is interested in the traffic flow offered by the sender). The *RESV* message winds its way back towards the sender, following the path established by the incoming *PATH* messages. As the *RESV* message progresses toward the sender, RSVP aware devices verify that they have the resources necessary to meet the QoS requirements requested. If a device can accommodate the resource request, it installs a classification state corresponding to the flow and allocates resources for the flow. The device then allows the *RESV* message to progress on up toward the sender. If a device cannot accommodate the resource request, the *RESV* message is rejected and a rejection is sent back to the receiver.

In addition, RSVP aware devices in the data path may extract policy information from *PATH* messages and/or *RESV* messages for verification against network policies. Devices may reject resource requests based on the results of these policy checks by preventing the message from continuing on its path and sending a rejection message.

When requests are not rejected for either resource availability or policy reasons, the incident *PATH* message is carried from sender to receiver, and a *RESV* message is carried in return. In this case, a reservation is said to be installed. An installed reservation indicates that RSVP aware devices in the traffic path have committed the requested resources to the appropriate flow and are prepared to allocate these resources to traffic belonging to the flow. This process of approving or rejecting RSVP messages is known as admission control and is a key QoS concept.

The Subnet Bandwidth Manager (SBM)

The SBM is based on an enhancement to the RSVP protocol, which extends its features to shared networks. In shared subnetworks or LANs (which may include a number of hosts and/or routers interconnected by a switch or hub), standard RSVP falls short. The problem arises because RSVP messages may pass through Layer 2 (RSVP unaware) devices in the shared network, implicitly admitting flows that require shared network resources. RSVP aware hosts and routers admit or reject flows based on availability of their private resources, but not based on availability of shared resources. As a result, RSVP requests destined for hosts on the shared subnet may result in the over commitment of resources in the shared subnet.

Basically using SBM, best-effort traffic generated from present day LANs relies on admission control and bandwidth management from SBM server to meet QoS specifications RFC2814.

The SBM solves this problem by enabling intelligent devices that reside on the shared network to volunteer their services as a "broker" for the shared network's resources. Eligible devices are (in increasing order of suitability)

▼ Attached SBM capable hosts

■ Attached SBM capable routers

▲ SBM capable switches which comprise the shared network

These devices automatically run an election protocol that results in the most suitable device(s) being appointed Designated SBMs (DSBM). When eligible switches participate in the election, they subdivide the shared network between themselves based on the Layer 2 network topology. Hosts and routers that send into the shared network discover the closest DSBM and route RSVP messages through the device. Thus, the DSBM sees all messages that will affect resources in the shared subnet and provides admission control on behalf of the subnet.

Policy Mechanisms and Protocols

Network administrators configure QoS mechanisms subject to certain policies. Policies determine which applications and users are entitled to varying amounts of resources in different parts of the network.

Policy components include:

▼ *Policy Data Store*: This contains the policy data itself. Things like user names, applications, and the network resources to which they're entitled.

■ *Policy decision points (PDP)*: These translate network wide higher layer policies into specific configuration information for individual network devices. PDPs also inspect resource requests carried in RSVP messages and accept or reject them based on a comparison against policy data.

■ *Policy enforcement points (PEP)*: PEPs act on the decisions made by PDPs. These are typically network devices that either do or do not grant resources to arriving traffic.

▲ Protocols between the data-store, PDPs, and PEPs.

Policy Data Store - Directory Services

Policy mechanisms rely on a set of data describing how resources in various parts of the network can be allocated to traffic that is associated with specific users and/or applications.

Policy *schemas* define the format of this information. Two general types of schemas are required:

▼ One that describes the resources that should be allocated in a *top-down* mechanism

▲ One that describes the resources that can be configured using end-to-end *signaling* mechanism

This information tends to be relatively static and (at least in part) needs to be distributed across the network. As a result, directories tend to be suitable data stores.

Policy Decision Points (PDP) and Policy Enforcement Points (PEP)

Policy Decision Points (PDP) interpret data stored in the *schemas* and control Policy Enforcement Points (PEP) accordingly. PEPs are the switches and routers through which the traffic passes. These devices have the ultimate control over which traffic is allocated resources and which is not.

In the case of *top-down* provisioned QoS, the PDP "pushes" policy information to PEPs in the form of classification information (IP addresses and ports), and the resources to which the classified packets are entitled.

In the case of *signaled* QoS, RSVP messages transit through the network along the data path. When an RSVP message arrives at a PEP, the device extracts a policy element from the message, as well as a description of the service type required and the traffic profile. The policy element generally contains authenticated user and/or application identification. The router then passes the relevant information from the RSVP message to the PDP for comparison of the resources requested against those allowable for the user and/or application (per policy in the data store). The PDP makes a decision regarding the admissibility of the resource request and returns an approval or denial to the PEP.

In certain cases, the PEP and the PDP can be in the same network device. In other cases, the PDP may be separated from the PEP in the form of a policy server. A single policy server may reside between the directory and multiple PEPs. Although many policy decisions can be made trivially by co-locating the PDP and the PEP, certain advantages can be realized by the use of a policy server.

Use of Policy Protocols

When RSVP messages transit RSVP aware network devices, they cause the configuration of traffic handling mechanisms in PEPs, including classifiers and queuing mechanisms that provide Intserv services. However, in many cases, RSVP cannot be used to configure these mechanisms. Instead, more traditional, *top-down* mechanisms must be used.

These protocols include Simple Network Management Protocol (SNMP), command line interface (CLI), Common Open Protocol Services (COPS), and others.

SNMP has been in use for many years, primarily for the purpose of monitoring network device functionality from a central console. It can also be used to set or configure device functionality.

CLI is a protocol used initially to configure and monitor Cisco network equipment. Due to its popularity, a number of other network vendors provide CLI type configuration interfaces to their equipment.

COPS is a protocol that has been developed in recent years in the context of QoS. It was initially targeted as an RSVP related policy protocol but has recently been pressed into service as a general Diffserv configuration protocol.

All these protocols are considered *top-down* because, traditionally, a higher level management console uses them to push configuration information down to a set of network devices.

In the case of *signaled* QoS (as opposed to *top-down* QoS), detailed configuration information is generally carried to the PEP in the form of RSVP signaling messages. However, the PEP must outsource the decision whether or not to honor the configuration request to the PDP. COPS was initially developed to pass the relevant information contained in the RSVP message from the PEP to the PDP, and to pass a policy decision in response. Obviously, when PEP and PDP are co-located, no such protocol is required.

A protocol is also required for communication between the PDP and the policy data store. Since the data store tends to take the form of a distributed directory, LDAP is commonly used for this purpose.

TRAFFIC ENGINEERING

Working hand in hand with the technologies and protocols that were outlined in the last section are the traffic shaping and conditioning mechanisms which can be deployed in network nodes in order to deliver priority services. This section looks briefly at some of the main algorithms and mechanisms that are used on a per-hop basis in order to deliver QoS solutions. Some of these approaches are for the purpose of limiting, smoothing, or shaping the traffic flows in some way; others provide a means of distributing resources most efficiently for the traffic patterns and services which are being supported on the network. Some of these mechanisms allow fine tuning of bandwidth allocation or router resources, based on the QoS requirements of each data packet's contents.

Committed Access Rate (CAR)

CAR allows the manager to limit the amount of bandwidth given to a particular class or type of traffic entering or leaving the network. Packets will be dropped if the limit for that type of traffic is crossed. This ensures adequate bandwidth on that interface for other types of traffic. CAR also allows packets to be classified (with IP precedence) on entering a network.

First In First Out (FIFO)

FIFO is the simplest form of queue, generally considered the default behavior for a router, and requires the least computational overhead of the queuing techniques. It is sometimes known as First Come First Served (FCFS).

Priority Queuing (PQ)

PQ Packets are shuffled towards the front of a FIFO queue depending on some prede-fined criteria. This approach has high overheads (in packet examination and queue re-shuffling) and is now considered far less efficient than some of the alternatives de-scribed in this section.

Class Based Queuing (CBQ)

A more sophisticated version of Priority Queuing in which a number of queues are main-tained, and system resources are spread out between them. The manager defines the cri-teria for each queue, the amount of preference, or weighting, each queue should be given, and the amount of traffic drained from each queue at each servicing. CBQ provides a weighted (but fair) distribution of resources.

Random Early Detection (RED)

RED is a queuing mechanism that monitors queue length and drops packets randomly as the queue grows. The closer to the maximum length, the more packets that are dropped. This mechanism relies on the Transmission Control Protocol (TCP) reacting to the dropped packets in a coherent and predictable manner by reducing the frequency of packets that it is sending to the router(s) that is experiencing congestion. User Datagram Protocol (UDP) packets will be lost when dropped (because of the lack of a control proto-col), and this will include those containing RTP packets with accompanying audio/video data. The RED algorithm has to drop packets in a random manner in order to prevent a synchronized response from affected end-stations. There is no traffic re-ordering or queue management, making this an efficient and fair algorithm.

Weighted Fair Queuing (WFQ)

Rather than have a simple universal FIFO queue, weighted fair queuing sets up multiple single-flow based queues, then services these in turn. WFQ tends to give low volume flows preferential treatment in order to stop them from being squeezed out of system re-sources by higher volume queues. The precise nature of the weighting is vendor specific, and may use the IP Precedence bits in order to weight the queues. The higher the bit value, the higher the amount of queue servicing resources given to that particular flow.

Class Based Weighted Fair Queuing (CBWFQ)

CBWFQ extends the WFQ mechanism to allow for user defined traffic classes. Allows specification of capped amounts of bandwidth for particular classes of traffic.

Weighted Random Early Detection (WRED)

WRED introduces an element of unfairness into the RED approach. When it is necessary to drop packets in order to avoid the buildup of congestion, IP precedence (set on ingress to the network) is examined. The higher the precedence, the lower the drop probability.

The lower the precedence of the packet, the higher the probability of it being dropped. When there is no congestion control necessary, packets are treated with parity.

Round Robin

A number of queues are maintained, and each of these is serviced in turn, in a fair or weighted fashion.

Deficit Round Robin (DRR)

Each out flowing queue is serviced as per Round Robin, but with the addition of a quantum. A quantum is a credit for service by the router for each queue. The queue is serviced until the credit reaches 0 or less. Any unused credit is held until the next pass. This approach has a low computational overhead and is simple to implement. It allows for different flow characteristics (packet size, burstiness, greed) that might otherwise lead to unfairness.

Modified Deficit Round Robin (MDRR)

Same as DRR but with the addition of a Low Latency High Priority (LLHP) queue which gets preferential treatment. The LLHP is inspected between each other queue as the round-robin progresses. If the LLHP has packets, they can be either serviced until there are no more packets on that queue (strict priority), or for a configurable number of bytes (alternate priority). Strict priority ensures a better QoS for LLHP packets at the expense of best effort traffic, the latter is fairer to best effort traffic but at the expense of the possibility of increased latency in the LLHP stream. Each queue can have WRED configured individually.

Generic Traffic Shaping (GTS)

GTS is a traffic shaping mechanism to control the flow of outward bound traffic on a particular interface. Traffic (or sub-sets of the traffic adhering to a particular profile) flowing downstream can be limited, smoothed, or shaped to meet downstream requirements.

SUMMARY

In this chapter we learned the concept of *Quality of Service* (QoS). We then covered the different QoS protocols used to govern how data travels over individual networks or any number of networks up to and including the global Internet. In the next chapter, we will introduce you to Traffic Engineering and how we not only decide what services and data get preferential treatment, but which paths they should take.

CHAPTER 21

Traffic Engineering—
Movement of Data

In the last chapter, we talked about the concept of *Quality of Service* (QoS). We then talked about the different QoS methods and why it is necessary and so important that QoS be integrated into our future networks and the Internet. We finished by providing you a very high-level view of the many standards and protocols used to accomplish QoS.

In this chapter, we will be talking about traffic engineering. Traffic engineering is a kind of "traffic cop" for data; it can tell it where to go and how to get there. More specifically, *traffic engineering* is the method of mapping traffic flows onto an existing physical topology.

ROUTING TO SWITCHING TO ROUTING?

If the title of this section sounds confusing, it is. The best way to learn about traffic engineering is to examine just how all of the ISPs had to learn about it. ISPs started their traffic engineering methods with routing, merged to switching, and are now heading back to routing. All of this will be made less confusing very soon. Before we get to that, let's talk about what has become even more confusing, the definitions of *switching* and *routing*.

Is That a Switch or Is It a Router?

By now, we've all been exposed to terms like *routing* and *switching* and the *routers* and *switches* that accompany them. We have also talked about how the companies that make these devices have blurred many of the terms, as they relate to the OSI model, that describe their products. But the beating that the terms *routing* and *switching* have gone through has never been more evident. There have been many advertisements for "Layer 3 switches," "switches that can route," and so on. This has led to a never-ending cycle of debate among systems engineers and to a trail of confusion for the rest of us.

Instead of feeding the flames of that fire, in this chapter, we will instead talk about the way data is actually moved from one point to another. After all, this is really what routing and switching devices do—move data. This is also the point where you become acquainted with the subject of this final section of the book: MPLS. You will start to see why MPLS is one of the biggest ideas for the Internet that has been created.

"Routing" is a term loosely used to describe the actions taken by the network to move packets through it. Many times you may hear of packets being "routed" from point A to point B, or of packets being routed "over" a network or "through" the Internet. There may be many routers in a network connected in some arbitrary fashion. Packets progress through the network by being sent from one machine to another toward their destination.

Technically, (maybe we should say, "historically") routing is a Layer 3 function. As such, its job is to get the data being sent from the originating computer on one network to the final destination computer on another network, regardless of how many other networks it has to pass through to get there.

Technically, "switching" is a Layer 2 function and, just like routing, its job is to also get the data being sent from the originating computer to the final destination computer. The difference is that those two computers need to be on the same network or data link, hence, the name Data Link layer.

For switches and routers to do these tasks, they need to unwrap the packet to see what the address of the destination computer is. Unfortunately for the router, it has to unwrap that same packet even further. Imagine two kids in a race to open their presents, and some mean parent wrapped his kid's present twice as a joke (hmmm, I'll have to remember that one…I'm mean to my kids, just ask 'em). The kid who had to unwrap his present twice would lose. The same principle holds true for switches and routers—at least it *did*. (We'll get into *did* soon.)

Routing protocols enable each machine to understand which other machine is the "next hop" that a packet should take toward its destination. Routers use the routing protocols to construct routing tables. Some of the more common routing protocols you may have heard of are

▼ Routing Information Protocol (RIP)

■ Open Shortest Path First (OSPF)

■ Interior Gateway Protocol (IGP)

▲ Border Gateway Protocol (BGP)

Whenever a router receives a packet and has to make a forwarding decision, the router will look at its routing table using the Destination IP address to determine the best interface to forward the packet to on to its "next hop" along the way to its next destination. The construction of the tables and the process of checking the table to decide which interface to send the packet are separate logical operations.

The best analogy might be comparing the routing process to those of a major airline. Let's say you are in Boston. You go to a counter and tell them you want to go to San Francisco. The salesperson looks at a list, gives you a ticket, and tells you to go and have a seat over in "Gate 29, Houston." You will wait there until your plane arrives and then walk through the "jet-bridge" into the plane.

In our analogy:

▼ The "airline" would be the *ISP* (humor me).

■ "You with a ticket to San Francisco" would be the *packet*.

■ The "list" would be the *routing table*.

■ The "salesperson" would be the *router*.

■ "Gate 29, Houston" would be the *queue*.

■ The "jet-bridge" at Gate 29 would be the *interface*.

▲ The "plane" would be the *next hop*.

There are a couple of other things in our analogy that relate to the routing process. When you purchased your ticket to San Francisco, the salesperson had to make a decision (the *algorithm*). The salesperson thinks that you can't fly to San Francisco from Boston. There is a list (the *routing table*) that says they only fly to Houston, Atlanta, Chicago, and Denver. The salesperson knows that planes in Chicago and Houston fly to San Francisco, so she sells you a ticket to San Francisco via Houston, because she knows Houston is a 1000 miles closer to San Francisco than Chicago (the *metric*).

Hey! What if the planes in Chicago are ten times as fast? What if from Houston you had to fly to Las Vegas and then fly to San Francisco? Those are different metrics that the salesperson could use when deciding what ticket to sell. Each routing algorithm has its own way of reasoning which way is best. If Boston could send you to any city in the world, but only had planes that went to the four cities in our example, the reality is that the only thing that the salesperson has to decide is which of those four cities she thinks is the best way to go.

If you were an airline, you would want to keep customers happy and sustain a high rate of growth. You do that by getting the passenger to his destination as fast and with as few inconveniences as possible. After all, what could possibly be more important than "Quality of Service"? The main challenge facing ISPs is keeping customers happy and sustaining high rates of growth as well.

TRAFFIC ENGINEERING 101

At a fundamental level, meeting these challenges requires that an ISP provision a number of circuits of various bandwidths over a geographic area. In other words, the ISP must deploy a physical topology that meets the needs of the customers connected to its network.

After the network is deployed, the ISP must map customer traffic flows onto the physical topology. In the early 1990s, mapping traffic flows onto a physical topology (traffic engineering) was not approached in a particularly scientific way. Instead, the mapping occurred as a byproduct of the technology of the time. Back then, they approached it based on what the routing protocols thought was best. That meant the traffic flows simply followed the shortest path calculated by the ISP's Interior Gateway Protocol (IGP).

In the early 1990s, ISP networks were composed of routers interconnected by leased lines like those used for multiplexed voice traffic, namely T1 (1.5-Mbps) and T3 (45-Mbps) links. As the Internet started to grow, the demand for bandwidth increased faster than the speed of individual network links. ISPs responded to this challenge by simply provisioning more links to provide additional bandwidth. At this point, traffic engineering became increasingly important to ISPs so that they could efficiently use aggregate network bandwidth when multiple parallel or alternate paths were available. Back then, the limitations of this haphazard mapping were easily resolved by over-provisioning bandwidth as individual links began to experience congestion.

Existing IGPs can actually contribute to network congestion, because they do not take bandwidth availability and traffic characteristics into account when building their forwarding tables. That was becoming painfully obvious to the ISPs.

Figure 21-1 illustrates how metric-based traffic engineering works. Let's say that Network A sends a large amount of traffic to Network C and Network D. With the metrics in Figure 21-1, Links 1 and 2 could become congested because both the Network A to Network C and the Network A to Network D flows go over those links. If the metric on Link 4 were changed to 2, the Network A to Network D flow would be moved to Link 4, but the

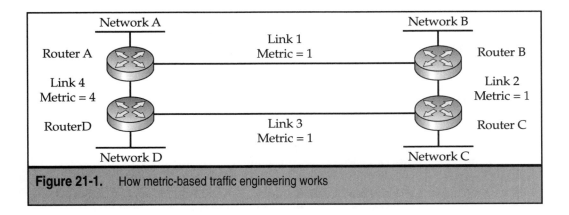

Figure 21-1.   How metric-based traffic engineering works

Network A to Network C flow would stay on Links 1 and 2. This would fix the "hot spot" without breaking anything else on the network.

The first attempt at traffic engineering was achieved by simply manipulating routing metrics. Metric-based control was adequate because Internet backbones were much smaller in terms of the number of routers, number of links, and amount of traffic.

Metric-based traffic controls provided an adequate traffic engineering solution until around 1994 or 1995. At this point, some ISPs began to reach a size at which they did not feel comfortable moving forward with metric-based traffic controls. They also did not feel comfortable moving forward with router-based cores (their backbone). It became increasingly difficult to ensure that a metric adjustment in one part of a huge network did not create a new problem in another part of the network. Now we will start to get into the *did* sentence at the end of our router/switch definitions in the "Routing to Switching to Routing?" section.

Limitations of the "Traditional" Routed Core

At the time, router-based cores did not offer the high-speed interfaces or deterministic performance that ISPs required as they planned to grow their core networks. Traditional routed cores were also starting to have problems with scalability. There were a number of reasons for this:

▼   The traditional routers made their path selections using the internal software to do the calculations. This could potentially cause traffic bottlenecks under heavy load, because their aggregate bandwidth and packet-processing capabilities were limited.

■   Traffic engineering based on metric manipulation was also a problem. That problem came to be known as the "N-squared problem." You will recall from Chapter 2 that the calculation for determining the number of routers required in a *Mesh Topology* is the total number of locations squared minus one (the original location). As ISP networks expanded, redundancy required the use of a Mesh

Topology. If they didn't adjust the metric properly in one part of the network, it would cause problems in another part of the network. Traffic engineering based on metric manipulation offered a trial-and-error approach rather than a scientific solution to an increasingly complex problem.

▲ The IGP protocol's route calculation was topology driven and based on a simple metric such as hop counts or manually entered administrative values. The IGP routing protocol did not distribute information such as bandwidth availability or traffic characteristics. This meant that the traffic load on the network was not taken into account when the IGP calculated its forwarding table. Thus, traffic was not evenly distributed across the network's links, causing inefficient use of expensive resources. Some of the links could become congested, while other links remained underutilized. They would now have to be able to control the paths that the traffic took in order to spread the load on each of the links relatively equally.

This would mean they would have to implement a solution that would allow them to control the "quality of service" on these links. Many of these ISPs turned to ATM to solve these dilemmas.

ATM was pretty much designed with QoS in mind. ATM allows point-to-point and point-to-multipoint virtual circuits (VCs) to be requested with prespecified QoS. ATM provides a rich set of QoS mechanisms with a wide variety of service categories, or *QoS descriptors*, providing very fine control capabilities and signaling mechanisms. This would give all the control they would ever need and more.

Given the need to incorporate QoS in IP, it would be wise to take full benefit of any inherent support for QoS in an underlying layer. This is especially true if that underlying layer is ATM, and many of the ISPs already had ATM in their backbones. This is the impetus behind IP over ATM (IPOA).

IP over ATM

In 1994, the volume of Internet traffic reached a point that ISPs were required to migrate their networks to support trunks that were larger than T3 (45 Mbps). Fortunately, at this time Optical Carrier 3 (OC-3) ATM interfaces (155 Mbps) became available for switches and routers. This meant that the ISPs would have to redesign their networks so that they could make use of the higher speeds supported by a switched (ATM) core.

Some ISPs transitioned from a network of T3 point-to-point links to routers with OC-3 ATM *Segmentation And Reassembly* (SAR) interfaces at the edges of their networks and OC-3 ATM switches in the core. A SAR interface is where an ATM switch connects to a router interface. It is at this point in the network that all of the variable IP packets are *segmented* into 53-byte cells. It is also where the 53-byte cells are reassembled into IP packets.

It wasn't long before the links between core ATM switches were upgraded to OC-12 (622 Mbps). You can see an illustration of this new physical topology of IP operating over an ATM core in Figure 21-2.

When IP runs over an ATM network, routers surround the edge of the ATM cloud. On the edge of the core network, each router communicates with every other router by a

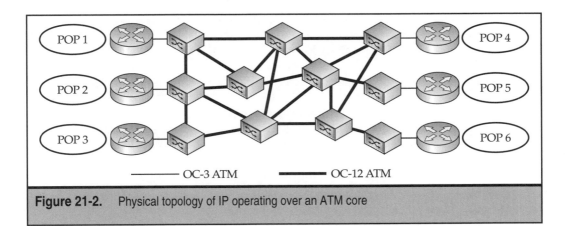

Figure 21-2.   Physical topology of IP operating over an ATM core

set of Permanent Virtual Circuits (PVCs) that are configured across the ATM physical to-pology. The PVCs function as logical circuits between edge routers. The routers do not have direct access to information describing the physical topology of the underlying ATM infrastructure supporting the PVCs. The routers have knowledge only of the indi-vidual PVCs that appear to them as simple point-to-point circuits between two routers. Figure 21-3 illustrates how the physical topology of an ATM core differs from the logical IP overlay topology.

For large ISPs, the ATM core is completely owned and operated by the ISP and is ded-icated to supporting Internet backbone service. This core infrastructure is entirely sepa-rate from the carrier's other private data services. Because the network is fully owned by the ISP and dedicated to IP service, all traffic flows across the ATM core utilizing the un-specified bit rate (UBR) ATM class of service—there is no policing, no traffic shaping, no peak cell rate, and no sustained cell rate. ISPs simply use the ATM switched infrastruc-

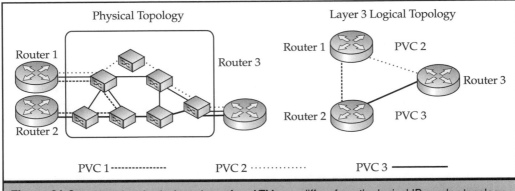

Figure 21-3.   How the physical topology of an ATM core differs from the logical IP overlay topology

ture as a high-speed transport without relying on ATM's traffic and congestion control mechanisms. There is little reason for them to use these advanced features, because each ISP owns its own backbone, and they do not need to police themselves.

The physical paths for the PVC overlay are typically calculated by an offline configuration utility on an as-needed basis when congestion occurs, a new trunk is added, or a new Point of Presence (POP) is deployed. The PVC paths and attributes are globally optimized by the offline configuration utility based on link capacity and historical traffic patterns. The offline configuration utility can also calculate a set of secondary PVCs that is ready to respond to failure conditions. Once the globally optimized PVC mesh has been calculated, the supporting configurations are downloaded to the routers and ATM switches to implement the single or double full-mesh logical topology. Figure 21-4 shows the logical IP topology over the ATM core network.

The distinct ATM Layer 2 and IP Layer 3 networks meet when the ATM PVCs are mapped to router subinterfaces. Subinterfaces on a router are associated with ATM PVCs, and then the routing protocol works to associate IP prefixes (routes) with the subinterfaces. In practice, the offline configuration utility generates both router and switch configurations, making sure that the PVC numbers are consistent and the proper mappings occur.

Finally, ATM PVCs are integrated into the IP network by running the IGP protocol across each of the PVCs to establish peer relationships and to exchange routing information. Between any two routers, the IGP metric for the primary PVC is set such that it is more preferred than the backup PVC. This guarantees that the backup PVC is used only when the primary PVC is unavailable. Also, if the primary PVC becomes available after an outage, traffic is automatically returned to the primary PVC from the backup.

Benefits of IP over ATM (IPOA)

In the mid-1990s, when ISPs required more bandwidth to handle ever-increasing traffic loads, ATM switches offered a solution. The ISPs who decided to migrate to

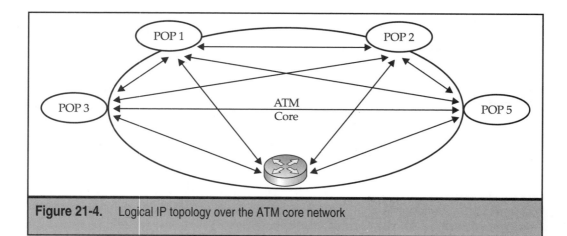

Figure 21-4. Logical IP topology over the ATM core network

ATM-based cores continued to experience growth and, in the process, discovered that ATM PVCs provided a tool that offered precise control over traffic as it flowed across their networks. ISPs have come to rely on the high-speed interfaces, deterministic performance, and PVC functionality that ATM switches provide to successfully manage the operation of their networks.

When compared with the traditional software-based routers, ATM switches provided higher-speed interfaces and significantly greater aggregate bandwidth, thereby eliminating the potential for router bottlenecks in the core of the network. Together, the speed and bandwidth provided deterministic performance for ISPs at a time when router performance was unpredictable.

The ATM-based core fully supported traffic engineering, because it could explicitly route PVCs. Routing PVCs is done by creating an arbitrary virtual topology on top of the network's physical topology, in which PVCs are routed to precisely distribute traffic across all links so that they are evenly utilized. This approach eliminates the traffic magnet effect of least-cost routing, which results in overutilized and underutilized links. The traffic engineering capabilities supported by ATM PVCs made the ISPs more competitive within their market, permitting them to provide lower costs and better service to their customers.

Though this new approach to traffic engineering required the operation and maintenance of two different networks, and the added expert personnel those networks need, it began to look like ATM was going to be the wave of the future, the end-all and be-all solution to the world's networking problems.

Limitations of IP over ATM (IPOA)

Over the past few years, ATM switches have empowered ISPs, allowing them to expand market share and increase their profitability. In the mid-1990s, ATM switches were selected for their unparalleled ability to provide high-speed interfaces, deterministic performance, and traffic engineering through the use of explicitly routed PVCs.

Today, however, all of the once-unique features of ATM switches are now supported by the newest generation of Internet backbone routers. The latest advances in routing technology are causing ISPs to reevaluate their willingness to continue tolerating the limitations of maintaining an IP network on top of an ATM network model, the administrative expense, equipment expense, operational stability, and scale.

One of the fundamental limitations of an ATM-based core is that it requires the management of two different networks: an ATM infrastructure and a logical IP overlay. By running an IP network over an ATM network, an ISP not only increases the complexity of its network, but also doubles its overhead, because it must manage and coordinate the operation of two separate networks. Also, routing and traffic engineering occur on different sets of systems. The routing is done on the routers, and the traffic engineering is run on the ATM switches. This makes it very difficult to fully integrate traffic engineering with routing.

Recent technological advances allow Internet backbone routers to provide the high-speed links and deterministic performance formerly found only in ATM switches. When considering a future migration to OC-48 (2.48 Gbps) speeds, ISPs must determine whether it is in their best interest to continue with a costly and complex design when the same functionality can now be achieved with a single set of equipment in a fully integrated router-based core.

ATM SAR router interfaces have not kept pace with the latest increases in optical bandwidth. The fastest commercially available ATM SAR router interface is OC-12. OC-48 and OC-192 (9.92 Gbps) SONET/SDH interfaces are available today for new-generation routers. OC-48 ATM SAR router interfaces are not likely to be available in the near future, and the reason for that is due to fact that *nobody* likes taxes…of any kind!

The ATM Cell Tax Problem

In 1998, the Congress passed the dreaded "ATM Cell Tax Bill" to pay for all the bandwidth it was taking from…(You're not buying this, are you?) I'm sorry, I lied about the Congress thing…but I didn't lie about the cell tax.

The OC-192 ATM SAR router interfaces might never be commercially available because of the expense and complexity of implementing the cell Segmentation and Reassembly (SAR) function at these high speeds. The limits in SAR scaling mean that ISPs attempting to increase the speed of their networks using the IP-over-ATM model will have to purchase large ATM switches and routers with a large number of slower ATM interfaces. This will dramatically increase the expense and complexity of growing the network, which will only compound as ISPs consider a future migration to OC-192. Here is why.

A *cell tax* is introduced when packet-oriented protocols such as IP are carried over an ATM infrastructure. This is because every ATM cell has a 5-byte header preceding it. Thus, almost 20 percent of *all* data traveling over the network is thrown away. That didn't use to matter at lower speeds; CPUs could run through that calculation easily and quickly. All of that has now changed, and that is just the beginning.

Assuming a 20 percent overhead for ATM when accounting for framing and realistic distribution of packets sizes, on a 2.488-Gbps OC-48 link, 1.99 Gbps is available for customer data, and 498 Mbps is required for the ATM overhead (cell tax). That is almost a full OC-12…so what?! If you are getting tired of your DSL line at home, go ahead and give your favorite ISP a jingle. Ask them what it would cost you to upgrade to an OC-12. (Oh, come on, it'll be fun! Everybody deserves a good laugh now and then.)

If you think they find *that* funny, just wait until 10-Gbps OC-192 interfaces become available. Now, 1.99 Gbps of the link's capacity will be consumed by the ATM cell tax. That is almost a full OC-48! Don't expect AOL to offer that to their customers anytime soon.

When faced with the challenge of migrating their networks to OC-48 and OC-192 speeds, ISPs must determine whether continuing to pay the ATM cell tax puts them at a competitive disadvantage when router-based cores offer the alternative of using the wasted overhead capacity for customer traffic. That isn't all they need to consider; their problems are about to become N^2 worse.

The N-Squared Problem

A network that deploys a full mesh of ATM PVCs exhibits the traditional N-squared problem. For relatively small or moderately sized networks, the N-squared problem is not a major issue. But for core ISPs with hundreds of attached routers, the challenge can be quite significant.

Let's say you want to expand your five-router network by adding one more router to connect to your new location. When changing a small network from five to six routers, an ISP is required to increase the number of simplex PVCs from 20 to 30. Now let's say you are a big company with 200 offices, and you want to add your new office, a new point of presence (POP), to your network. Remember, we are only adding one more router this time as well. Increasing the number of attached routers from 200 to 201 requires the addition of 400 new simplex PVCs. That is an increase from 39,800 to 40,200 PVCs. That doesn't even include backup PVCs or additional PVCs for networks running multiple services. (Those services require more than one PVC between any two routers.) A number of operational challenges are caused by the N-squared problem:

▼   New PVCs need to be mapped over the physical topology.

■   The new PVCs must be tuned so that they have minimal impact on existing PVCs.

■   The large number of PVCs might exceed the configuration and implementation capabilities of the ATM switches.

▲   The configuration of each switch and router in the core must be modified.

In the mid-1990s, a full mesh of PVCs was required to eliminate interior router hops because of their slow speed interfaces and lack of deterministic performance. The rules have changed. The new-generation Internet backbone routers have now eliminated all of those limitations, leaving the N-squared problem to ISPs that keep the IP-over-ATM architecture. Their problems do not end here, however; there is also the IGP stress problem to deal with.

Internet Gateway Protocol (IGP) Stress Problem

Deploying a full mesh of PVCs also stresses the Interior Gateway Protocol (IGP). This stress results from the number of peer relationships that must be maintained, the challenge of processing link-state updates in the event of a failure, and the complexity of performing the Dijkstra algorithm calculation over a topology containing a significant number of logical links. Anytime the topology results in a full mesh, the impact of the IGP problem is not only subpar, it is also extremely difficult to maintain. Once you reach this point, it can only get worse. As an ATM core expands, the N-squared problem and the stress it puts on the IGP will only compound your problems.

Again, with a new-generation router-based core, the N-squared stress on the IGP is eliminated. As with the cell tax, IGP stress is a heritage of the IP-over-ATM model.

Traffic engineering based on the overlay model requires the presence of a Layer 2 technology that supports switching and PVCs. On a mixed-media network, the dependency on a specific Layer 2 technology like ATM to support traffic engineering can severely affect any possible solutions. If an ISP wants to perform traffic engineering across these newer optical networks, the Layer 2 transport cannot perform traffic engineering, because it has been eliminated. The growth of mixed-media networks and the goal of reducing the number of layers between IP and the glass require that traffic engineering capabilities be supported at Layer 3 to provide an integrated approach. As ISPs continue to build out their networks based on the optical internetworking model, the limitations of the IP-over-ATM architecture become significantly more restrictive.

SUMMARY

In summary, the fundamental assumptions supporting the original deployment of ATM-based cores are no longer valid. There are numerous disadvantages for continuing to follow the IP over ATM (IPOA) model when other alternatives are available.

High-speed interfaces, deterministic performance, and traffic engineering using PVCs no longer distinguish ATM switches from Internet backbone routers. Instead, you are left with:

▼ The complexity and expense of maintaining two sets of network equipment

■ The complexity and expense of coordinating two sets of administrative personnel

■ The bandwidth limitations of ATM SAR interfaces

■ The cell tax

■ The N-squared PVC problem

■ The IGP stress

■ The limitation of not being able to operate over a mixed-media infrastructure

▲ The disadvantages of not being able to seamlessly integrate Layer 2 and Layer 3

Migrating back to a new-generation router-based core solves a number of inherent problems with the ATM model. Router to Switch to Router....the more things change, the more they stay the same. Enter MPLS.

CHAPTER 22

MPLS Background

When compared with the traditional software-based routers, ATM switches provided higher-speed interfaces and significantly improved the bandwidth problems that were facing the service providers. By overlaying IP onto ATM's QoS-rich infrastructure, ISPs were able to eliminate the potential for router bottlenecks in the core of the network. Together, the speed and bandwidth provided deterministic performance for ISPs at a time when router performance was unpredictable.

The ATM-based core fully supported traffic engineering, because it could explicitly route PVCs. Routing PVCs is done by creating an arbitrary virtual topology on top of the network's physical topology in which PVCs are routed to precisely distribute traffic across all links so that they are evenly utilized.

However, while ATM has its claims to providing QoS assurances, it still cannot do anything above Layer 2. That means that all Layer 3 flows that have been aggregated cannot be differentiated by ATM. Because of this, they end up competing against one another for the same resources. Today, we need a way to implement finer granularity in the control of traffic flow, and this is best done in Layer 3.

Several approaches were made by a number of companies over the last couple of years to find some way to apply Layer 2 switching technology to Layer 3 routing. Four companies in particular were having some success: Ipsilon, Cisco, IBM, and Toshiba. They all had their strengths, but then again, they were all proprietary.

This led the IETF to form the MPLS Working Group. The goal of this group is to develop a standard for the integration of Layer 2 switching with Layer 3 routing in order to improve performance and scalability of network layer routing while providing greater flexibility and QoS.

WHAT IS MPLS?

Multi-Protocol Label Switching (MPLS) was originally presented as a way of improving the forwarding speed of routers, but is now emerging as a technology that offers much more than that. *Traffic engineering,* the ability of network operators to dictate the path that traffic takes through their network, and VPN support are two examples where MPLS is proving superior to any currently available IP technology.

MPLS generates a short fixed-length label that acts as a shorthand representation of an IP packet's header. This is much the same as the way a ZIP code acts as shorthand for the house, street, and city in a postal address and is used to make forwarding decisions about the letter. IP packets have a field in their header that contains the address to which the packet is to be routed. Traditional routed networks process this information at every router in a packet's path through the network in a hop-by-hop approach.

In MPLS, the IP packets are encapsulated with these labels by the first MPLS device (called an *edge router*) they encounter as they enter (*ingress*) the network. The MPLS edge router analyzes the contents of the IP header and selects an appropriate label with which to encapsulate the packet.

One of the things that makes MPLS so powerful, in contrast to conventional IP routing, is that this analysis can be based on more than just the Destination Address carried in the IP header. At all the subsequent nodes within the network, it is the MPLS label, not the IP header, that is used to make the forwarding decision for the packet. Finally, as MPLS-labeled packets leave the network (*egress*), another edge router removes the labels and forwards it normally as an IP packet.

In MPLS terminology, the packet-handling nodes or routers are called *Label Switched Routers* (LSR). MPLS routers forward packets by making switching decisions based on the MPLS label. This is another key concept in MPLS. Conventional IP routers contain routing tables that are looked up using the IP header from a packet to decide how to forward that packet. These tables are built by IP routing protocols that carry around IP reachability information in the form of IP addresses.

In practice, we find that the *forwarding plane* (IP header lookup) and *control planes* (generation of the routing tables) are tightly coupled. The beautiful thing about MPLS is that since forwarding is based on the labels, you can cleanly separate the (label-based) *forwarding plane* from the routing protocol *control plane*.

This is awesome, because by separating the two, each can be modified independently. With such a separation, we don't need to change the forwarding machinery, for example, to migrate a new routing strategy into the network!

Two broad categories of LSR exist. At the edge of the network, we require high-performance packet classifiers that can apply and remove the MPLS labels. These are called the MPLS *edge routers*. The Core LSRs need only process the labels on these packets, and this allows them to do this at the highest bandwidths available today.

WHY DO WE NEED MPLS?

MPLS is the key to the future of large-scale IP networks. MPLS has applications in:

- ▼ Deployment of IP networks across ATM-based WANs
- ■ Providing traffic engineering capabilities to packet-based networks
- ■ Providing IP QoS capabilities
- ▲ Bulletproofing Virtual Private Networks (VPNs)

MPLS will also make it possible to address all of the IP-over-ATM issues facing ISPs that we talked about in the last chapter:

- ▼ The complexity and expense of maintaining two sets of network equipment
- ■ The complexity and expense of coordinating two sets of administrative personnel
- ■ The bandwidth limitations of ATM SAR interfaces
- ■ The "cell tax"

- The N-squared PVC problem
- The IGP stress
- The limitation of not being able to operate over a mixed media infrastructure
- ▲ The disadvantages of not being able to seamlessly integrate Layer 2 and Layer 3

What's more, MPLS technology will allow them to make all of those changes without having to throw away expensive ATM equipment, or Frame Relay for that matter, as you will soon see. It almost sounds too good to be true, but fortunately, it's not. MPLS, on its way to becoming the premier packet-switching technique for backbone networking, could feasibly even make ATM, Frame Relay, and IP routing-based networks a thing of the past in ISP backbones and large enterprise WANs as well.

ATM and Frame-Relay Forwarding Methods

We tend to love what we know. MPLS is still relatively new, and because it was designed to do pretty much everything, people tend to think it is just another flavor of the technology they have been working with all along. Because MPLS can use ATM switches, the "ATM people" think that MPLS is really just tunneled through ATM virtual circuits. And, because MPLS is connection-oriented and can use variable-length frames at Layer 2, the "Frame Relay people" think that MPLS is simply some kind of a "Super Frame Relay" technology with a clever twist.

The key thing that they are missing is what we talked about in the last section. What sets MPLS apart from everything is the separation of the packet forwarding and control planes. All switches and routers perform these two functions, whether they are connectionless or connection oriented, or use frames or cells.

Within an IP router, the control function involves mostly route calculation. The router uses whatever routing protocol is installed to process some metric to update the routing tables in the CPU.

Frame forwarding is a whole different function. The router simply looks at the Destination Address in the packet's header and consults the routing table to determine the best interface for that packet's "next hop." In the new-generation routers, forwarding is performed using specialized hardware, Application Specific Integrated Circuits (ASICs), giving it "switch-like" speeds.

In ATM switches, the control function is based on the Private Network-to-Network Interface (PNNI) protocol. This is ATM's link state routing protocol and is conceptually similar to what happens in a router. The big difference here is that since ATM is connection-oriented, more parameters are involved in calculating the path, including consideration of available bandwidth at the time of virtual circuit setup. Also, instead of forwarding variable-length packets, it forwards only fixed-length 53-byte cells based upon the already determined VC paths. Since it only needs to look at Layer 2 information, the forwarding can be done using specialized, hardware switching ASICs that are optimized for high performance and separate from the processors that control the virtual circuits and calculate their paths.

MPLS Forwarding Methods

MPLS provides a third mechanism for control and path calculation. As with IP routers, MPLS nodes or Label Switched Routers (LSRs) also use routing protocols to calculate network paths and establish reachability.

Unlike IP, MPLS paths are called Label Switched Paths (LSPs) and, like ATM and Frame Relay, these LSPs are connection-oriented rather than connectionless and can be provisioned manually. This should sound familiar from what we learned in the last chapter. These LSPs are like the ATM PVCs we talked about that were used to solve the old IP traffic engineering problems.

These are built using CR-LDP or RSVP-TE protocols, which are essentially modifications of some of the QoS protocols we talked about in Chapter 20. There is a huge debate as to which one is best, but we won't touch that. Suffice it to say, essentially what these protocols do is set up the proper use of MPLS labels along each path.

Now for the cool part. Because MPLS path calculation and control are separate from forwarding (just as with IP routers or ATM switches), MPLS developers decided to make the control plane independent of the type of forwarding that is used. That means that LSRs can forward frames *or* cells. It's a good thing they can, because this is what makes it possible for ATM switches or IP routers to become MPLS LSRs.

If that isn't neat, I don't know what is. This means that all the ATM switch or IP router products that are scattered throughout all the ISPs can be converted by reprogramming their control processors. In either case, the forwarding hardware continues to operate exactly as it did before, and continues to forward either frames or cells. Of course, they will be locked into being a "frame forwarder" or "cell forwarder" forever. Pure MPLS LSRs can do both.

When cells are used for MPLS transport, what was formerly the ATM VPI/VCI field becomes the location for the MPLS label. When frames are used, an additional "shim" header is added between PPP and IP headers to carry the MPLS label. Router and cell switch LSRs can be mixed in the same MPLS network, and a single LSP can cross both cell- or frame-based links, because the LSPs are independent of underlying link layer protocols. Most frame-based MPLS links will use PPP, while the cell-based MPLS links will use "pseudo-ATM" 53-byte cells.

An MPLS network is not an ATM network, a Frame Relay network, or an IP routed network. It is an *MPLS* network, with its own unique control plane operating in Label Switched Routers.

HISTORY BEHIND MPLS

To find out how this incredible technology came to be, we can go back to where we left off in Chapter 21. As ISPs continued migrating to the IP-over-ATM model, a number of technical, marketing, and financial trends began to influence the development of new technologies designed for the Internet. IP quickly became the only protocol that mattered, winning out over IPX, AppleTalk, OSI, and SNA.

IP was now becoming the foundation not only for the Internet but also for many organizations' private wide and local area networks. For ISPs the future was clear, and so began the big push for convergence of voice, data, multimedia, and everything else onto the Internet. All of these new networks, computers, and even "Internet appliances" will be based on IP protocols, leading to the need for technical and operational improvements.

The notion of IP convergence needed to deliver a solution that provided the price and performance of an ATM switch and the control of an IP router, while eliminating the complex mapping required by the IP-over-ATM model. Toward the end of 1996, a number of vendors were starting to develop proprietary multilayer switching solutions that integrated ATM switching and IP routing, most notably:

▼ IP Switching designed by Ipsilon/Nokia

■ Tag Switching developed by Cisco Systems

■ Aggregate Route based IP Switching (ARIS) designed by IBM Corporation

▲ Cell Switching Router (CSR) developed by Toshiba

IP Switching

IP Switching, developed by Ipsilon (now Nokia), was announced in 1996 and has been delivered in commercial products. IP Switching enables a device with the performance of an ATM switch to act as a router, thereby overcoming the limited packet throughput of traditional routers.

The basic goal of IP Switching is to integrate ATM switches and IP routing in a simple and efficient way by eliminating the ATM control plane. IP Switching uses the presence of data traffic (flows) to drive the establishment of a label.

A label-binding protocol called the *Ipsilon Flow Management Protocol* (IFMP) and a switch-management protocol called the *General Switch Management Protocol* (GSMP) are defined. GSMP is used solely to control an ATM switch and the virtual circuits made across it.

Tag Switching

Tag Switching is the label-switching approach developed by Cisco Systems. In contrast to CSR and IP Switching, Tag Switching is a control-driven technique that does not depend on the flow of data to stimulate setup of label-forwarding tables in the router.

A Tag Switching network consists of *Tag Edge Routers* (TER) and *Tag Switching Routers* (TSR), with packet tagging being the responsibility of the edge router. Standard IP routing protocols are used to determine the next hop for traffic. Tag Switching binds the tags to routes in a routing table. It then distributes those bindings to peers using the *Tag Distribution Protocol* (TDP).

Aggregate Route based IP Switching (ARIS)

Aggregate Route based IP Switching (ARIS) is IBM's contribution to label switching. ARIS binds labels to aggregate routes (groups of address prefixes) rather than to flows, unlike CSR or IP Switching. Label bindings and *Label Switched Paths* (LSP) are set up in response to control traffic, such as routing updates, rather than data flows, with the egress router generally being the initiator.

Routers that are ARIS capable are called *Integrated Switch Routers* (ISR). ARIS was designed focusing on ATM as the data link layer of choice. One of the features provides loop prevention mechanisms that are not available in ATM. The ARIS protocol is a peer-to-peer protocol that runs between ISRs directly over IP and provides a means to establish neighbors and to exchange label bindings. A key concept in ARIS is the *egress identifier*. Label distribution begins at the egress router and propagates in an orderly fashion toward the ingress router.

Cell Switching Router (CSR)

The *Cell Switching Router* (CSR) approach was developed by Toshiba in 1994. It was one of the earliest public proposals for using IP protocols to control an ATM switching fabric. CSR is designed to function as a router for connecting logical IP subnets in a classical *IP-over-ATM* (IPOA) environment.

Label-switching devices communicate over standard ATM virtual circuits (VCs). CSR labeling is *data driven.* That means the labels are assigned on the basis of flows that are identified locally. The *Flow Attribute Notification Protocol* (FANP) is used to identify the dedicated VCs between CSRs and to establish the association between individual flows and individual dedicated VCs.

The objective of the CSR is to allow *cut through* forwarding of these flows. Cut through switches the ATM cell flow that constitutes the packet rather than reassembling it and making an IP level forwarding decision on it.

Which Multilayer Switching Is Best?

Although these approaches had a number of characteristics in common, they did not work and play well with each other. Each company's solution relied on different technologies to combine IP routing and ATM switching into an integrated solution. However, by early 1997, many in the Internet community were so impressed with the simplicity and elegance of these solutions that they began to view multilayer switching as the future technology for the design of large ISP backbone networks.

Each company's multilayer switching solutions sought to combine the best properties of IP routing and ATM switching. The basic idea was to take the control software from an IP router, integrate it with the forwarding performance of a label swapping ATM switch, and create an extremely fast and cost-efficient IP router. You can get an idea of what we mean in Figure 22-1.

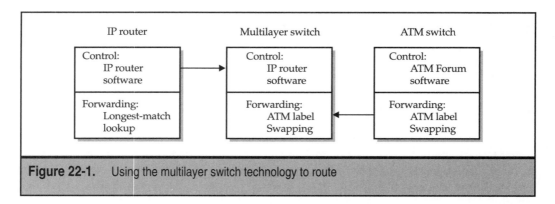

Figure 22-1. Using the multilayer switch technology to route

For the control component, each multilayer switch would run standard IP routing protocols and a proprietary label-binding mechanism. The routing protocols would permit the multilayer switches to exchange Layer 3 network reachability information. The label-binding mechanism mapped Layer 3 routes to these labels and distributed them to neighbors to establish Label Switched Paths (LSP) across the core of the network. Using routing protocols on core devices, as opposed to the IP-over-ATM model which ran the routing protocols on just the edge devices, provided a number of benefits and solutions to their current overlay models:

▼ Eliminated the IP-over-ATM model's N-squared PVC scaling problem.

■ Reduced the IGP stress by dramatically decreasing the number of peers that each router had to maintain.

▲ Allowed information about the core's actual physical topology to be made available to Network layer routing procedures.

For the *forwarding* component, multilayer switches used conventional ATM switching hardware and label swapping to forward cells across the core of the network. You can see what we mean in Figure 22-2.

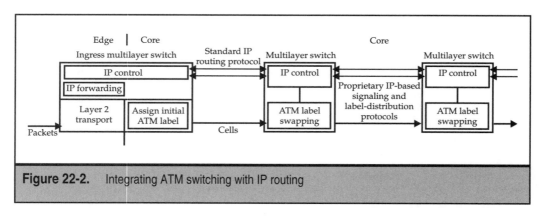

Figure 22-2. Integrating ATM switching with IP routing

However, the *control* procedures that assigned the labels to routes, distributed the labels among the multilayer switches, and created forwarding tables were managed by proprietary IP-based protocols and not the ATM protocols. ATM label-swapping in the core of the network provided a number of benefits:

▼   Label-swapping optimized network performance, because it was able to use the ATM switches' fast ASIC (hardware-based) forwarding.

■   Label-swapping made explicit routing practical. An explicit route is a preconfigured sequence of hops that establishes the path that traffic should take across a service provider's network. Instead of simplistic paths created by routing protocols, explicit paths could now give the ISPs precise control over traffic flows, making it possible to support traffic engineering and QoS.

▲   Label-swapping provided an instrument to extend control beyond the limitations of conventional destination-based routing.

By avoiding ATM's routing and signaling protocols, multilayer switching reduced operational complexity by eliminating the need to coordinate and map between the IP and ATM's different protocol architectures.

However, if you recall the list at the beginning of this chapter, you will see that this still left one glaring omission. The majority of these proprietary, multilayer switching solutions was that they were still restricted to running over a cell-based ATM infrastructure. What about the cell tax? Those companies lost sight of the original intention of trying to emulate the future of the Internet and IP, which are packet oriented.

Also, fundamental differences existed between the companies' multilayer switching solutions. While the various multilayer switching solutions had numerous features in common, they relied on two fundamentally different approaches to initiate the assignment and distribution of label bindings to establish LSPs:

▼   A data-driven model

▲   A control-driven model

Data-Driven Model

In the *data-driven model*, label bindings are created when user data packets arrive. A *flow* is a sequence of packets that have the same source and destination IP addresses and TCP or UDP port numbers. A multilayer switch can either create a label binding as soon as it sees the first packet in a traffic flow, or it can wait until it has seen a number of packets in the flow.

The benefit of waiting for a number of packets ensures that the flow is long enough to merit the overhead of assigning and distributing a label. Multilayer switching solutions that implemented the data-driven approach were IP Switching by Ipsilon and the Cell Switching Router by Toshiba. (MPLS does not support the data-driven model.)

The advantage of the data-driven model is that a label binding is created only when there is a traffic flow that uses the label binding. However, this model has a number of limitations for deployment in the core of a large ISP network, where there can be an enormous number of individual traffic flows:

▼ Each multilayer switch must provide sophisticated and high-performance packet classification capabilities to identify traffic flows.

■ There is a latency between the recognition of a flow and the assignment of a label to the flow. This means that each multilayer switch must also support longest-match IP forwarding during the setup phase so packets that have not been assigned to a flow can be forwarded and not dropped.

■ The amount of control traffic needed to distribute label bindings is directly proportional to the number of traffic flows.

▲ The presence of a significant number of relatively short-lived flows can impose a heavy burden on network operations. Conventional wisdom dictates that the data-driven model does not have the scaling properties required for application in the core of the Internet.

Control-Driven Model

In the *control-driven model,* label bindings are created when control information arrives. Labels are assigned in response to the normal processing of routing protocol traffic, control traffic such as RSVP traffic, or in response to static configuration. The multilayer switching solutions that implemented the control-driven model were Tag Switching by Cisco Systems and ARIS by IBM. (MPLS uses the control-driven model.)

The control-driven model has a number of benefits for deployment in the core of a large ISP network:

▼ Labels are assigned and distributed before the arrival of user data traffic. This means that if a route exists in the IP forwarding table, a label has already been allocated for the route, so traffic arriving at a multilayer switch can be label-swapped immediately.

■ Scalability is significantly better than in the data-driven model, because the number of Label Switched Paths is proportional to the number of entries in the IP forwarding table, not to the number of individual traffic flows.

■ Label assignments are based on prefixes rather than individual flows.

■ In a stable topology, the label assignment and distribution overhead is lower than in the data-driven model, because Label Switched Paths are established only after a topology change or the arrival of control traffic, not with the arrival of each new traffic flow.

■ Every packet in a flow is label switched, not just the tail end of the flow as in the data-driven model.

▲ Each multilayer switching solution maintained the IP control component and used ATM label swapping as the forwarding component. The challenge facing the ISP community was that each solution was proprietary and therefore not interoperable. Also, the majority of multilayer switching solutions required an ATM transport, because they could not operate over mixed media infrastructures like Frame Relay, PPP, SONET, and Ethernet. If multilayer switching was to be widely deployed by ISPs, there had to be a compatible standard that could run over any link layer technology.

Multi-Protocol Label Switching (MPLS)

MPLS is the latest step in the evolution of multilayer switching in the Internet. MPLS is an IETF standards-based approach built on the efforts of all the companies' proprietary multilayer switching solutions you have just read about. In early 1997, the IETF established the MPLS working group to produce a unified and interoperable multilayer switching standard.

The charter of the MPLS working group is to standardize a base technology that combines the use of label swapping in the forwarding component with Network layer routing in the control component. To achieve its objectives, the MPLS working group has to deliver a solution that satisfies a number of requirements, including:

▼ MPLS must run over any link layer technology, not just ATM.

■ MPLS core technologies must support the forwarding of both unicast and multicast traffic flows.

■ MPLS must be compatible with the IETF Integrated Services Model, including RSVP.

■ MPLS must scale to support constant Internet growth.

▲ MPLS must support operations, administration, and maintenance facilities at least as extensive as those supported in current IP networks.

MPLS defines new standard-based IP signaling and label distribution protocols, as well as extensions to some of the existing protocols to support multivendor interoperability. MPLS does not use or require any of the ATM signaling or routing protocols, so the complexity of coordinating two different protocol architectures is eliminated. In this way, MPLS brings significant benefits to a packet-oriented Internet.

SUMMARY

Clearly, these were all big advances, and since multiple proprietary solutions for label-based switching is counterproductive to global interoperation of Internet devices, it was recognized that standards were needed, and an IETF working group was formed. A charter was agreed to in the IETF in early 1997, and the inaugural meeting of the working group was held in April 1997. After much deliberation, the term *Multi-Protocol Label Switching* (MPLS) was selected to refer to the standard whose goal is to "integrate the label swapping forwarding paradigm with network layer routing" with an initial focus on IPv4 and IPv6. MPLS provides the mechanisms, and these can be applied in various ways according to the network's needs.

CHAPTER 23

MPLS Components
and Concepts

Τhis brings us to our last chapter, where we will talk about the components and concepts and how they are used to forward information through an MPLS network. Finally, we will also see why MPLS is the ultimate solution for VPNs.

MPLS COMPONENTS AND CONCEPTS

You have learned some of the components and concepts mentioned in the last couple of chapters, but we will quickly go over what each of them is and how they work together from ingress to egress.

The Control Component

The *control component* builds and maintains a forwarding table for the node to use. It works with the control components of other nodes to distribute routing information consistently and accurately. The control component also ensures that consistent local procedures are used to create the forwarding tables. Standard routing protocols are used to exchange routing information among the control components. The control component must react when network changes such as a link failure occur, but it is not involved in the processing of individual packets.

MPLS uses the control-driven model to initiate the assignment and distribution of label bindings for the establishment of LSPs. LSPs are simple in nature: traffic flows in one direction from the ingress LSR to the egress LSR. Duplex traffic requires that two LSPs be created, one LSP to carry traffic in each direction. The MPLS control component centers around IP and is similar to proprietary multilayer switching solutions we have been talking about with one important exception. MPLS cleanly separates the (label-based) *forwarding plane* from the routing protocol *control plane.* Take a look at Figure 23-1 to get an idea of how this is done.

The Forwarding Component

The *forwarding component* performs the actual packet forwarding. It uses information from the forwarding table in the router. Forwarding decisions are made based on the information in the packet itself and a set of local procedures. A conventional router uses a *longest address match* algorithm that compares the destination address in the packet with entries in the forwarding table until it finds the best available match. More importantly, the full decision-making process has to be repeated at each node along the path from source to destination. On the other hand, the forwarding component of a Label Switching Router (LSR) uses an *exact match label swapping* algorithm that uses the label in the packet and a label-based forwarding table to obtain a "new" label and output interface for the packet.

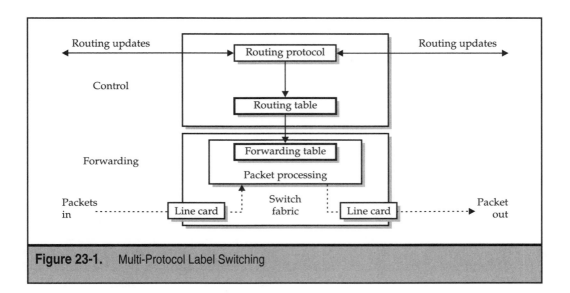

Figure 23-1. Multi-Protocol Label Switching

The Forwarding Table

A *forwarding table* is the set of entries in a table that provides information to help the forwarding component perform its switching function. The forwarding table must associate each packet with an entry. When a packet containing a label arrives at an LSR, the LSR examines the label and uses it as an index into its MPLS forwarding table. Each entry in the forwarding table contains an interface-inbound label pair that is mapped to a set of forwarding information that is applied to all packets arriving on the specific interface with the same inbound label. Figure 23-2 shows an example of an MPLS forwarding table.

MPLS Forwarding Table

In Interface	In Label	Out Interface	Out Label
⋮	⋮	⋮	⋮
3	21	4	18
3	56	6	135
⋮	⋮	⋮	⋮

Figure 23-2. MPLS forwarding table

Forwarding Equivalence Class (FEC)

A *Forwarding Equivalence Class* (FEC) is defined as any group of packets that can be treated in an equivalent manner for purposes of forwarding. An example of an FEC would be a set of unicast packets whose Destination Addresses match a particular IP address prefix. Another FEC would be a set of packets whose Source Address and Destination Address are the same. FECs can be defined at different levels of granularity. For example, all packets matching a given address prefix is a coarser granularity than all packets from a given source going to a specific destination application port.

MPLS Label

A *label* is a relatively short, fixed-length, unstructured identifier that can be used to assist in the forwarding process. Labels are associated with an FEC through a binding process. Labels are normally local to a single data link and have no global significance (as would an address). Labels are analogous to the DLCIs used in a Frame Relay network or the VPI/VCIs used in an ATM environment. Labels are bound to an FEC as a result of some event that indicates a need for the binding making them meaningful. *Control-driven bindings* are established as a result of control plane activity and are independent of the data. Label bindings might be established in response to routing updates or receipt of RSVP messages.

If the Layer 2 technology supports a label field such as the ATM VPI/VCI or the Frame Relay DLCI fields, the native label field encapsulates the MPLS label. However, if the Layer 2 technology does not support a label field, the MPLS label is encapsulated in a standardized MPLS header that is inserted between the Layer 2 and IP headers. The MPLS header permits any link layer technology to carry an MPLS label so it can benefit from label swapping across an LSP. You can see how this is done in Figure 23-3.

The 32-bit MPLS header contains the following fields:

▼ **Label**   20 bits. This carries the actual value of the MPLS label.

■ **CoS**   3 bits. This can affect the queuing and discard algorithms applied to the packet as it is transmitted through the network.

■ **Stack (S)**   1 bit. This is to support a hierarchical label stack.

▲ **TTL (time-to-live)**   8 bits. This provides conventional IP TTL functionality.

Label-Swapping Forwarding Algorithm

The forwarding component of virtually all multilayer switching solutions and MPLS is based on a label-swapping forwarding algorithm. This is the same algorithm used to forward data in ATM and Frame Relay switches.

Recall that a label is a short, fixed-length value carried in the packet's header to identify a Forwarding Equivalence Class (FEC). A label is analogous to a connection identifier, such as an ATM VPI/VCI or a Frame Relay DLCI, because it has only link-local significance, does not encode information from the network layer header, and maps traf-

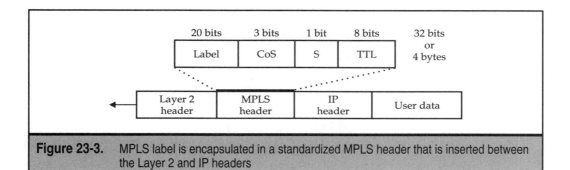

Figure 23-3.   MPLS label is encapsulated in a standardized MPLS header that is inserted between the Layer 2 and IP headers

fic to a specific FEC. Also, as we mentioned earlier, an FEC is a set of packets that are forwarded over the same path through a network even if their ultimate destinations are different. For example, in conventional longest-match IP routing, the set of unicast packets whose destination addresses map to a given IP address prefix is an example of an FEC.

The label-swapping forwarding algorithm requires packet classification at the ingress edge of the network to assign an initial label to each packet. In Figure 23-4, the ingress label switch receives an unlabeled packet with a destination address of 192.4.2.1. The label switch performs a longest-match routing table lookup and maps the packet to an FEC, 192.4/16. The ingress label switch then assigns a label, in this case with a value of 5, to the packet and forwards it to the next hop in the Label Switched Path (LSP).

Label Switched Path (LSP)

An LSP is functionally equivalent to a virtual circuit because it defines an ingress-to-egress path through a network that is followed by all packets assigned to a specific FEC. The first label switch in an LSP is called the *ingress*, or head-end, label switch. The last label switch in an LSP is called the *egress*, or tail-end, label switch.

Label swapping provides a significant number of operational benefits when compared with conventional hop-by-hop network layer routing:

▼ Label swapping gives a service provider tremendous flexibility in the way that it assigns packets to FECs. For example, to simulate conventional IP forwarding, the ingress label switch can be configured to assign a packet to an FEC based on its destination address. However, packets can also be assigned to an FEC based on an unlimited number of policy-based considerations—the source address alone, the application type, the point of entry into the label-swapping network, the point of exit from the label-swapping network, the CoS conveyed in the packet header, or any combination of the above.

■ Service providers can construct customized LSPs that support specific application requirements. LSPs can be designed to minimize the number of hops, meet certain bandwidth requirements, support precise performance requirements, bypass potential points of congestion, direct traffic away from the default path selected by the IGP, or simply force traffic across certain links or nodes in the network.

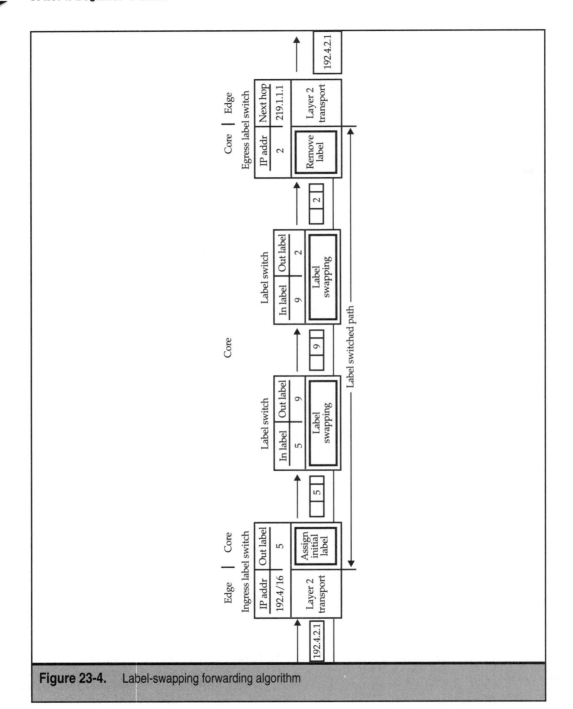

Figure 23-4.   Label-swapping forwarding algorithm

▲ The most important benefit of the label-swapping forwarding algorithm is its ability to take any type of user traffic, associate it with an FEC, and map the FEC to an LSP that has been specifically designed to satisfy the FEC's requirements. The deployment of technologies based on label-swapping forwarding techniques offers ISPs precise control over the flow of traffic in their networks. This unprecedented level of control results in a network that operates more efficiently and provides more predictable service.

Label Switching Routers (LSR)

Label switching is an advanced form of packet forwarding that replaces conventional longest-address-match forwarding with a more efficient label-swapping algorithm. There are three important distinctions between label switching and conventional routing. These distinctions are listed in Table 23-1.

A Label Switching Router (LSR) is any device that supports both a label-swapping forwarding component and a standard IP control component, such as standard routing protocols, RSVP, and so on. Figure 23-5 shows a simple label-switching network and illustrates the edge LSRs providing the ingress and egress functions and core LSRs providing high-speed switching. A label-switching network serves the same purpose as any conventional routed network: it delivers traffic to one or more destinations. The addition of label-based forwarding complements conventional routing but does not replace it.

Action	Conventional Routing	Label Switching
Full IP header analysis	Occurs at every node	Occurs only once at the network edge when the label is assigned
Unicast and multicast support	Requires multiple complex forwarding algorithms	Requires only one forwarding algorithm
Routing decisions	Based on Address only	Can be based on any number of parameters such as QoS, VPN Identifiers, and so on

Table 23-1. Distinctions Between Label Switching and Conventional Routing

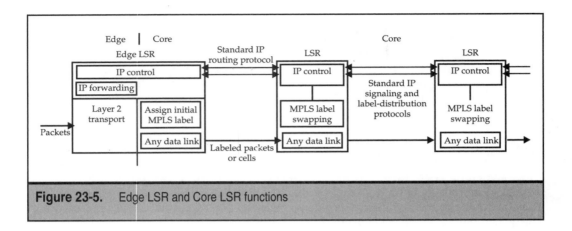

Figure 23-5.   Edge LSR and Core LSR functions

The Role of the Core LSR

In the core of the network, LSRs ignore the packet's network layer header and simply forward the packet using the label-swapping algorithm. When a labeled packet arrives at a switch, the forwarding component uses the input port number and label to perform an exact match search of its forwarding table. When a match is found, the forwarding component retrieves the outgoing label, the outgoing interface, and the next-hop address from the forwarding table. The forwarding component then swaps (or replaces) the incoming label with the outgoing label and directs the packet to the outbound interface for transmission to the next hop in the LSP.

The Role of the Edge LSR

At the boundary of an MPLS network, the ingress-edge LSRs make classification and forwarding decisions by examining the IP header in the unlabeled packets. The result is that appropriate labels are applied to the packets, and they are then forwarded to an LSR that serves as the next hop toward the ultimate destination.

The LSR-generated, fixed-length "label" acts as a shorthand representation for the IP packet's header, thereby reducing the processing complexity at all subsequent nodes in the path. The label is generated during header processing at the LSR node. All subsequent nodes in the network use the label for their forwarding decisions. Of course, the value of the label may, and usually does, change at each LSR in the path through the network. As packets emerge from the core of an MPLS network, the edge LSRs that find they have to forward packets onto an unlabeled interface can simply remove any label encapsulation before doing so.

When the labeled packet arrives at the egress edge LSR, the forwarding component searches its forwarding table. If the next hop is not a label switch, the egress edge LSR discards the label and forwards the packet using conventional longest-match IP forwarding.

When a core LSR receives a labeled packet, the label is first extracted and used as an index into the forwarding table that resides in the LSR. When the entry indexed by the incoming label is found, the outgoing label is extracted and added to the packet, and the packet is then sent out the outgoing interface to the next hop that is specified in the entry. (Multicast involves multiple outgoing packets and therefore multiple interfaces.) Label-switching forwarding tables may be implemented at the node level (a single table per node) or at the interface level (one table per interface).

What is most important about label-based forwarding is that only a single forwarding algorithm is needed for all types of switching, and this can be implemented in hardware for extra speed.

The MPLS Forwarding Component of the LSR

All multilayer switching solutions, including MPLS, are composed of two distinct functional components: a *control* component and a *forwarding* component. The control component uses standard routing protocols to exchange information with other routers to build and maintain a forwarding table. When packets arrive, the forwarding component searches the forwarding table maintained by the control component to make a routing decision for each packet. Specifically, the forwarding component examines information contained in the packet's header, searches the forwarding table for a match, and directs the packet from the input interface to the output interface across the system's switching fabric. Take a look at Figure 23-6 to see how this works.

By completely separating the control component from the forwarding component, each component can be independently developed and modified. The only requirement is that the control component continue to communicate with the forwarding component by managing the packet-forwarding table. We will see that the deployment of an extremely simple forwarding algorithm, such as label swapping, can provide the extended forwarding capabilities needed to support new revenue-generating customer services.

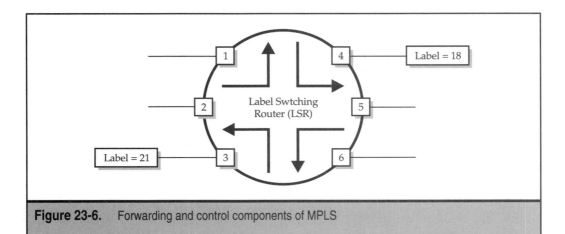

Figure 23-6.   Forwarding and control components of MPLS

The MPLS Control Component of the LSR

Labels are attached to the packets by an upstream LSR. The downstream LSR that receives these labeled packets must know, or find out, what to do with them. It is the responsibility of the label-switching control component to handle this task. It uses the contents of an entry in the label-switching forwarding table as its guide.

The establishment and maintenance of table entries are essential functions and must be performed by each LSR. The label-switching control component is responsible for distributing routing information among the LSRs in a consistent fashion and for executing the procedures that are used by the LSRs to convert this information into a forwarding table.

The label-switching control component includes all the conventional routing protocols. These routing protocols provide the LSRs with the mapping between the FEC and the next hop addresses. In addition, the LSR must:

▼ Create the bindings between the labels and the FECs

■ Distribute those bindings to other LSRs

▲ Construct its own label forwarding table

The binding between a label and an FEC is control driven, meaning that it is directed by the topology as represented in routing updates or other control messages.

DISTRIBUTION OF LABEL INFORMATION

A label-switching forwarding table entry provides, at a minimum, information about the outgoing interface and a new label, but may also contain other information. It might, for example, indicate the output queuing decision to be applied to the packet. The incoming label uniquely identifies a single entry in this table.

Every label that is distributed must be bound to an entry in the forwarding table. This binding may be performed in the local LSR or be supplied by a remote LSR. MPLS uses downstream binding in which locally bound labels are used as incoming labels and remotely bound labels are used as outgoing labels. For MPLS, the entries in the forwarding table are established as follows:

▼ The *Next Hop* is provided by the routing protocols. This is the FEC to next hop mapping.

■ The *Incoming Label* is provided by creating a local binding between an FEC and the label.

▲ The *Outgoing Label* is provided by a remote binding between the FEC and the label.

The MPLS architecture uses both *local control* and *egress control*. With *local control*, the LSR can decide to create and advertise a binding without waiting to receive a binding from a neighbor for the same FEC. With *egress control*, the LSR waits for a binding from its downstream neighbor before allocating a label and advertising it upstream.

In order that LSPs can be used, the forwarding tables at each LSR must be populated with the mappings from [*incoming interface, label value*] to [*outgoing interface, label value*]. This process is called *LSP setup,* or the Label Distribution process.

Several different approaches to label distribution can be used depending on the requirements of the hardware that forms the MPLS network and the administrative policies used on the network. The underlying principles are that an LSP is set up either in response to a request from the ingress LSR, known as *Downstream-on-Demand,* or preemptively by LSRs in the network, including the egress LSR, known as *Downstream-Unsolicited.* It is possible for both to take place at once and for the LSP to meet in the middle.

Some possible options for controlling how LSPs are set up, and the protocols that can be used to achieve them, are described next.

▼   *Hop-by-hop label assignment* is the process by which the LSP setup requests are routed according to the next-hop routing toward the destination of the data. LSP setup could be initiated by updates to the routing table, or in response to a new traffic flow. The IETF MPLS Working Group has specified (but not mandated) LDP as a protocol for hop-by-hop label assignment. RSVP and CR-LDP can also be used. (We'll talk about these two shortly.)

■   In *Downstream-Unsolicited* label distribution, the egress LSR distributes the label to be used to reach a particular host. The trigger for this will usually be new routing information received at the egress node. Additionally, if the label distribution method is Ordered Control, each upstream LSR distributes a label further upstream. This effectively builds a tree of LSPs rooted at each egress LSR. LDP is currently the only protocol suitable for this mode of label distribution.

■   Once LSPs have been established across the network, they can be used to support new routes as they become available. As the routing protocols distribute the new routing information upstream, they can also indicate which label (which LSP) should be used to reach the destinations to which the route refers.

■   If an ingress LSR wants to set up an LSP that does not follow the next-hop routing path, it must use a *Label Distribution Protocol* (LDP) that allows specification of an Explicit Route. This requires Downstream-on-Demand label distribution. CR-LDP and RSVP are two protocols that provide this function.

▲   An ingress LSR may also want to set up an LSP that provides a particular level of service by, for example, reserving resources at each intermediate LSR along the path. In this case, the route of the LSP may be constrained by the availability of resources and the ability of nodes to fulfill the QoS requirements. CR-LDP and RSVP are two protocols that allow *Downstream-on-Demand* label distribution to include requests for specific QoS guarantees.

Explicit Routes

An *Explicit Route* is most simply understood as a precise sequence of steps from ingress to egress. An LSP in MPLS can be set up to follow an explicit path that is essentially a list of IP addresses. However, it does not need to be specified this fully. For example, the route could specify only the first few hops. After the last explicitly specified hop has been reached, routing of the LSP proceeds using hop-by-hop routing.

A component of an explicit route may also be less precisely specified. A collection of nodes, known as an *Abstract Node,* may be presented as a single step in the route, for example, by using an IP prefix rather than a precise address. The LSP must be routed to some node within this Abstract Node as the next hop. The route may contain several hops within the Abstract Node before emerging to the next hop specified in the Explicit Route.

An Explicit Route may also contain the identifier of an *Autonomous System* (AS). This allows the LSP to be routed through an area of the network that is out of the administrative control of the initiator of the LSP. The route may contain several hops within the AS before emerging to the next hop specified in the Explicit Route.

An Explicit Route may be classified as "strict" or "loose." A *strict* route must contain only those nodes, Abstract Nodes, or ASes specified in the Explicit Route and must use them in the order specified. A *loose* route must include all of the hops specified and must maintain the order, but it may also include additional hops as necessary to reach the hops specified. Once a loose route has been established, it can be modified as a hop-by-hop route could be, or it can be "pinned" so that it does not change.

Explicit routing is particularly useful to force an LSP down a path that differs from the one offered by the routing protocol. It can be used to distribute traffic in a busy network, to route around network failures or hot spots, or to provide preallocated backup LSPs to protect against network failures.

Constraint-Based Routing

The route that an LSP may take can be constrained by many requirements selected at the ingress LSR. An Explicit Route is an example of a constrained route where the constraint is the order in which intermediate LSRs may be reached. Other constraints can be imposed by a description of the traffic that is to flow and may include bandwidth, delay, resource class, and priority.

One approach is for the ingress LSR to calculate the entire route based on the constraints and information that it has about the current state of the network. This leads it to produce an Explicit Route that satisfies the constraints.

The other approach is a variation on hop-by-hop routing where, at each LSR, the next hop is calculated using information held at that LSR about local resource availability.

The two approaches are combined if information about part of the route is unavailable, such as when it traverses an AS. In this case, the route may be loosely specified in part and explicitly routed using the constraints where necessary.

Resource Reservation

To secure promised services, it is not sufficient simply to select a route that can provide the correct resources. These resources must be reserved to ensure that they are not "shared" or "stolen" by another LSP. The traffic requirements can be passed during LSP setup just as with constraint-based routing. They are used at each LSR to reserve the resources required, or to fail the setup if the resources are not available.

Label Distribution Protocols (LDP)

MPLS allows the use of just one set of protocols in the network. Using MPLS to meet the aims described in the previous three sections while avoiding the problems described earlier requires an LDP that supports Explicit Routes and constraint-based routing. There are currently two label-distribution protocols that meet this definition:

▼   CR-LDP

▲   Extended RSVP

There is a debate about which of these protocols is preferable, which is most suitable for particular scenarios, and whether it is necessary to implement both of the protocols in an MPLS network.

CR-LDP

CR-LDP is a set of extensions to LDP specifically designed to facilitate constraint-based routing of LSPs. Like LDP, it uses TCP sessions between LSR peers and sends label-distribution messages along the sessions. This allows it to assume reliable distribution of control messages. Setting up an LSP using CR-LDP is shown in Figure 23-7, followed by the list of steps.

1. The Ingress LSR, LSR A, determines that it needs to set up a new LSP to LSR C. The traffic parameters required for the session or administrative policies for the network enable LSR A to determine that the route for the new LSP should go through LSR B, which might not be the same as the hop-by-hop route to LSR C. LSR A builds a *LABEL_REQUEST* message with an explicit route of (B, C) and details of the traffic parameters requested for the new route. LSR A reserves the resources it needs for the new LSP and then forwards the LABEL_REQUEST to LSR B on the TCP session.

2. LSR B receives the LABEL_REQUEST message, determines that it is not the egress for this LSP, and forwards the request along the route specified in the message. It reserves the resources requested for the new LSP, modifies the explicit route in the LABEL_REQUEST message, and passes the message to LSR C. If necessary, LSR B may reduce the reservation it makes for the new LSP if the appropriate parameters were marked as negotiable in the LABEL_REQUEST.

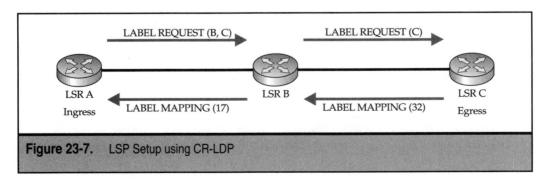

Figure 23-7.   LSP Setup using CR-LDP

3. LSR C determines that it is the egress for this new LSP. It performs any final negotiation on the resources and makes the reservation for the LSP. It allocates a label to the new LSP and distributes the label to LSR B in a *LABEL_MAPPING* message, which contains details of the final traffic parameters reserved for the LSP.

4. LSR B receives the LABEL_MAPPING and matches it to the original request using the LSP ID contained in both the LABEL_REQUEST and LABEL_MAPPING messages. It finalizes the reservation, allocates a label for the LSP, sets up the forwarding table entry, and passes the new label to LSR A in a LABEL_MAPPING message.

5. The processing at LSR A is similar, but it does not have to allocate a label and forward it to an upstream LSR, because it is the ingress LSR for the new LSP.

Extended RSVP

Generic RSVP uses a message exchange to reserve resources across a network for IP flows. The Extensions to RSVP for LSP Tunnels enhances generic RSVP so that it can be used to distribute MPLS labels. RSVP is a separate protocol at the IP level. It uses IP datagrams (or UDP at the margins of the network) to communicate between LSR peers. It does not require the maintenance of TCP sessions, but as a consequence of this, it must handle the loss of control messages. Setting up an LSP using RSVP for LSP Tunnels is shown in Figure 23-8 followed by the list of steps.

1. The Ingress LSR, LSR A, determines that it needs to set up a new LSP to LSR C. The traffic parameters required for the session or administrative policies for the network enable LSR A to determine that the route for the new LSP should go through LSR B, which might not be the same as the hop-by-hop route to LSR C. LSR A builds a *PATH* message with an explicit route of (B, C) and details of the traffic parameters requested for the new route. LSR A then forwards the PATH to LSR B as an IP datagram.

2. LSR B receives the PATH request, determines that it is not the egress for this LSP, and forwards the request along the route specified in the request. It modifies the explicit route in the PATH message and passes the message to LSR C.

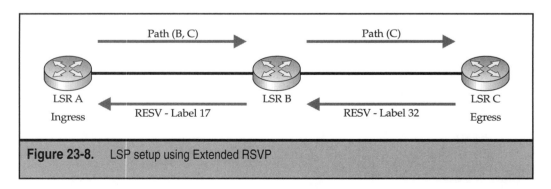

Figure 23-8.   LSP setup using Extended RSVP

3. LSR C determines that it is the egress for this new LSP, determines from the requested traffic parameters what bandwidth it needs to reserve, and allocates the resources required. It selects a label for the new LSP and distributes the label to LSR B in a *RESV* message, which also contains actual details of the reservation required for the LSP.

4. LSR B receives the RESV message and matches it to the original request using the *LSP ID* contained in both the PATH and RESV messages. It determines what resources to reserve from the details in the RESV message, allocates a label for the LSP, sets up the forwarding table, and passes the new label to LSR A in a RESV message.

5. The processing at LSR A is similar, but it does not have to allocate a new label and forward this to an upstream LSR, because it is the ingress LSR for the new LSP.

MPLS VIRTUAL PRIVATE NETWORKS (MPLS VPNS)

In the past, private WANs were created using dedicated leased line connections, each line providing a point-to-point connection between two customer sites. Such networks are expensive to put in place, especially if the connections between sites need to support some level of redundancy.

MPLS VPNs are a method of interconnecting multiple sites belonging to a customer using an ISP backbone network in place of dedicated leased lines. Each customer site is directly connected to the ISP backbone. The ISP can offer a VPN service more economically than if dedicated private WANs are built by each individual customer because the ISP can share the same backbone network resources (bandwidth, redundant links) between many customers. The customer also gains by outsourcing the complex task of planning and managing these VPNs to the ISP.

Currently, not all VPN solutions using dedicated commercial equipment at customer locations are interoperable, and some may lock you into one equipment vendor and in many cases a single ISP. This has created strong interest in MPLS VPNs running over the public Internet using standards-based interoperable implementations that work across multiple ISPs.

MPLS forwards data using labels that are attached to each data packet. Intermediate MPLS nodes do not need to look at the content of the data in each packet. In particular, the destination IP addresses in the packets are not examined, which enables MPLS to offer an efficient encapsulation mechanism for private data traffic traversing the ISP backbone.

MPLS is the ultimate solution for VPNs for a number of reasons:

▼ **Connectionless service** A significant technical advantage of MPLS VPNs is that they are connectionless. This means that no prior action is necessary to establish communication between hosts, making it easy for two parties to communicate. To establish privacy in a connectionless IP environment, current VPN solutions impose a connection-oriented, point-to-point overlay on the network. Even if it runs over a connectionless network, a VPN cannot take advantage of the ease of connectivity and multiple services available in connectionless networks. When you create a connectionless VPN, you do not need tunnels and encryption for network privacy.

▲ **Centralized service**   Building VPNs in Layer 3 allows delivery of targeted services to a group of users represented by a VPN. A VPN must give service providers more than a mechanism for privately connecting users to intranet services. It must also provide a way to flexibly deliver value-added services to targeted customers. Scalability is critical, because customers want to use services privately in their intranets and extranets. Because MPLS VPNs are seen as private intranets, you may use new IP services such as:

- Multicast
- Quality of Service (QoS)
- IP telephony support within a VPN
- Centralized services including content and web hosting to a VPN

You can customize several combinations of specialized services for individual customers. For example, a service that combines IP multicast with a low-latency service class enables videoconferencing within an intranet.

▼ **Scalability**   If you create a VPN using connection-oriented, point-to-point overlays, Frame Relay, or ATM virtual connections (VCs), the VPN's key deficiency is scalability. Specifically, connection-oriented VPNs without fully meshed connections between customer sites are not optimal. MPLS-based VPNs instead use the peer model and Layer 3 connectionless architecture to leverage a highly scalable VPN solution. The peer model requires a customer site to only be a peer with one provider edge (PE) router as opposed to all other CPE or customer edge (CE) routers that are members of the VPN. The connectionless architecture allows the creation of VPNs in Layer 3, eliminating the need for tunnels or VCs. Other scalability issues of MPLS VPNs are due to the partitioning of VPN routes between PE routers and the further partitioning of VPN and IGP routes between PE routers and provider (P) routers in a core network.

- PE routers must maintain VPN routes for those VPNs who are members.
▲ P routers do not maintain any VPN routes.

This increases the scalability of the provider's core and ensures that no one device is a scalability bottleneck.

▼ **Security**   Historically, VPN implementations based on ATM or Frame Relay VCs have provided this security by virtue of the connection-oriented nature of the physical network. However, the connectionless public IP network cannot provide the same protection, and standard VPNs have relied on cryptographic means to provide security and authentication. MPLS brings to IP security benefits similar to leased line VCs. This means that the customer equipment connected to the VPN does not need to run IPSec or other cryptographic software. This results in a considerable saving for the customer in terms of equipment expense and management complexity. MPLS VPN security is achieved as described next.

- ■ At the ingress SP edge router, all data for a VPN is assigned a label stack that is unique to the VPN destination. This ensures that the data is delivered only to that destination and cannot "leak" out of the VPN.

- ■ Any other packet entering the ISP network is either routed without the use of MPLS or is assigned a different label stack, so a malicious third-party cannot "spoof" or insert data into the VPN from outside the ISP network.

- ▲ ISP routers can simply use hash algorithms such as MD5 or similar techniques to protect against insertion of fake labels or LSRs into the label-distribution protocols.

There are two situations when a customer may still require the use of cryptographic security measures even when using an MPLS VPN solution:

- ▼ If the customer data is considered sufficiently sensitive that it must be protected against snooping even from within the ISP network. In this case, IPSec or similar cryptographic techniques would just be applied to the VPN data *before* it enters the ISP network. Then, the customer retains control and responsibility for distributing the cryptographic keys.

- ■ When a VPN is served by more than one ISP, the ISPs may choose to use IPSec-based tunnels to carry the VPN traffic between their networks on the public IP network if a direct MPLS connection between the ISPs is not available. In this case, the ISPs are responsible for distributing cryptographic keys.

- ■ **Easy to create**   To take full advantage of VPNs, it must be easy for customers to create new VPNs and user communities. Because MPLS VPNs are connectionless, no specific point-to-point connection maps or topologies are required. You can add sites to intranets and extranets and form closed user groups. When you manage VPNs in this manner, it enables membership of any given site in multiple VPNs, maximizing flexibility in building intranets and extranets.

- ■ **Flexible addressing**   To make a VPN service more accessible, customers of a service provider can design their own addressing plan, independent of addressing plans for other service provider customers. This means that the time and expense of converting to public IP addresses to enable intranet connectivity is eliminated. MPLS VPNs allow customers to continue to use their present address spaces without resorting to Network Address Translation (NAT) by providing a public and private view of the address. A NAT is required only if two VPNs with overlapping address spaces want to communicate. This enables customers to use their own unregistered private addresses and to communicate freely across a public IP network.

- **Integrated Class of Service (CoS)**   CoS is an important requirement for many IP VPN customers. It provides the ability to address two fundamental VPN requirements:

  - Predictable performance and policy implementation

  - Support for multiple levels of service in a MPLS VPN Network traffic is classified and labeled at the edge of the network before traffic is aggregated according to policies defined by subscribers and implemented by the provider and transported across the provider core. Traffic at the edge and core of the network can then be differentiated into different classes by drop probability or delay.

▲ **Easy migration**   With MPLS it is easy for ISPs to quickly deploy VPN services. MPLS VPNs are unique because you can build them over multiple network architectures, including IP, ATM, Frame Relay, and hybrid networks. Migration for the end customer is simplified because there is no requirement to support MPLS on the customer edge (CE) router, and no modifications are required to a customer's intranet. Figure 23-9 shows an example of a VPN with a service provider (P) backbone network, service provider edge routers (PE), and customer edge routers (CE).

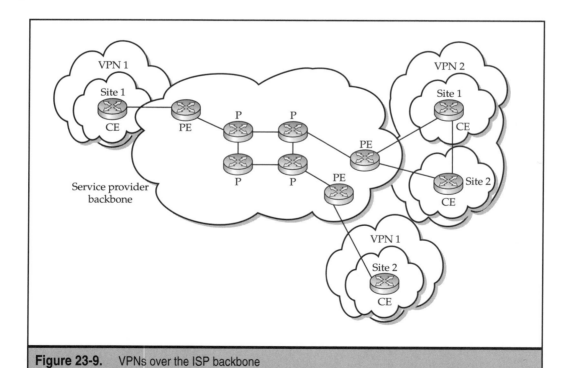

Figure 23-9.   VPNs over the ISP backbone

A VPN contains customer devices attached to the CE routers. These customer devices use VPNs to exchange information between devices. Only the PE routers are aware of the VPNs. Figure 23-10 shows five customer sites communicating within three VPNs. The VPNs can communicate with the following sites:

▼ **VPN1**   Sites 2 and 4

■ **VPN2**   Sites 1, 3, and 4

▲ **VPN3**   Sites 1, 3, and 5

SUMMARY

MPLS will provide a new technical foundation for the next generation of multiuser, multiservice internetworks. MPLS will provide higher performance, unlimited scalability, bulletproof connectionless VPN solutions, and QoS capabilities. While the expansion of the Internet has been a major driver for development of label switching, it is not the only, or even the most important, factor.

Label switching provides significant improvements in the packet-forwarding process, simplifying the processing requirements by avoiding the need to duplicate header processing at every step in the path. IP will finally have a way of operating in an environment that can support controlled QoS and traffic engineering.

MPLS is a new technology that is just beginning to be recognized as beneficial. MPLS is starting to see widespread deployment in ISP backbones and over time will be seen in many large enterprise networks, finally paving the way for true convergence of telephony, video, VPN, and computing services.

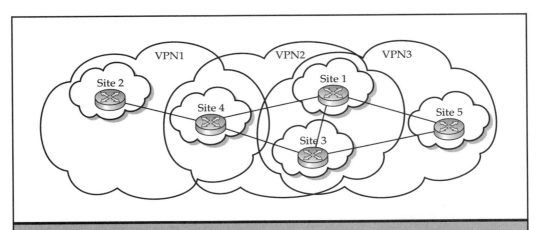

Figure 23-10.   Different sites within VPNs

Index

B

O

INTERNATIONAL CONTACT INFORMATION

AUSTRALIA
McGraw-Hill Book Company Australia Pty. Ltd.
TEL +61-2-9417-9899
FAX +61-2-9417-5687
http://www.mcgraw-hill.com.au
books-it_sydney@mcgraw-hill.com

CANADA
McGraw-Hill Ryerson Ltd.
TEL +905-430-5000
FAX +905-430-5020
http://www.mcgrawhill.ca

GREECE, MIDDLE EAST, NORTHERN AFRICA
McGraw-Hill Hellas
TEL +30-1-656-0990-3-4
FAX +30-1-654-5525

MEXICO (Also serving Latin America)
McGraw-Hill Interamericana Editores S.A. de C.V.
TEL +525-117-1583
FAX +525-117-1589
http://www.mcgraw-hill.com.mx
fernando_castellanos@mcgraw-hill.com

SINGAPORE (Serving Asia)
McGraw-Hill Book Company
TEL +65-863-1580
FAX +65-862-3354
http://www.mcgraw-hill.com.sg
mghasia@mcgraw-hill.com

SOUTH AFRICA
McGraw-Hill South Africa
TEL +27-11-622-7512
FAX +27-11-622-9045
robyn_swanepoel@mcgraw-hill.com

UNITED KINGDOM & EUROPE (Excluding Southern Europe)
McGraw-Hill Education Europe
TEL +44-1-628-502500
FAX +44-1-628-770224
http://www.mcgraw-hill.co.uk
computing_neurope@mcgraw-hill.com

ALL OTHER INQUIRIES Contact:
Osborne/McGraw-Hill
TEL +1-510-549-6600
FAX +1-510-883-7600
http://www.osborne.com
omg_international@mcgraw-hill.com

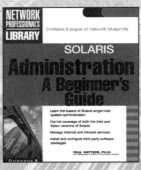

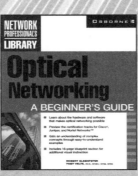